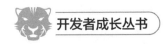

开发者成长丛书

PySpark原理深入与编程实战

微课视频版

辛立伟 辛雨桐◎编著

清华大学出版社

北京

内 容 简 介

本书系统讲述 Apache Spark/PySpark 大数据计算平台的原理，以及如何将 Apache PySpark 应用于大数据的实时流处理、批处理等场景。通过对原理的深入讲解和对实践示例、案例的讲解，使读者了解并掌握 Apache Spark/PySpark 的基本原理和技能，拉近理论与实践的距离。

全书共分为 8 章，主要内容包括 Spark 架构原理与集群搭建、开发和部署 PySpark 应用程序、PySpark 核心编程、PySpark SQL、PySpark 结构化流、PySpark 大数据分析综合案例。本书源码全部在 Apache Spark 3.1.2 上调试成功，所有示例和案例均基于 Python 3.x 语言。

为降低读者学习大数据技术的门槛，本书除提供了丰富的上机实践操作和详细的范例程序讲解之外，还提供了搭建好的 Hadoop、Hive 数据仓库和 PySpark 大数据开发和学习环境。读者既可参照本书的讲解自行搭建 Hadoop 和 PySpark 环境，也可直接使用本书提供的开发和学习环境，快速开始大数据和 PySpark 的学习。

本书内容全面、实例丰富、可操作性强，做到了理论与实践相结合。本书适合大数据学习爱好者、想要入门 Apache Spark/PySpark 的读者作为入门和提高的技术参考书，也适合用作高等院校大数据专业相关课程的教材或教学参考书。

本书封面贴有清华大学出版社防伪标签，无标签者不得销售。
版权所有，侵权必究。举报：010-62782989，beiqinquan@tup.tsinghua.edu.cn。

图书在版编目(CIP)数据

PySpark 原理深入与编程实战：微课视频版/辛立伟，辛雨桐编著. —北京：清华大学出版社，2023.7
（开发者成长丛书）
ISBN 978-7-302-62597-1

Ⅰ.①P… Ⅱ.①辛… ②辛… Ⅲ.①数据处理 Ⅳ.①TP274

中国国家版本馆 CIP 数据核字(2023)第 022854 号

责任编辑：赵佳霓
封面设计：刘　键
责任校对：时翠兰
责任印制：丛怀宇

出版发行：清华大学出版社
　　　　　网　　址：http://www.tup.com.cn，http://www.wqbook.com
　　　　　地　　址：北京清华大学学研大厦 A 座　　邮　编：100084
　　　　　社 总 机：010-83470000　　　　　　　　　邮　购：010-62786544
　　　　　投稿与读者服务：010-62776969，c-service@tup.tsinghua.edu.cn
　　　　　质 量 反 馈：010-62772015，zhiliang@tup.tsinghua.edu.cn
　　　　　课 件 下 载：http://www.tup.com.cn，010-83470236
印 装 者：三河市铭诚印务有限公司
经　　销：全国新华书店
开　　本：186mm×240mm　　　印　张：31.5　　　字　数：688 千字
版　　次：2023 年 8 月第 1 版　　　　　　　　印　次：2023 年 8 月第 1 次印刷
印　　数：1～2000
定　　价：119.00 元

产品编号：097802-01

前 言
PREFACE

大数据分析一直是一个热门话题，需要大数据分析的场景也越来越多。Apache Spark 是一个用于快速、通用、大规模数据处理的开源项目。现在，Apache Spark 已经成为一个统一的大数据处理平台，拥有一个快速的统一分析引擎，可用于大数据的批处理、实时流处理、机器学习和图计算。

2009 年，Spark 诞生于伯克利大学 AMP 实验室，最初属于伯克利大学的研究性项目。它于 2010 年被正式开源，于 2013 年被转交给 Apache 软件基金会，并于 2014 年成为 Apache 基金的顶级项目，整个过程不到五年时间。Apache Spark 诞生以后，迅速发展成为大数据处理技术中的佼佼者，目前已经成为大数据处理领域炙手可热的技术，其发展势头非常强劲。

自 2010 年首次发布以来，Apache Spark 已经成为极为活跃的大数据开源项目之一。如今，Apache Spark 实际上已经是大数据处理、数据科学、机器学习和数据分析工作负载的统一引擎，是从业人员及希望进入大数据行业人员必须学习和掌握的大数据技术之一。

Apache Spark 支持 Java、Scala、Python 和 R 语言，并提供了相应的 API，而在数据科学领域，Python 是应用得非常普遍的数据处理语言，但是作为大数据的初学者，即使精通 Python 语言，在学习 PySpark 时通常也会遇到以下几个难题：

（1）缺少面向零基础读者的 PySpark 入门教程。
（2）缺少系统化的 PySpark 大数据教程。
（3）现有的 PySpark 资料、教程或图书陈旧或者碎片化。
（4）官方全英文文档难以阅读和理解。
（5）缺少必要的数据集、可运行的实验案例及学习平台。

特别是 Spark 3 发布以后，性能得到了极大提升。为此，一方面是为了编者自己能更系统、更及时地跟进 Spark/PySpark 的演进和迭代；另一方面也是为了降低面向零基础读者学习 PySpark（及其他大数据技术）的入门难度，编写了这本《PySpark 原理深入与编程实战（微课视频版）》。编者以为，本书具有以下几个特点：

（1）面向零基础读者，知识点深浅适当，代码完整易懂。
（2）内容全面系统，包括架构原理、开发环境及程序部署、流和批计算、综合项目案例等。

（3）版本较新，所有代码均基于 Spark 3.1.2 和 Python 3.7。

（4）全书包含大量的示例代码讲解和完整项目案例。

扫描目录上方二维码可获取本书配套资源。

本书特别适合想要入门并深入掌握 Apache Spark/PySpark 大数据开发和大数据分析、大数据 OLAP 引擎、流计算的读者，希望拥有大数据系统参考教材的教师，及想要了解最新 Spark/PySpark 技术应用的从业人员。

本书第 1～3 章由辛雨桐执笔编写，第 4～8 章由辛立伟执笔编写。

由于编者水平所限，行文及内容难免有疏漏之处，请读者见谅并予以反馈，笔者会在后续的版本重构中不断提升内容质量。

编　者

2023 年 4 月

目录
CONTENTS

程序源码

教学课件(PPT)

第1章 Spark 架构原理与集群搭建(▶78min) ··· 1
- 1.1 Spark 简介 ··· 1
- 1.2 Spark 技术栈 ··· 3
  - 1.2.1 Spark Core ··· 4
  - 1.2.2 Spark SQL ··· 4
  - 1.2.3 Spark Streaming 和 Structured Streaming ··· 5
  - 1.2.4 Spark MLlib ··· 5
  - 1.2.5 Spark GraphX ··· 6
  - 1.2.6 SparkR ··· 6
- 1.3 Spark 和 PySpark 架构原理 ··· 7
  - 1.3.1 Spark 集群和资源管理系统 ··· 7
  - 1.3.2 Spark 应用程序 ··· 7
  - 1.3.3 Spark Driver 和 Executor ··· 9
  - 1.3.4 PySpark 架构 ··· 10
- 1.4 Spark 程序部署模式 ··· 11
- 1.5 安装和配置 Spark 集群 ··· 12
  - 1.5.1 安装 Spark ··· 12
  - 1.5.2 了解 Spark 目录结构 ··· 13
  - 1.5.3 配置 Spark/PySpark 集群 ··· 14
  - 1.5.4 验证 PySpark 安装 ··· 15
- 1.6 配置 Spark 历史服务器 ··· 17
  - 1.6.1 历史服务器配置 ··· 17
  - 1.6.2 启动 Spark 历史服务器 ··· 19
- 1.7 使用 PySpark Shell 进行交互式分析 ··· 20
  - 1.7.1 运行模式--master ··· 20
  - 1.7.2 启动和退出 PySpark Shell ··· 21
  - 1.7.3 PySpark Shell 常用命令 ··· 23
  - 1.7.4 SparkContext 和 SparkSession ··· 23
  - 1.7.5 Spark Web UI ··· 25
- 1.8 使用 spark-submit 提交 PySpark 应用程序 ··· 28
  - 1.8.1 spark-submit 指令的各种参数说明 ··· 28

1.8.2　提交 pi.ipynb 程序，计算圆周率 π 值 ·· 31
　　　1.8.3　将 PySpark 程序提交到 YARN 集群上执行 ··· 32

第 2 章　开发和部署 PySpark 应用程序(▶80min) ··· 34

2.1　使用 PyCharm 开发 PySpark 应用程序 ·· 34
　　2.1.1　准备数据文件 ·· 34
　　2.1.2　安装 PyCharm ·· 35
　　2.1.3　创建一个新的 PyCharm 项目 ·· 35
　　2.1.4　安装 PySpark 包 ··· 36
　　2.1.5　创建 PySpark 应用程序 ·· 39
　　2.1.6　部署到集群中运行 ·· 40
2.2　使用 Zeppelin 进行交互式分析 ··· 41
　　2.2.1　下载 Zeppelin 安装包 ··· 42
　　2.2.2　安装和配置 Zeppelin ·· 42
　　2.2.3　配置 Spark/PySpark 解释器 ··· 44
　　2.2.4　创建和执行 Notebook 文件 ··· 45
2.3　使用 Jupyter Notebook 进行交互式分析 ··· 46
　　2.3.1　配置 PySpark Driver 使用 Jupyter Notebook ·· 47
　　2.3.2　使用 findSpark 包 ·· 49

第 3 章　PySpark 核心编程(▶212min) ··· 51

3.1　理解数据抽象 RDD ·· 51
3.2　RDD 编程模型 ·· 53
　　3.2.1　单词计数应用程序 ·· 53
　　3.2.2　理解 SparkSession ·· 56
　　3.2.3　理解 SparkContext ·· 57
3.3　创建 RDD ·· 58
　　3.3.1　将现有的集合并行化以创建 RDD ·· 58
　　3.3.2　从存储系统读取数据集以创建 RDD ··· 60
　　3.3.3　从已有的 RDD 转换得到新的 RDD ·· 60
　　3.3.4　创建 RDD 时指定分区数量 ··· 60
3.4　操作 RDD ·· 61
　　3.4.1　RDD 上的 Transformation 和 Action ··· 62
　　3.4.2　RDD Transformation 操作 ·· 64
　　3.4.3　RDD Action 操作 ··· 70
　　3.4.4　RDD 上的描述性统计操作 ·· 73
3.5　Key-Value Pair RDD ·· 74
　　3.5.1　创建 Pair RDD ··· 75
　　3.5.2　操作 Pair RDD ··· 76
　　3.5.3　关于 reduceByKey() 操作 ··· 81
　　3.5.4　关于 aggregateByKey() 操作 ··· 83
　　3.5.5　关于 combineByKey() 操作 ·· 87

- 3.6 持久化 RDD ... 90
  - 3.6.1 缓存 RDD ... 90
  - 3.6.2 RDD 缓存策略 ... 92
  - 3.6.3 检查点 RDD ... 93
- 3.7 数据分区 ... 94
  - 3.7.1 获取和指定 RDD 分区数 ... 95
  - 3.7.2 调整 RDD 分区数 ... 96
  - 3.7.3 内置数据分区器 ... 97
  - 3.7.4 自定义数据分区器 ... 101
  - 3.7.5 避免不必要的 shuffling ... 102
  - 3.7.6 基于数据分区的操作 ... 104
- 3.8 使用共享变量 ... 108
  - 3.8.1 广播变量 ... 109
  - 3.8.2 累加器 ... 114
- 3.9 PySpark RDD 可视化 ... 116
- 3.10 PySpark RDD 编程案例 ... 117
  - 3.10.1 合并小文件 ... 117
  - 3.10.2 二次排序实现 ... 119
  - 3.10.3 Top N 实现 ... 121
  - 3.10.4 数据聚合计算 ... 125

第 4 章 PySpark SQL(初级) (▶163min) ... 127
- 4.1 PySpark SQL 数据抽象 ... 127
- 4.2 PySpark SQL 编程模型 ... 129
- 4.3 程序入口 SparkSession ... 132
- 4.4 PySpark SQL 中的模式和对象 ... 134
  - 4.4.1 模式 ... 134
  - 4.4.2 列对象和行对象 ... 135
- 4.5 简单构造 DataFrame ... 136
  - 4.5.1 简单创建单列和多列 DataFrame ... 137
  - 4.5.2 从 RDD 创建 DataFrame ... 140
  - 4.5.3 读取外部数据源创建 DataFrame ... 144
- 4.6 操作 DataFrame ... 166
  - 4.6.1 列的多种引用方式 ... 167
  - 4.6.2 对 DataFrame 执行 Transformation 转换操作 ... 170
  - 4.6.3 对 DataFrame 执行 Action 操作 ... 184
  - 4.6.4 对 DataFrame 执行描述性统计操作 ... 185
  - 4.6.5 提取 DataFrame Row 中特定字段 ... 188
  - 4.6.6 操作 DataFrame 示例 ... 189
- 4.7 存储 DataFrame ... 191
  - 4.7.1 写出 DataFrame ... 191

4.7.2 存储模式 … 194
   4.7.3 控制 DataFrame 的输出文件数量 … 195
   4.7.4 控制 DataFrame 实现分区存储 … 199
 4.8 临时视图与 SQL 查询 … 201
   4.8.1 在 PySpark 程序中执行 SQL 语句 … 201
   4.8.2 注册临时视图并执行 SQL 查询 … 203
   4.8.3 使用全局临时视图 … 206
   4.8.4 直接使用数据源注册临时视图 … 208
   4.8.5 查看和管理表目录 … 209
 4.9 缓存 DataFrame … 211
   4.9.1 缓存方法 … 211
   4.9.2 缓存策略 … 213
   4.9.3 缓存表 … 214
 4.10 PySpark SQL 可视化 … 214
   4.10.1 PySpark DataFrame 转换到 Pandas … 214
   4.10.2 PySpark SQL DataFrame 可视化 … 218
 4.11 PySpark SQL 编程案例 … 220
   4.11.1 实现单词计数 … 220
   4.11.2 用户数据集分析 … 222
   4.11.3 航空公司航班数据集分析 … 224

第 5 章 PySpark SQL(高级) (▶115min) … 234
 5.1 PySpark SQL 函数 … 234
 5.2 内置标量函数 … 234
   5.2.1 日期时间函数 … 235
   5.2.2 字符串函数 … 239
   5.2.3 数学计算函数 … 243
   5.2.4 集合元素处理函数 … 244
   5.2.5 其他函数 … 248
   5.2.6 函数应用示例 … 252
   5.2.7 PySpark 3 数组函数 … 255
 5.3 聚合与透视函数 … 264
   5.3.1 聚合函数 … 264
   5.3.2 分组聚合 … 271
   5.3.3 数据透视 … 274
 5.4 高级分析函数 … 277
   5.4.1 使用多维聚合函数 … 277
   5.4.2 使用时间窗口聚合 … 281
   5.4.3 使用窗口分析函数 … 286
 5.5 用户自定义函数（UDF） … 296
   5.5.1 内部原理 … 296

   5.5.2 创建和使用 UDF ·········297
   5.5.3 特殊处理 ·········303
  5.6 数据集的 join 连接 ·········305
   5.6.1 join 表达式和 join 类型 ·········306
   5.6.2 执行 join 连接 ·········307
   5.6.3 处理重复列名 ·········314
   5.6.4 join 连接策略 ·········317
  5.7 读写 Hive 表 ·········319
   5.7.1 PySpark SQL 的 Hive 配置 ·········320
   5.7.2 PySpark SQL 读写 Hive 表 ·········322
   5.7.3 分桶、分区和排序 ·········332
  5.8 PySpark SQL 编程案例 ·········334
   5.8.1 电商订单数据分析 ·········334
   5.8.2 电影评分数据集分析 ·········344

第 6 章 PySpark 结构化流(初级) (▷195min) ·········349
  6.1 PySpark DStream 流简介 ·········349
  6.2 PySpark 结构化流简介 ·········354
  6.3 PySpark 结构化流编程模型 ·········356
  6.4 PySpark 结构化流核心概念 ·········360
   6.4.1 数据源 ·········360
   6.4.2 输出模式 ·········361
   6.4.3 触发器类型 ·········362
   6.4.4 数据接收器 ·········362
   6.4.5 水印 ·········363
  6.5 使用各种流数据源 ·········363
   6.5.1 使用 Socket 数据源 ·········364
   6.5.2 使用 Rate 数据源 ·········365
   6.5.3 使用 File 数据源 ·········368
   6.5.4 使用 Kafka 数据源 ·········372
  6.6 流 DataFrame 操作 ·········379
   6.6.1 选择、投影和聚合操作 ·········379
   6.6.2 执行 join 连接操作 ·········385
  6.7 使用数据接收器 ·········388
   6.7.1 使用 File Data Sink ·········389
   6.7.2 使用 Kafka Data Sink ·········391
   6.7.3 使用 Foreach Data Sink ·········394
   6.7.4 使用 Console Data Sink ·········399
   6.7.5 使用 Memory Data Sink ·········401
   6.7.6 Data Sink 与输出模式 ·········402
  6.8 深入研究输出模式 ·········402
   6.8.1 无状态流查询 ·········403

6.8.2 有状态流查询 ... 404
6.9 深入研究触发器 ... 409
　　6.9.1 固定间隔触发器 ... 410
　　6.9.2 一次性的触发器 ... 412
　　6.9.3 连续性的触发器 ... 413

第7章 PySpark 结构化流(高级) (▶62min) ... 416
7.1 事件时间和窗口聚合 ... 416
　　7.1.1 固定窗口聚合 ... 416
　　7.1.2 滑动窗口聚合 ... 421
7.2 水印 ... 426
　　7.2.1 限制维护的聚合状态数量 ... 426
　　7.2.2 处理迟到的数据 ... 429
7.3 处理重复数据 ... 436
7.4 容错 ... 439
7.5 流查询度量指标 ... 441
7.6 结构化流案例：运输公司车辆超速实时监测 ... 443
　　7.6.1 实现技术剖析 ... 443
　　7.6.2 完整实现代码 ... 449
　　7.6.3 执行步骤演示 ... 451

第8章 PySpark 大数据分析综合案例 (▶21min) ... 455
8.1 项目需求说明 ... 455
8.2 项目架构设计 ... 456
8.3 项目实现：数据采集 ... 457
　　8.3.1 爬虫程序实现：使用 requests 库 ... 457
　　8.3.2 爬虫程序实现：使用 Scrapy 框架 ... 460
8.4 项目实现：数据集成 ... 466
　　8.4.1 Flume 简介 ... 466
　　8.4.2 安装和配置 Flume ... 467
　　8.4.3 实现数据集成 ... 468
8.5 项目实现：数据 ELT ... 469
8.6 项目实现：数据清洗与整理 ... 472
8.7 项目实现：数据分析 ... 476
8.8 项目实现：分析结果导出 ... 479
8.9 项目实现：数据可视化 ... 480
　　8.9.1 Flask 框架简介 ... 480
　　8.9.2 ECharts 图表库介绍 ... 481
　　8.9.3 Flask Web 程序开发 ... 483
　　8.9.4 前端 ECharts 组件开发 ... 485

第 1 章 Spark 架构原理与集群搭建

CHAPTER 1

25min

Apache Spark 是一个用于快速、通用、大规模数据处理的开源项目。它类似于 Hadoop 的 MapReduce，但对于执行批处理来讲速度更快、更高效。Apache Spark 可以部署在大量廉价的硬件设备上，以创建大数据并行处理计算集群。

Apache Spark 作为一个用于大数据处理的内存并行计算框架，它利用内存缓存和优化执行来获得更高的性能，并且支持以任何格式读取/写入 Hadoop 数据，同时保证了高容错性和可扩展性。现在，Apache Spark 已经成为一个统一的大数据处理平台，拥有一个快速的统一分析引擎，可用于大数据的批处理、实时流处理、机器学习和图计算。

自 2010 年首次发布以来，Apache Spark 已经成为最活跃的大数据开源项目之一。如今，Apache Spark 实际上已经是大数据处理、数据科学、机器学习和数据分析工作负载的统一引擎。

1.1 Spark 简介

2009 年，Spark 诞生于伯克利大学的 AMP 实验室，最初属于伯克利大学的研究性项目。它于 2010 年被正式开源，于 2013 年被转交给 Apache 软件基金会，并于 2014 年成为 Apache 软件基金会的顶级项目，整个过程不到五年时间。Apache Spark 诞生以后，迅速发展成为大数据处理技术中的佼佼者，目前已经成为大数据处理领域炙手可热的技术，其发展势头非常强劲。

Spark 的内存计算模型如图 1-1 所示。

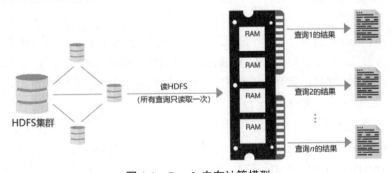

图 1-1　Spark 内存计算模型

在图 1-1 中，Spark 一次性从 HDFS 中读取所有的数据，并以分布式的方式缓存在计算机集群的各节点的内存中。

Spark 用于迭代算法的内存数据共享表示如图 1-2 所示。

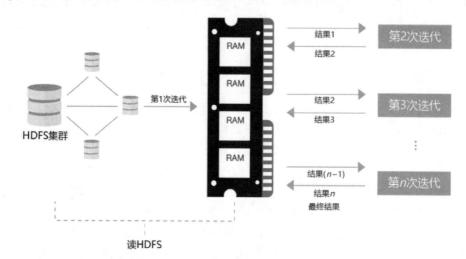

图 1-2　Spark 用于迭代算法的内存数据共享表示

Spark 与其他分布式计算平台相比有许多独特的优势：
（1）用于迭代机器学习和交互式数据分析的更快的执行平台。
（2）用于批处理、SQL 查询、实时流处理、图处理和复杂数据分析的单一技术栈。
（3）通过隐藏分布式编程的复杂性，提供高级 API 来供用户开发各种分布式应用程序。
（4）对各种数据源的无缝支持，如 RDBMS、HBase、Cassandra、Parquet、MongoDB、HDFS、Amazon S3 等。

Spark 隐藏了编写核心 MapReduce 作业的复杂性，并通过简单的函数调用提供了大部分功能。由于它的简单性，受到了用户的广泛应用和认同，例如数据科学家、数据工程师、统计学家，以及 R/Python/Scala/Java 开发人员。由于 Spark 采用了内存计算，并采用函数式编程，提供了大量高阶函数和算子，因此它具有以下 3 个显著特性：速度、易用性和灵活性。

在 2014 年，Spark 赢得了 Daytona GraySort 竞赛，该竞赛是对 100 TB 数据进行排序的行业基准（1 万亿条记录）。来自 Databricks 的提交声称 Spark 能够以比之前的 Hadoop MapReduce 所创造的世界纪录快三倍的速度对 100 TB 的数据进行排序，并且使用的资源减少了约 90%，如图 1-3 所示。

Spark 可以连接到许多不同的数据源，包括文件（如 CSV、JSON、Parquet、AVRO）、MySQL、MongoDB、HBase 和 Cassandra。此外，它还可以连接到特殊用途的引擎和数据源，如 Elasticsearch、Apache Kafka 和 Redis。这些引擎支持 Spark 应用程序中的特定功能，如搜索、流、缓存等。Spark 提供了 DataSource API 以支持各种数据源（包括自定义数据源）的 Spark 连接，如图 1-4 所示。

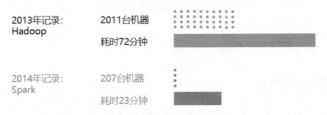

图 1-3　与 Hadoop MapReduce 相比，Spark 计算速度更快，使用资源更少

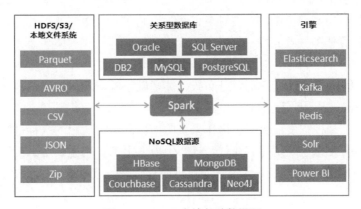

图 1-4　Spark 支持各种数据源

Spark 提供了 4 种编程语言接口，分别是 Java、Scala、Python 和 R。因为 Apache Spark 本身是用 Scala 构建的，所以 Scala 是首选语言。由于 Spark 内置了对 Scala、Java、R 和 Python 的支持，因此大多数的开发人员和数据工程师能够利用整个 Spark 栈应用于不同的应用场景。

1.2　Spark 技术栈

Spark 提供了一个统一的数据处理引擎，称为 Spark 栈。Spark 栈的基础是其核心模块（称为 Spark Core）。Spark Core 提供了管理和运行分布式应用程序的所有必要功能，如调度、协调和容错。此外，它还为数据处理提供了强大的通用编程抽象，称为弹性分布式数据集（Resilient Distributed Datasets，RDD）。

在 Spark Core 之上是一个组件集合，其中每个组件都是为特定的数据处理工作而设计的，它们建立在 Spark Core 的强大基础引擎之上。Spark 技术栈如图 1-5 所示。

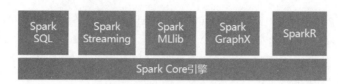

图 1-5 Spark 技术栈

下面分别了解 Spark Core 引擎和各个功能组件。

1.2.1 Spark Core

Spark Core 由两部分组成：分布式计算基础设施和 RDD 编程抽象，其中分布式计算基础设施的职责包括以下几点：

（1）负责集群中多节点上的计算任务的分发、协调和调度。

（2）处理计算任务失败。

（3）高效地跨节点传输数据（数据传输 shuffling）。

Spark 的高级用户需要对 Spark 分布式计算基础设施有深入的了解，从而能够有效地设计高性能的 Spark 应用程序。

Spark Core 在某种程度上类似于操作系统的内核。它是通用的执行引擎，它既快速又容错。整个 Spark 生态系统是建立在这个核心引擎之上的。它主要用于工作调度、任务分配和跨 worker 节点的作业监控。此外它还负责内存管理，与各种异构存储系统交互，以及各种其他操作。

Spark Core 的主要编程抽象是弹性分布式数据集（RDD），RDD 是一个不可变的、容错的对象集合，它可以在一个集群中进行分区，因此可以并行操作。本质上，RDD 为 Spark 应用程序开发人员提供了一组 APIs，使这些开发人员能够轻松高效地执行大规模的数据处理，而不必担心数据驻留在集群上的什么位置或处理机器故障。

Spark 可以从各种数据源创建 RDD，如 HDFS、本地文件系统、Amazon S3、其他 RDD、NoSQL 数据存储等。RDD 适应性很强，会在失败时自动重建。RDD 是通过惰性并行转换构建的，它们可能被缓存和分区，可能会也可能不会被具体化。

1.2.2 Spark SQL

Spark SQL 是构建在 Spark Core 之上的组件，被设计用来在结构化数据上执行查询、分析操作。因为 Spark SQL 具有灵活性、易用性和良好性能，现在它是 Spark 技术栈中最受欢迎和应用最多的组件。

Spark SQL 提供了一种名为 DataFrame 的分布式编程抽象。DataFrame 是分布式二维表集合，类似于 SQL 表或 Python 的 Pandas 库中的 DataFrame。可以从各种数据源构造

DataFrame，如 Hive、Parquet、JSON、关系型数据库（如 MySQL 等）及 Spark RDD。这些数据源可以具有各种模式。

Spark SQL 可以用于不同格式的 ETL 处理，然后进行即时查询分析。Spark SQL 附带一个名为 Catalyst 的优化器框架，它能解析 SQL 查询并自动进行优化以提高效率。Spark SQL 利用 Catalyst 优化器来执行许多分析数据库引擎中常见的优化类型。Spark SQL 的座右铭是 write less code, read less data, and let the optimizer do the hard work。

1.2.3　Spark Streaming 和 Structured Streaming

为了解决企业的数据实时处理需求，Spark 提供了流处理组件，它具有容错能力和可扩展性。Spark 支持实时数据流的实时数据分析。因为具有统一的 Spark 技术栈，所以在 Spark 中可以很容易地将批处理和交互式查询及流处理结合起来。

目前的 Spark 流处理模块实际上包含两代流处理引擎，分别是第 1 代的 Spark Streaming 和第 2 代的 Spark Structured Streaming，其中 Spark Streaming 是基于 RDD 的，而 Spark Structured Streaming 是基于 DataFrame 的。

Spark Streaming 和 Spark Structured Streaming 模块能够以高吞吐量和容错的方式处理来自各种数据源的实时流数据。数据可以从像 Kafka、Flume、Kinesis、Twitter、HDFS 或 TCP 套接字等资源中摄取。

在第 1 代 Spark Streaming 处理引擎中，主要的数据抽象是离散化流（DStream），它通过将输入数据分割成小批量（基于时间间隔）实现增量流处理模型，该模型可以定期地组合当前的处理状态以产生新的结果。换句话说，一旦传入的数据被分成微批，每批数据都将被视为一个 RDD，并将其复制到集群中，这样它们就可以被作为基本的 RDD 进行处理。通过在 DStreams 上应用一些更高级别的操作，可以产生其他的 DStream。Spark 流的最终结果可以被写回 Spark 所支持的各种数据存储，或者可以被推送到任何仪表盘进行可视化。

从 Spark 2.1 开始，Spark 引入了一个新的可扩展和容错的流处理引擎，称为结构化流（Structured Streaming）。结构化流构建在 Spark SQL 引擎之上，它进一步简化了流处理应用程序开发，处理流计算就像在静态数据上表示批计算一样。随着新的流数据的持续到来，结构化流引擎将自动地、增量地、持续地执行流处理逻辑。结构化流提供的一个新的重要特性是基于事件时间（Event Time）处理输入流数据的能力。在结构化流引擎中还支持端到端的、精确一次性保证。

1.2.4　Spark MLlib

MLlib 是 Spark 栈中内置的机器学习库，它的目标是使机器学习变得可扩展并且更容易。MLlib 提供了执行各种统计分析的必要功能，如相关性、抽样、假设检验等。该组件还开箱即用地提供了常用的机器学习算法，如分类、回归、聚类和协同过滤。

Spark 机器学习库实际上包含两种,分别是基于 RDD 的第 1 代机器学习库(在 Spark 0.8 引入)和基于 DataFrame 的第 2 代机器学习库(在 Spark 2.0 引入)。目前基于 RDD 的机器学习库已经处于维护模式,而第 2 代机器学习库受益于 Spark SQL 引擎中的 Catalyst 优化器和 Tungsten 项目,以及这些组件所提供的许多优化。

机器学习工作流程包括收集和预处理数据、构建和部署模型、评估结果和改进模型。在现实世界中,预处理步骤需要付出很大的努力。这些都是典型的多阶段工作流,涉及昂贵的中间读/写操作。通常,这些处理步骤可以在一段时间内多次执行。Spark 机器学习库引入了一个名为 ML 管道的新概念,以简化这些预处理步骤。管道是一个转换序列,其中一个阶段的输出是另一个阶段的输入,形成工作流链。

除了提供超过 50 种常见的机器学习算法之外,Spark MLlib 库提供了一些功能抽象,用于管理和简化许多机器学习模型构建任务,如特征化,用于构建、评估和调优模型的管道,以及模型的持久性(以帮助将模型从开发转移到生产环境)。

1.2.5　Spark GraphX

GraphX 是 Spark 的统一图分析框架。它被设计成一个通用的分布式数据流框架,取代了专门的图处理框架。它具有容错特性,并且利用内存进行计算。

GraphX 是一种嵌入式图处理 API,用于操纵图(例如,社交网络)和执行图并行计算(例如,谷歌的 Pregel)。它结合了 Spark 栈上的图并行和数据并行系统的优点,以统一探索性数据分析、迭代图计算和 ETL 处理。它扩展了 RDD 抽象来引入弹性分布式图(Resilient Distributed Graph,RDG),这是一个有向图,具有与每个顶点和边相关联的属性。

GraphX 组件包括一组通用图处理算法,包括 PageRank、K-Core、三角计数、LDA、连接组件、最短路径等。

目前的 Spark GraphX 组件是基于 RDD 的,社区正在构建基于 DataFrame(及其底层的 Catalyst 优化器和 Tungsten 项目)的 DataFrame 版本,称为 GraphFrames,目前还没有集成到 Spark 发行版中,但已经得到了广泛的应用。

1.2.6　SparkR

SparkR 项目将 R 的统计分析和机器学习能力与 Spark 的可扩展性集成在一起。它解决了 R 的局限性,即它处理单个机器内存中所需要的大量数据的能力。R 程序现在可以通过 SparkR 在分布式环境中进行扩展。

SparkR 实际上是一个 R 包,它提供了一个 R Shell 来利用 Spark 的分布式计算引擎。有了 R 丰富的用于数据分析的内置包,数据科学家可以交互式地分析大型数据集。

1.3 Spark 和 PySpark 架构原理

在深入了解 Spark/PySpark 的架构之前,一定要对 Spark 的核心概念和各种核心组件有深入的理解。这些核心概念和组件包括以下几方面:
(1) Spark 集群。
(2) 资源管理系统。
(3) Spark 应用程序。
(4) Spark Driver。
(5) Spark Executor。

1.3.1 Spark 集群和资源管理系统

Spark 本质上是一个分布式系统,设计的目的是用来高效、快速地处理海量数据。这个分布式系统通常部署在一个计算机集合上,称为 Spark 集群。为了高效和智能地管理这个集群,通常依赖于一个资源管理系统,如 Apache YARN 或 Apache Mesos。

资源管理系统内部有两个主要组件:集群管理器(Cluster Manager)和工作节点(Worker)。它有点像主从架构,其中集群管理器充当主节点,工作节点充当集群中的从节点。集群管理器跟踪与工作节点及其当前状态相关的所有信息。集群管理器维护的信息包括以下几方面:
(1) Worker 节点的状态(busy/available)。
(2) Worker 节点位置。
(3) Worker 节点内存。
(4) Worker 节点的总 CPU 核数。

集群管理器知道 Worker 节点的位置,其内存大小,以及每个 Worker 的 CPU 核数量。集群管理器的主要职责之一是管理 Worker 节点并根据 Worker 节点的可用性和容量为它们分配任务(Task)。每个 Worker 节点都向集群管理器提供自己可用的资源(如内存、CPU 等),并负责执行集群管理器分配的任务,如图 1-6 所示。

1.3.2 Spark 应用程序

Spark 应用程序也采用了主从架构,其中 Spark Driver 是 master,Spark Executors 是 slave。每个组件都作为一个独立的 JVM 进程运行在 Spark 集群上。Spark 应用程序由一个且只有一个 Spark Driver 和一个或多个 Spark Executors 组成,如图 1-7 所示。

Spark 应用程序由两部分组成,分别如下:
(1) 应用程序数据处理逻辑,使用 Spark API 表示。
(2) Spark 驱动程序(Spark Driver)。

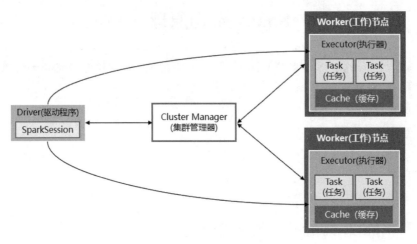

图 1-6　Spark 集群架构

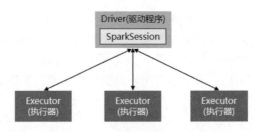

图 1-7　Spark 程序组成

应用程序数据处理逻辑(Task)是用 Java、Scala、Python 或 R 这几种语言编写的数据处理逻辑代码。它可以简单到几行代码来执行一些数据处理操作，也可以复杂到训练一个大型机器学习模型（这个模型需要多次迭代，可能要运行很多个小时才能完成）。

Spark 驱动程序用于运行应用程序 main()函数并创建 SparkSession 的进程。它是 Spark 应用程序的主控制器，负责组织和监控一个 Spark 应用程序的执行。它与集群管理器进行交互，以确定哪台机器来运行数据处理逻辑。Driver 及其子组件（Spark Session 和 Scheduler）负责的职责如下：

（1）向集群管理器请求内存和 CPU 资源。

（2）将应用程序逻辑分解为阶段（Stage）和任务（Task）。

（3）请求集群管理器启动名为 Executor 的进程（在运行 Task 的节点上）。

（4）向 Executor 发送 Tasks（应用程序数据处理逻辑），每个 Task 都在一个单独的 CPU Core 上执行。

（5）与每个 Executor 协调以收集计算结果并将它们合并在一起。

Spark 应用程序的入口点是通过一个名为 SparkSession 的类实现的。一旦 Driver 程序被启动之后，它将立即启动并配置 SparkSession 的一个实例。SparkSession 是访问 Spark

运行时的主要接口。SparkSession 对象连接到一个集群管理器，并提供了设置配置的工具，以及用于表示数据处理逻辑的 API。

除此之外，还需要一个客户端组件。客户端进程负责启动 Driver 程序。客户端进程可以是一个用于运行程序的 spark-submit 脚本，也可以是一个 spark-Shell 脚本或一个使用 Spark API 的自定义应用程序。客户端进程为 Spark 程序准备 class path 和所有配置选项，并将应用程序参数（如果有）传递给运行在 Driver 中的程序。

1.3.3 Spark Driver 和 Executor

每个 Spark 应用程序都有一个 Driver 进程。Spark Driver 包含多个组件，负责将用户代码转换为在集群上执行的实际作业，如图 1-8 所示。

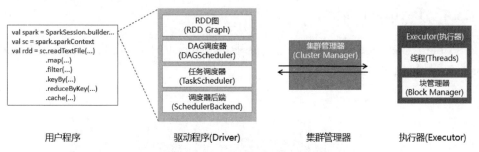

图 1-8 Spark 程序从代码到执行的转换过程

Spark Driver 中各个组件的功能如下。

（1）SparkContext：表示到 Spark 集群的连接，可用于在该集群上创建 RDD、累加器和广播变量。

（2）DAGScheduler：计算每个作业的 stages 的 DAG，并将它们提交给 TaskScheduler，确定任务的首选位置（基于缓存状态或 shuffle 文件位置），并找到运行作业的最优调度。

（3）TaskScheduler：负责将任务（Tasks）发送到集群，然后运行它们，在出现故障时重试，并减少掉队的情况。

（4）SchedulerBackend：用于调度系统的后端接口，允许插入不同的实现（Mesos、YARN、单机、本地）。

（5）BlockManager：提供用于在本地和远程将 block 块放入和检索到各种存储（如内存、磁盘和非堆）中的接口。

每个 Spark 应用程序都有一组 Executor 进程。每个 Executor 都是一个 JVM 进程，扮演 slave 角色，专门分配给特定的 Spark 应用程序，以便执行命令，并以任务的形式执行数据处理逻辑。每个任务在一个单独的 CPU 核心上执行。

Executors 驻留在 Worker 节点上，一旦集群管理器建立连接，就可以直接与 Driver 通信，接受来自 Driver 的任务，执行这些任务，并将结果返回给 Driver。每个 Executor 都有几个并行运行任务的任务槽。可以将任务槽的数量设置为 CPU 核心数量的 2 倍或 3 倍。尽管这些任务槽通常被称为 Spark 中的 CPU Cores，但它们是作为线程实现的，并且不需要与机器上的物理 CPU Cores 数量相对应。另外，每个 Spark Executor 都有一个 Block Manager 组件，Block Manager 负责管理数据块。这些块可以缓存 RDD 数据、中间处理的数据或广播数据。当可用内存不足时，它会自动将一些数据块移动到磁盘。Block Manager 还有一个职责，即执行跨节点的数据复制。

在启动一个 Spark 应用程序时，可以向资源管理器请求该应用程序所需的 Executor 数量，以及每个 Executor 应该拥有的内存大小和 CPU 核数。要计算出适当数量的 Executor、内存大小和 CPU 数量，需要了解将要处理的数据量、数据处理逻辑的复杂性及 Spark 应用程序完成处理逻辑所需的时间。

1.3.4　PySpark 架构

PySpark 构建在 Spark 的 Java API 之上。数据在 Python 中处理，在 JVM 中缓存和 Shuffle。PySpark 架构如图 1-9 所示。

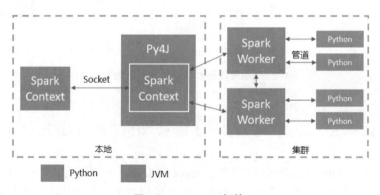

图 1-9　PySpark 架构

在 Python 驱动程序中，当 PySpark 的 Python 解释器启动时，SparkContext 使用 Py4J 启动 JVM 并创建 JavaSparkContext，并通过 Socket 套接字与之通信。JVM 作为实际的 Spark 驱动程序运行，并通过 JavaSparkContext 与集群中的 Spark Executor 进行通信。Py4J 只在驱动程序上用于 Python 和 JavaSparkContext 对象之间的本地通信（大型数据传输是通过一种不同的机制执行的）。

对 SparkContext 对象的 Python API 调用被转换为对 JavaSparkContext 的 Java API 调用。例如，PySpark 的 sc.textFile() 的实现分派为对 JavaSparkContext 的 .textFile() 方法的调用，该方法最终与 Spark Executor JVM 通信，以便从 HDFS 加载文本数据。

集群上的 Spark Executor 为每个 core 启动一个 Python 解释器，当它们需要执行用户代码时，通过管道与该解释器通信，便可以发送用户代码和要处理的数据。

本地 PySpark 客户端中的 PythonRDD 对应于本地 JVM 中的 PythonRDD 对象。与此 RDD 关联的数据实际上作为 Java 对象存在于 Spark JVM 中。例如，在 Python 解释器中运行 sc.textFile()将调用 JavaSparkContext 的 textFile()方法，该方法将数据作为 Java 字符串对象加载到集群中。类似地，使用 newAPIHadoopFile 加载 Parquet/Avro 文件将以 Java Avro 对象的形式加载对象。

PySpark 目前使用 Python cPickle 序列化器序列化数据。PySpark 使用 cPickle 序列化数据，因为它相当快，并且支持几乎任何 Python 数据结构。当在 Python RDD 上进行 API 调用时，任何相关的代码（例如 Python lambda 函数）都会通过 cloudpickle（一个由 PiCloud 构建的定制模块）进行序列化，并分发给执行器，然后将数据从 Java 对象转换为与 Python 兼容的表示（例如 pickle 对象），并通过管道流向与 executor 相关的 Python 解释器。

任何必要的 Python 处理都在解释器中执行，结果数据作为 RDD（默认情况下作为 pickle 对象）存储回 JVM 中。

1.4　Spark 程序部署模式

Spark Driver 程序的运行有两种基本的方式：集群部署模式和客户端部署模式。

在集群部署模式下，Driver 进程作为一个单独的 JVM 进程运行在集群中，集群负责管理其资源(主要是 JVM 堆内存)，如图 1-10 所示。

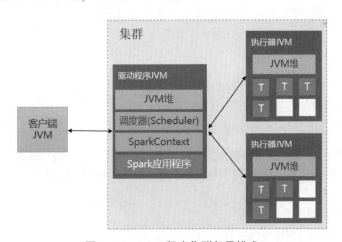

图 1-10　Spark 程序集群部署模式

在客户端部署模式下，Driver 进程运行在客户端的 JVM 进程中，并与受集群管理的 Executors 进行通信，如图 1-11 所示。

选择不同的部署模式将影响如何配置 Spark 和客户端 JVM 的资源需求。通常使用客户端部署模式，在这种模式下，用户可以在客户端获取并显示作业执行情况。

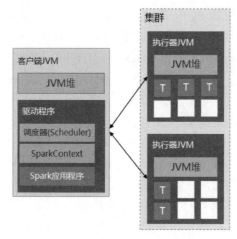

图 1-11　Spark 程序客户端部署模式

1.5　安装和配置 Spark 集群

为了学习 Spark，最好在自己的计算机上本地安装 Spark。通过这种方式，用户可以轻松地尝试 Spark 特性或使用小型数据集测试数据处理逻辑。

Apache Spark 是用 Scala 编程语言编写的，而 Scala 需要运行在 JVM 上，因此，在安装 Spark 之前，应确保已经在自己的计算机上安装了 Java（JDK 8）。

1.5.1　安装 Spark

要在自己的计算机上本地安装 Spark，建议按以下步骤操作。

（1）将预先打包的二进制文件下载到~/software 目录下，它包含运行 Spark 所需的 JAR 文件。下载网址为 http://spark.apache.org/downloads.html，下载界面如图 1-12 所示。

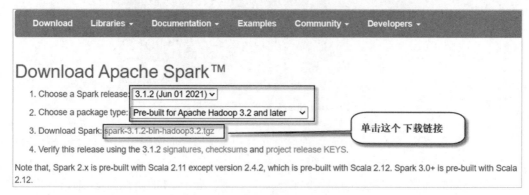

图 1-12　选择要下载的 Spark 版本

（2）将下载的压缩包解压缩到~/bigdata/目录下，并重命名为 spark-3.1.2，命令如下：

```
$ cd ~/bigdata
$ tar -zxvf ~/software/spark-3.1.2-bin-hadoop3.2.tgz
$ mv spark-3.1.2-bin-hadoop3.2 spark-3.1.2
```

（3）配置环境变量。使用任意编辑器（本书使用的编辑器是 nano）打开/etc/profile 文件，命令如下：

```
$ cd
$ sudo nano /etc/profile
```

在文件最后，添加以下内容：

```
export SPARK_HOME=/home/hduser/bigdata/spark-3.1.2
export PATH=$SPARK_HOME/bin:$PATH
```

保存文件并关闭。

（4）执行/etc/profile 文件使配置生效，命令如下：

```
$ source /etc/profile
```

1.5.2　了解 Spark 目录结构

查看解压缩后的 Spark 安装目录，会发现其中包含多个目录，如图 1-13 所示。

名称	修改日期	类型	大小
bin	2021-08-05 16:44	文件夹	
conf	2021-08-05 16:44	文件夹	
data	2021-08-05 16:44	文件夹	
examples	2021-08-05 16:44	文件夹	
jars	2021-08-05 16:44	文件夹	
kubernetes	2021-08-05 16:44	文件夹	
licenses	2021-08-05 16:44	文件夹	
python	2021-08-05 16:44	文件夹	
R	2021-08-05 16:44	文件夹	
sbin	2021-08-05 16:44	文件夹	
yarn	2021-08-05 16:44	文件夹	
LICENSE	2021-05-24 12:45	文件	23 KB
NOTICE	2021-05-24 12:45	文件	57 KB
README.md	2021-05-24 12:45	MD 文件	5 KB
RELEASE	2021-05-24 12:45	文件	1 KB

图 1-13　Spark 安装目录

其中几个主要目录的作用见表 1-1。

表 1-1 Spark 安装目录说明

目录	描述
bin	包含各种可执行文件,以启动 Scala 或 Python 中的 Spark Shell、提交 Spark 应用程序和运行 Spark 示例
conf	包含用于 Spark 的各种配置文件
data	包含用于各种 Spark 示例的小示例数据文件
examples	包含所有 Spark 示例的源代码和二进制文件
jars	包含运行 Spark 所需的二进制文件
sbin	包含管理 Spark 集群的可执行文件

1.5.3 配置 Spark/PySpark 集群

Spark 的配置文件位于 conf 目录下。conf 目录下会默认存放着几个 Spark 的配置示例模板文件,见表 1-2。

表 1-2 Spark conf 目录下的模板文件说明

文件名	说明
fairscheduler.xml.template	Hadoop 公平调度配置模板文件
log4j.properties.template	Spark Driver 节点的日志配置模板文件
metrics.properties.template	Metrics 系统性能监控工具的配置模板文件
spark-defaults.conf.template	Spark 运行时的属性配置模板文件
spark-env.sh.template	Spark 环境变量配置模板文件
workers.template	Spark 集群的 Worker 节点配置模板文件

这些模板文件均不会被 Spark 读取,需要将 .template 后缀去除,Spark 才会读取这些文件。在这些配置文件中,在 Spark 集群中主要需要关注的是 spark-env.sh、spark-defaults.conf 和 workers 这几个配置文件。

接下来,对 Spark 进行配置,包括其运行环境和集群配置参数。建议按以下步骤执行。

(1) 从模板文件 spark-env.sh.template 复制一份,并重命名为 spark-env.sh,命令如下:

```
$ cd ~/spark-3.1.2/conf/
$ cp spark-env.sh.template spark-env.sh
```

(2) 使用编辑器打开 spark-env.sh 进行编辑,命令如下:

```
$ nano spark-env.sh
```

加入以下内容:

```
export JAVA_HOME=/usr/local/jdk1.8.0_251
export HADOOP_CONF_DIR=/home/hduser/bigdata/hadoop-3.2.2/etc/hadoop
export YARN_CONF_DIR=/home/hduser/bigdata/hadoop-3.2.2/etc/hadoop
```

```
export SPARK_HOME=/home/hduser/bigdata/spark-3.1.2
export
SPARK_DIST_CLASSPATH=$(/home/hduser/bigdata/hadoop-3.2.2/bin/hadoop
classpath)

#PySpark 配置
export
PYTHONPATH=$SPARK_HOME/python:$SPARK_HOME/python/lib/py4j-0.10.7-src.zip
export PYSPARK_PYTHON=/opt/anaconda3/bin/python
#如果要使用 Jupyter Notebook，则应将下面这句注释掉
export PYSPARK_DRIVER_PYTHON=/opt/anaconda3/bin/python
```

注意，应将 JDK 和 Hadoop 修改为读者自己安装的版本和路径，然后保存并关闭文件。

（3）从模板文件 workers.template 复制一份，并重命名为 workers，命令如下：

```
$ cp workers.template workers
$ nano workers
```

然后将其中的 localhost 删除，将集群中所有 Worker 节点的机器名或 IP 地址填写进去，一个一行。例如，笔者的机器名是 xueai8，因此 workers 文件的内容修改如下：

```
xueai8
```

（4）将配置好的 Spark 目录复制到集群中其他的节点，放在相同的路径下。

1.5.4　验证 PySpark 安装

PySpark 配置完成后就可以直接使用了，不需要像 Hadoop 运行那样启动命令。下面通过运行 PySpark 自带的蒙特卡洛求圆周率 π 值示例，以验证 Spark 是否安装成功。

PySpark 支持以本地模式运行 PySpark 程序，或者以集群模式运行 PySpark 程序。

在本地（Local）模式下，进入 Spark 主目录，直接使用 spark-submit 命令来提交示例程序 pi.py 运行即可，命令如下：

```
$ cd ~/bigdata/spark-3.1.2
$ ./bin/spark-submit --master local[*] examples/src/main/python/pi.py
```

执行过程如图 1-14 所示。

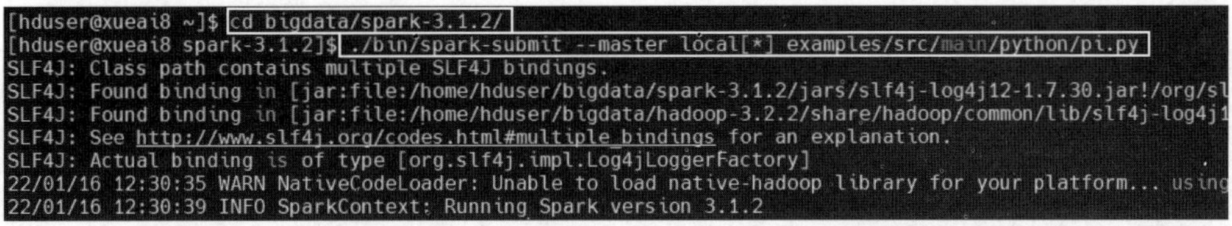

图 1-14　以本地模式运行 PySpark 自带的蒙特卡洛求圆周率 π 值程序

执行结果如图 1-15 中所示。

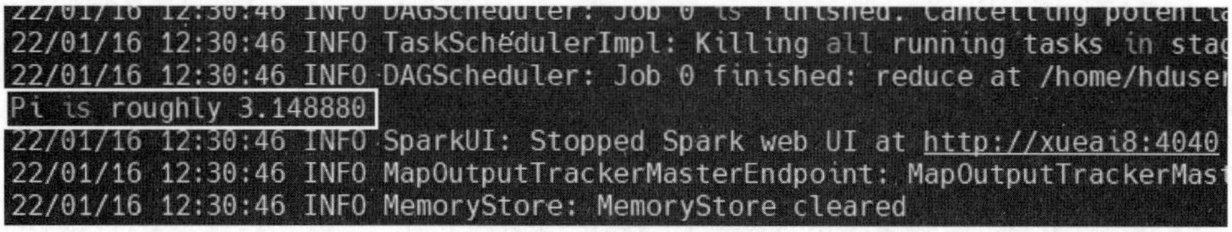

图 1-15 计算出的π值

或者，以 standalone 模式（需要先执行 ./sbin/start-all.sh 启动 Spark 集群）来执行，命令如下：

```
$ cd ~/bigdata/spark-3.1.2
$ ./sbin/start-all.sh

$ ./bin/spark-submit --master spark://xueai8:7077
examples/src/main/python/pi.py
```

说明：

（1）--master 参数用于指定要连接的集群管理器，这里是 standalone 模式。

（2）最后一个参数是所提交的 Python 程序。

执行过程如图 1-16 所示。

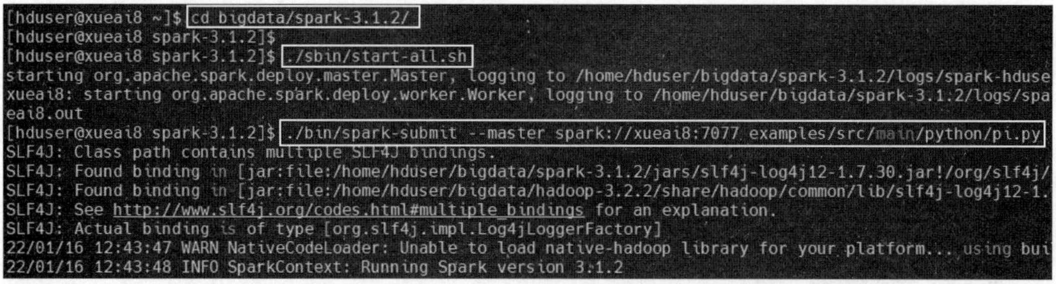

图 1-16 以 standalone 模式运行 PySpark 自带的蒙特卡洛求圆周率π值程序

执行结果如图 1-17 所示。

图 1-17 蒙特卡洛求圆周率π值程序执行结果

1.6 配置 Spark 历史服务器

7min

当用户提交一个 Spark 应用程序时，会创建一个 SparkContext，它提供了 Spark Web UI 来监视应用程序的执行。监控包括以下内容：

（1）Spark 使用的配置。
（2）Spark Jobs、stages 和 tasks 细节。
（3）DAG 执行。
（4）Driver 和 Executor 资源利用率。
（5）应用程序日志等。

当应用程序完成处理后，SparkContext 将终止，因此 Web UI 也将终止。如果用户还想看到已经完成的应用程序的监控信息，则用户就必须配置一个单独的 Spark 历史记录服务器。

Spark History Server（历史记录服务器）是一个用户界面，用于监控已完成的 Spark 应用程序的指标和性能。它是 Spark 的 Web UI 的扩展，保存了所有已完成的应用程序的历史（事件日志）及其运行时信息，允许用户稍后检查度量并及时监控应用程序。当用户试图改进应用程序的性能时，历史度量非常有用，用户可以将以前的运行度量与最近的运行度量进行比较。

Spark History Server 可以保存事件日志的历史信息，用于如下操作：

（1）所有通过 spark-submit 提交的应用程序。
（2）通过 REST API 提交的应用程序。
（3）运行的每个 spark-shell。
（4）通过 Notebook 提交的作业。

1.6.1 历史服务器配置

为了存储所有提交的应用程序的事件日志，首先，Spark 需要在应用程序运行时收集信息。默认情况下，Spark 不收集事件日志信息。用户可以通过在 spark-defaults.conf 文件中设置下面的配置来启用它：

（1）将配置项 spark.eventLog.enabled 设置为 true 来启用事件日志功能。
（2）使用 spark.history.fs.logDirectory 和 spark.eventLog.dir 指定存储事件日志历史的位置。默认位置为 file://tmp/spark-events。需要提前创建该目录。

建议按以下步骤操作。

（1）在 Spark 安装目录下，创建存储事件日志历史的文件夹，命令如下：

```
$ cd ~/bigdata/spark-3.1.2
$ mkdir spark-events
```

（2）在 spark-defaults.conf 文件中启用事件日志记录功能。首先从模板文件复制一份，并去掉 .template 后缀，得到 spark-defaults.conf 文件，命令如下：

```
$ cd ~/bigdata/spark-3.1.2/conf
$ cp spark-defaults.conf.template spark-defaults.conf

#编辑 spark-defaults.conf
$ nano spark-defaults.conf
```

将下面的配置项添加到文件的末尾：

```
#启用存储事件日志
spark.eventLog.enabled true

#存储事件日志的位置
spark.eventLog.dir file:///home/hduser/bigdata/spark-3.1.2/spark-events

#历史服务器读取事件日志的位置
spark.history.fs.logDirectory file:///home/hduser/bigdata/spark-3.1.2/spark-events

#日志记录周期
spark.history.fs.update.interval 10s

#历史服务器端口号
spark.history.ui.port 18080
```

Spark 通过为每个应用程序创建一个子目录来保存运行的每个应用程序的历史，并在该目录中记录与该应用程序相关的事件。

还可以设置如 HDFS 目录这样的位置，以便历史文件可以被历史服务器读取，示例代码如下：

```
spark.eventLog.dir hdfs://xueai8:8020/user/spark/spark-events
```

如果想要为 org.apache.spark.deploy.history 的日志记录器（logger）启用 INFO 日志记录级别，则可以将下面这行配置添加到 conf/log4j.properties 文件中：

```
log4j.logger.org.apache.spark.deploy.history=INFO
```

当启用事件日志记录时，默认的行为是保存所有日志，这会导致可用存储空间随时间的增长而变小。要启用自动清理功能，建议编辑 spark-defaults.conf 文件并添加以下选项：

```
#设置日志清除周期
spark.history.fs.cleaner.enabled true
spark.history.fs.cleaner.interval 1d
spark.history.fs.cleaner.maxAge 7d
```

对于这些设置，将启用自动清理，每天执行清理，并删除超过 7 天的日志。

1.6.2 启动 Spark 历史服务器

要启动 Spark 历史服务器，打开一个终端窗口，执行的命令如下：

```
$ SPARK_HOME/sbin/start-history-server.sh
```

如果未明确指定，start-history-server.sh 会使用默认配置文件 spark-defaults.conf。另外，它也可以接受--properties-file [propertiesFile]命令行选项，该选项用于指定带有自定义 Spark 属性的属性文件，命令如下：

```
$ SPARK_HOME/sbin/start-history-server.sh --properties-file
history.properties
```

使用更显式的 spark-class 方法来启动 Spark History Server，可以更容易地跟踪执行，因为可以看到日志被打印到标准输出（直接输出到终端），命令如下：

```
$ SPARK_HOME/bin/spark-class org.apache.spark.deploy.history.HistoryServer
```

如果在 Windows 系统上运行 Spark，则启动历史记录服务器的命令如下：

```
$ SPARK_HOME/bin/spark-class.cmd
org.apache.spark.deploy.history.HistoryServer
```

默认情况下，历史记录服务器监听 18080 端口，可以使用 http://localhost:18080/从浏览器访问它，如图 1-18 所示。

图 1-18 监控 Spark 应用程序

在每个 App ID 上单击，可以得到该 Spark 应用程序的 job、stage、task、executor 的详细环境信息。

要停止 Spark 历史服务器，则需要在终端窗口中执行的命令如下：

```
$ $SPARK_HOME/sbin/stop-history-server.sh
```

使用 Spark 历史服务器，用户可以跟踪所有已完成的应用程序，因此需要启用此功能。在进行性能调优时，这些指标会派上用场。

1.7 使用 PySpark Shell 进行交互式分析

在进行数据分析时，通常需要进行交互式数据探索和数据分析。为此，PySpark 提供了一个交互式的工具 PySpark Shell。通过 Spark Shell，用户可以和 PySpark 进行实时交互，以进行数据探索、数据清洗和整理及交互式数据分析等工作。

使用 PySpark Shell 的命令格式如下：

```
$ ./bin/pyspark [options]
```

要查看完整的参数选项列表，可以执行 pyspark --help 命令，命令如下：

```
$ pyspark --help
```

1.7.1 运行模式--master

Spark/PySpark 的运行模式取决于传递给 SparkContext 的 Master URL 的值。参数选项--master 表示当前的 PySpark Shell 要连接到哪个 master（告诉 Spark/PySpark 使用哪种集群类型）。

如果是 local[*]，则表示使用本地模式启动 PySpark Shell，其中，中括号内的星号(*)表示需要使用几个 CPU 核，也就是启动几个线程模拟 Spark 集群。如果不指定，则默认为 local。

当运行 PySpark Shell 命令时，可以像下面这样定义这个参数：

```
$ pyspark --master <master_connection_url>
```

<master_connection_url>根据所使用的集群的类型而变化。Master URL(--master 参数）的值见表 1-3。

表 1-3　Spark 本地运行时的 --master 参数值

部署模式	Master URL	说明
本地部署	local	使用一个 Worker 线程本地化运行 Spark（完全不并行），相当于 local[1]
	local[N]	使用 N 个 Worker 线程本地化运行 Spark（最好 N 等于机器的 CPU 核数）
	local[*]	使用逻辑 CPU 个数数量的线程来本地化运行 Spark

当集群部署时，Master URL(--master 参数）的值见表 1-4。

表 1-4　Spark 集群运行时的 --master 参数值

部署模式	集群管理器	Master URL	说　　明	示　　例
集群部署	standalone	spark://HOST:PORT	连接到指定的 Spark Master	spark://master:7077
	on YARN	yarn	连接到 YARN 集群	yarn
	on Mesos	mesos://HOST:PORT	连接到指定的 Mesos 集群	mesos://master:5050

1.7.2　启动和退出 PySpark Shell

以下操作均在终端窗口中进行。

（1）启动 PySpark Shell 方式一：local 模式。以 local 模式启动 PySpark Shell，命令如下：

```
$ cd ~/bigdata/spark-3.1.2
$ ./bin/pyspark
```

启动过程如图 1-19 所示。

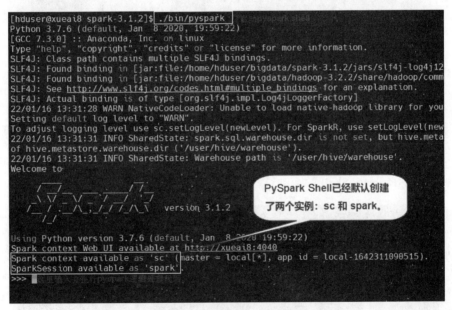

图 1-19　以 local 模式启动 PySpark Shell 过程

从图 1-19 中可以看出，PySpark Shell 在启动时，已经帮用户创建好了 SparkSession 对象的实例 spark（实际上也包括 SparkContext 对象的实例 sc），用户可以在 PySpark Shell 中直接使用 sc 和 spark 这两个对象。另外，默认情况下，启动的 PySpark Shell 采用 local 部署模式。

在创建 SparkContext 对象的实例 sc 之后，它将等待资源。一旦资源可用，sc 将设置内部服务并建立到 Spark 执行环境的连接。

退出 PySpark Shell，命令如下：

```
>>> exit()
```

（2）启动 PySpark Shell 方式二：standalone 模式。

首先要确保启动了 Spark 集群。启动 Spark 集群的命令如下：

```
$ cd ~/bigdata/spark-3.1.2
$ ./sbin/start-all.sh
```

使用 jps 命令查看启动的进程。如果有 master 和 worker 进程，则说明 Spark 集群已经启动，如图 1-20 所示。

图 1-20 使用 jps 命令查看 Spark 启动的进程

然后启动 PySpark Shell，并指定 --master spark://xueai8:7077 参数，以 standalone 模式运行，命令如下：

```
$ ./bin/pyspark --master spark://xueai8:7077
```

在 Master URL 中指定的 xueai8 是笔者当前的机器名，读者需要根据实际情况修改为自己的机器名。PySpark Shell 的启动过程如图 1-21 所示。

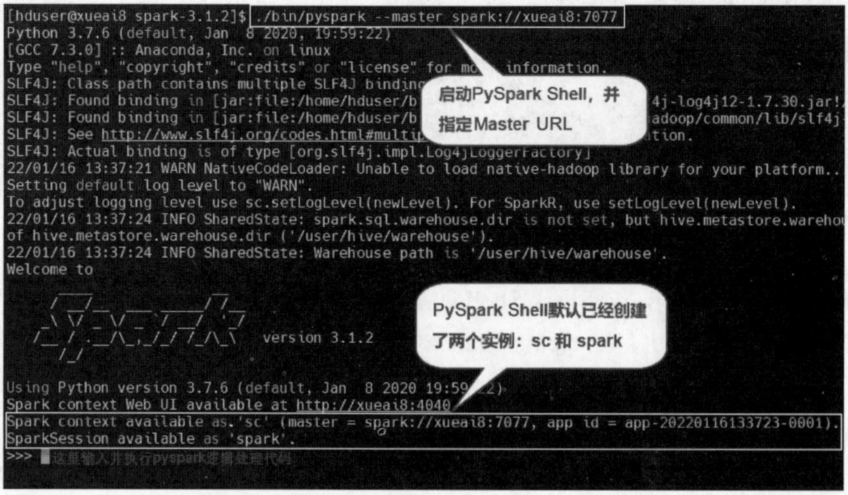

图 1-21 以集群模式启动 PySpark Shell 过程

1.7.3 PySpark Shell 常用命令

可以在 PySpark Shell 里面输入 Python 代码进行调试，如图 1-22 所示。

图 1-22　在 PySpark Shell 里面输入 Python 代码进行调试

可以在 PySpark Shell 中键入以下命令，查看 PySpark Shell 常用的命令，命令如下：

```
>>> help()
```

显示帮助内容，如图 1-23 所示。

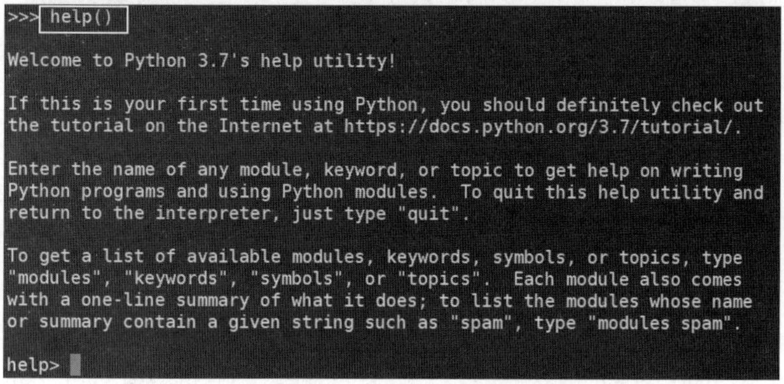

图 1-23　查看 PySpark Shell 帮助信息

可以在 help 模式下键入模块的名称，查看该模块的使用说明。例如，要查看 sql 模块的使用说明，命令如下：

```
help> pyspark.sql
```

使用说明的界面如图 1-24 所示。

可以按 q 键退出 help 帮助界面，也可以输入 exit() 命令，返回 Shell 命令行。

1.7.4 SparkContext 和 SparkSession

在 Spark 2.0 中引入了 SparkSession 类，以提供与底层 Spark 功能交互的单一入口点。这个类具有用于从非结构化文本文件及各种格式的结构化数据和二进制数据文件读取数据的 API，包括 JSON、CSV、Parquet、ORC 等。此外，SparkSession 还提供了检索和设置与 Spark 相关的配置的功能。

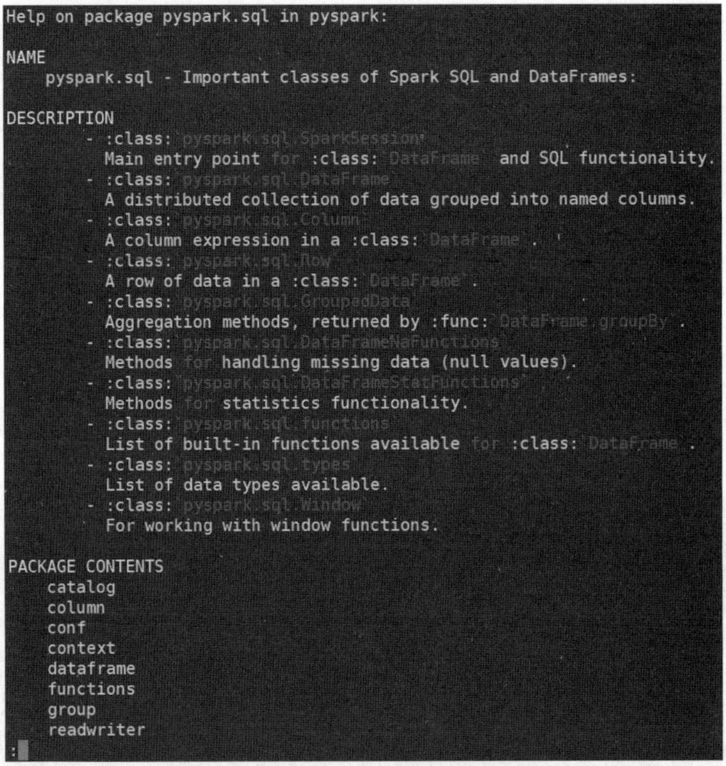

```
Help on package pyspark.sql in pyspark:

NAME
    pyspark.sql - Important classes of Spark SQL and DataFrames:

DESCRIPTION
        - :class: pyspark.sql.SparkSession
          Main entry point for :class: DataFrame and SQL functionality.
        - :class: pyspark.sql.DataFrame
          A distributed collection of data grouped into named columns.
        - :class: pyspark.sql.Column
          A column expression in a :class: DataFrame .
        - :class: pyspark.sql.Row
          A row of data in a :class: DataFrame .
        - :class: pyspark.sql.GroupedData
          Aggregation methods, returned by :func: DataFrame.groupBy .
        - :class: pyspark.sql.DataFrameNaFunctions
          Methods for handling missing data (null values).
        - :class: pyspark.sql.DataFrameStatFunctions
          Methods for statistics functionality.
        - :class: pyspark.sql.functions
          List of built-in functions available for :class: DataFrame .
        - :class: pyspark.sql.types
          List of data types available.
        - :class: pyspark.sql.Window
          For working with window functions.

PACKAGE CONTENTS
    catalog
    column
    conf
    context
    dataframe
    functions
    group
    readwriter
```

图 1-24 查看 **pyspark.sql** 模块的使用说明

SparkContext 在 Spark 2.0 中，成为 SparkSession 的一个属性对象。

一旦一个 PySpark Shell 成功启动，它就会初始化一个 SparkSession 类的实例（名为 spark），以及一个 SparkContext 类的实例（名为 sc）。这个 spark 变量和 sc 变量可以在 PySpark Shell 中直接使用。可以使用 type() 函数来验证这一点，命令如下：

```
>>> type(sc)
>>> type(spark)
```

执行过程如图 1-25 所示。

```
>>> type(sc)
<class 'pyspark.context.SparkContext'>
>>> type(spark)
<class 'pyspark.sql.session.SparkSession'>
>>>
```

图 1-25 验证 **sc** 和 **spark** 实例变量

查看当前使用的 Spark/PySpark 版本号，使用的命令如下：

```
>>> spark.version
>>> sc.version
```

执行过程如图 1-26 所示。

```
>>> spark.version
'2.4.6'
>>> sc.version
'2.4.6'
>>> spark.sparkContext.version
'2.4.6'
>>>
```

图 1-26　查看当前使用的 Spark/PySpark 版本号

要查看在 PySpark Shell 中配置的默认配置，可以访问 Spark 的 conf 变量。下面的命令可显示 PySpark Shell 中默认的配置信息：

```
>>> for item in spark.sparkContext.getConf().getAll():
...     print(item)
```

执行过程如图 1-27 所示。

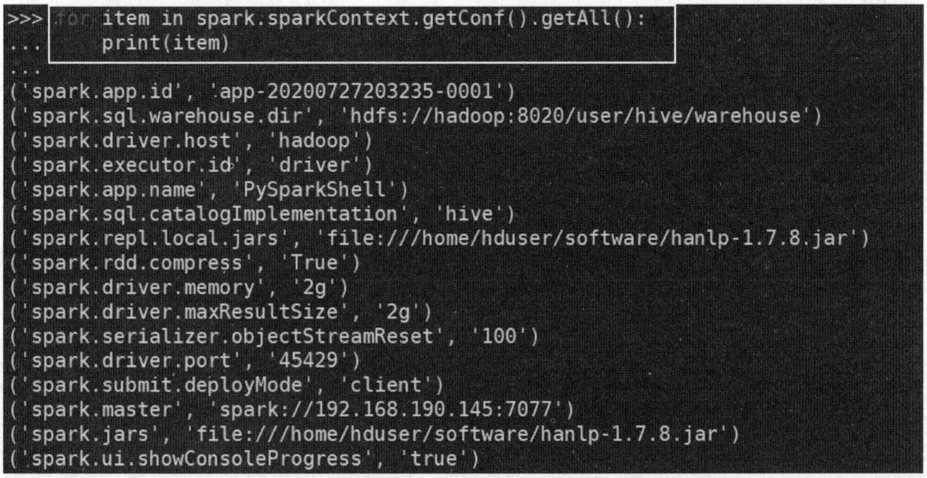

```
>>> for item in spark.sparkContext.getConf().getAll():
...     print(item)
...
('spark.app.id', 'app-20200727203235-0001')
('spark.sql.warehouse.dir', 'hdfs://hadoop:8020/user/hive/warehouse')
('spark.driver.host', 'hadoop')
('spark.executor.id', 'driver')
('spark.app.name', 'PySparkShell')
('spark.sql.catalogImplementation', 'hive')
('spark.repl.local.jars', 'file:///home/hduser/software/hanlp-1.7.8.jar')
('spark.rdd.compress', 'True')
('spark.driver.memory', '2g')
('spark.driver.maxResultSize', '2g')
('spark.serializer.objectStreamReset', '100')
('spark.driver.port', '45429')
('spark.submit.deployMode', 'client')
('spark.master', 'spark://192.168.190.145:7077')
('spark.jars', 'file:///home/hduser/software/hanlp-1.7.8.jar')
('spark.ui.showConsoleProgress', 'true')
```

图 1-27　查看在 PySpark Shell 中配置的默认配置

或者，通过另一种方式查看，命令如下：

```
>>> for item in sc.getConf().getAll():
...     print(item)
```

执行过程如图 1-28 所示。

1.7.5　Spark Web UI

每次初始化 SparkSession 对象时，Spark 都会启动一个 Web UI，提供关于 Spark 环境和作业执行统计信息的信息。Web UI 的默认端口是 4040，但是如果这个端口已经被占用

（例如，被另一个 Spark Web UI 占用），Spark 会增加该端口号值，直到找到一个空闲的端口号为止。

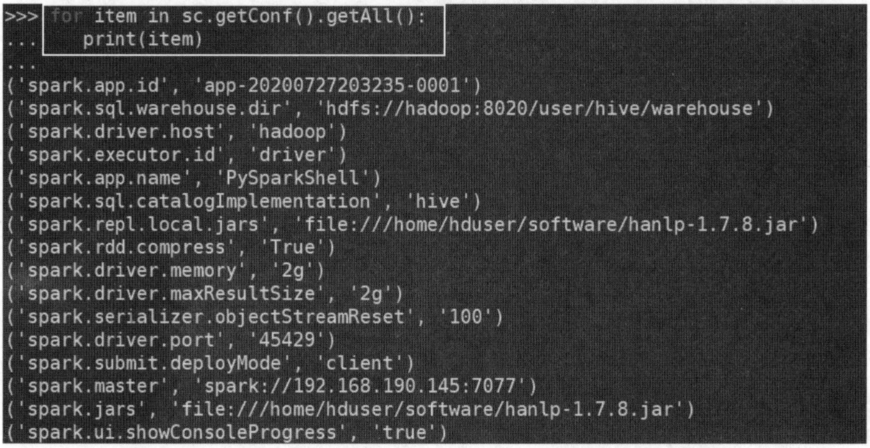

```
>>> for item in sc.getConf().getAll():
...     print(item)
...
('spark.app.id', 'app-20200727203235-0001')
('spark.sql.warehouse.dir', 'hdfs://hadoop:8020/user/hive/warehouse')
('spark.driver.host', 'hadoop')
('spark.executor.id', 'driver')
('spark.app.name', 'PySparkShell')
('spark.sql.catalogImplementation', 'hive')
('spark.repl.local.jars', 'file:///home/hduser/software/hanlp-1.7.8.jar')
('spark.rdd.compress', 'True')
('spark.driver.memory', '2g')
('spark.driver.maxResultSize', '2g')
('spark.serializer.objectStreamReset', '100')
('spark.driver.port', '45429')
('spark.submit.deployMode', 'client')
('spark.master', 'spark://192.168.190.145:7077')
('spark.jars', 'file:///home/hduser/software/hanlp-1.7.8.jar')
('spark.ui.showConsoleProgress', 'true')
```

图 1-28　另一种方式查看在 PySpark Shell 中配置的默认配置

在启动一个 PySpark Shell 时，将看到与此类似的输出行（除非关闭了 INFO log 消息）：

```
Spark context Web UI available at http://xueai8:4040
```

启动 PySpark Shell 时会输出行信息，如图 1-29 所示。

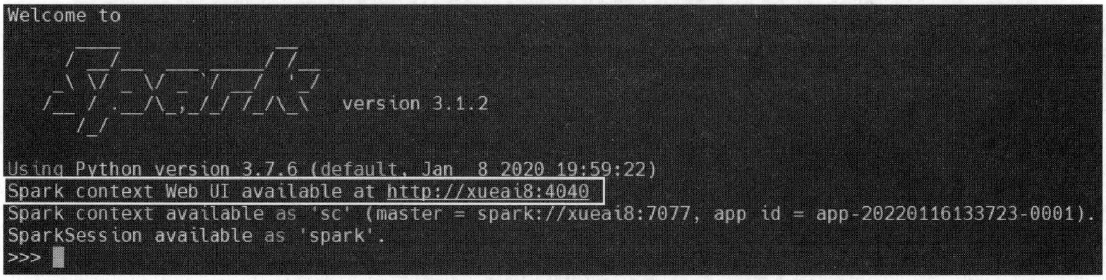

```
Welcome to
      ____              __
     / __/__  ___ _____/ /__
    _\ \/ _ \/ _ `/ __/  '_/
   /__ / .__/\_,_/_/ /_/\_\   version 3.1.2
      /_/

Using Python version 3.7.6 (default, Jan  8 2020 19:59:22)
Spark context Web UI available at http://xueai8:4040
Spark context available as 'sc' (master = spark://xueai8:7077, app id = app-20220116133723-0001).
SparkSession available as 'spark'.
>>>
```

图 1-29　初始化 SparkSession 对象时，Spark 都会启动一个 Web UI

注意：可以通过将 spark.ui.enabled 配置参数设为 false 来禁用 Spark Web UI。可以用 spark.ui.port 参数来改变它的端口。

Spark Web UI 欢迎页面的一个示例，如图 1-30 所示。

这个 Web UI 是从一个 PySpark Shell 启动的，所以它的名字被设置为 PySpark Shell，如图 1-30 右上角所示。

在运行 PySpark 命令时，也可以使用 --conf spark.app.name=<new_name> 在命令行上设置程序名称，但不能在启动 PySpark Shell 时更改应用程序名称。在这种情况下，它总是默认为 PySparkShell。

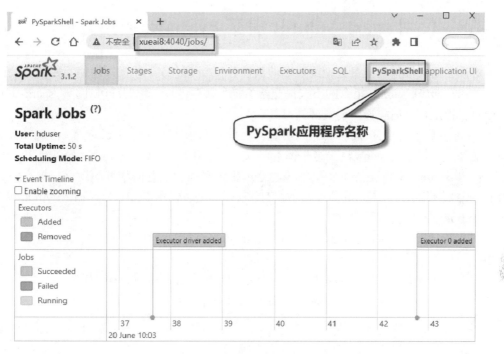

图 1-30　Spark Web UI 欢迎页面

在 Spark Web UI 的 Environment 页面，可以查看影响 PySpark 应用程序的配置参数的完整列表，如图 1-31 所示。

图 1-31　影响 PySpark 应用程序的配置参数

1.8 使用 spark-submit 提交 PySpark 应用程序

对于公司大数据的批量处理或周期性数据分析/处理任务，通常采用编写好的 Spark 程序，并通过 spark-submit 指令的方式提交给 Spark 集群进行具体的任务计算，spark-submit 指令可以指定一些向集群申请资源的参数。

Spark 安装包附带有 spark-submit.sh 脚本文件（适用于 Linux、Mac）和 spark-submit.cmd 命令文件（适用于 Windows）。这些脚本可以在 $SPARK_HOME/bin 目录下找到。

spark-submit 命令是一个实用程序，通过指定选项和配置向集群中运行或提交 Spark 或 PySpark 应用程序（或 job 作业），提交的应用程序可以用 Scala、Java 或 Python 编写。spark-submit 命令支持以下功能。

（1）在 YARN、Kubernetes、Mesos、Standalone 等不同的集群管理器上提交 Spark 应用。

（2）在 client 客户端部署模式或 cluster 集群部署模式下提交 Spark 应用。

下面是一个带有最常用命令选项的 spark-submit 命令：

```
./bin/spark-submit \
  --master <master-url> \
  --deploy-mode <deploy-mode> \
  --conf <key>=<value> \
  --driver-memory <value>g \
  --executor-memory <value>g \
  --executor-cores <number of cores> \
  --jars <comma separated dependencies> \
  --class <main-class> \
  <python file> \
  [application-arguments]
```

1.8.1 spark-submit 指令的各种参数说明

在 Linux 环境下，可通过 spark-submit --help 命令来了解 spark-submit 指令的各种参数说明，命令如下：

```
$ cd ~/bigdata/spark-3.1.2
$ ./bin/spark-submit --help
```

spark-submit 的完整语法如下：

```
$ ./bin/spark-submit [options] <app jar | python file> [app options]
```

其中，options 的主要标志参数说明如下。

（1）--master：指定使用哪个集群管理器来运行应用程序。Spark 目前支持 YARN、Mesos、Kubernetes、Standalone 和 local。

（2）--deploy-mode：是否要在本地("client")启动驱动程序，或者在集群中("cluster")的一台 worker 机器上启动驱动程序。在 client 模式下，驱动程序在调用 spark-submit 的机器上本地运行。客户端模式主要用于交互和调试目的。在 cluster 模式下，驱动程序会被发送到集群的一个 worker 节点上运行，该节点在应用程序的 Spark Web UI 上显示为 driver。集群模式用于运行生产作业。默认为 client 模式。

（3）--class：应用程序的主类（带有 main 方法的类），如果运行 Java 或 Scala 程序。

（4）--name：应用程序易读的名称，这将显示在 Spark 的 Web UI 上。

（5）--jars：一系列 JAR 文件的列表，会被上传并设置到应用程序的 classpath 上。如果应用程序依赖于少量的第三方 JAR 包，则可以将它们加到这里（逗号分隔）。

（6）--files：使用逗号分隔的文件。通常，这些文件可以来自 resource 文件夹。使用此选项，Spark 将所有这些文件提交到集群。这个标志参数可被用于想要分布到每个节点上的数据文件（注：使用--files 指定的文件会被上传到集群）。

（7）--py-files：一系列文件的列表，会被添加到应用程序的 PYTHONPATH。这可以包括.ipynb、.egg 或.zip 文件。

（8）--executor-memory：executor 使用的内存数量，以字节为单位。可以指定不同的后缀，如"512m"或"15g"。

（9）--driver-memory：driver 进程所使用的内存数量，以字节为单位。可以指定不同的后缀，如"512m"或"15g"。

（10）--verbose：显示详细信息。例如，将 spark 应用程序使用的所有配置写入日志文件。

（11）--config：用于指定应用程序配置、shuffle 参数、运行时配置。

关于 driver 和 executor 资源（如 CPU 核和内存）配置，需要深入了解一下。在提交应用程序时，用户可以指定需要为 driver 和 executor 提供多少内存和核数。这些资源相关的选项说明见表 1-5。

表 1-5 Spark 程序提交时的资源选项

选　　项	说　　明
--driver-memory	Spark driver 需要使用的内存
--driver-cores	Spark driver 需要使用的 CPU 内核数
--num-executors	要使用的执行器 executor 总数
--executor-memory	executor 进程使用的内存量
--executor-cores	executor 进程使用的 CPU 核数
--total-executor-cores	要使用的执行器 executor 内核总数

spark-submit 使用--config 支持几种配置，这些配置用于指定应用程序配置、shuffle 参数、运行时配置。这些配置对于用 Java、Scala 和 Python 编写的 Spark 应用程序（PySpark）来讲是相同的。几种常用的配置 key 及其说明见表 1-6。

表1-6 spark-submit 使用 --config 支持几种配置

选 项	说 明
spark.sql.shuffle.partitions	为宽 shuffle 转换(join 连接和聚合)创建的分区数
spark.executor.memoryOverhead	在集群模式下为每个 executor 进程分配的额外内存量，通常是用于 JVM 开销的内存（PySpark 不支持）
spark.serializer	org.apache.spark.serializer. JavaSerializer (default) org.apache.spark.serializer.KryoSerializer
spark.sql.files.maxPartitionBytes	读取文件时为每个分区使用的最大字节数。默认为 128 MB
spark.dynamicAllocation.enabled	指定是否根据工作负载动态增加或减少 executor 的数量。默认值为 true
spark.dynamicAllocation.minExecutors	启用动态分配时使用的最小 executor 数量
spark.dynamicAllocation.maxExecutors	启用动态分配时使用的最大 executor 数量
spark.extraJavaOptions	指定 JVM 选项

无论使用哪种语言，大多数选项是相同的，但是也有少数选项是特定于某种语言的。

1）用于 Scala 或 Java 程序的参数

例如，要运行用 Scala 或 Java 语言编写的 Spark 应用程序，需要使用的额外选项见表 1-7。

表1-7 用于运行 Scala 或 Java 语言编写的 Spark 应用程序的额外选项

选 项	说 明
--jars	如果在一个文件夹中有所有的依赖 JAR，则可以使用 spark-submit --jars 选项传递所有这些 JAR。所有的 JAR 文件都应该用逗号分隔。例如，--jars jar1.jar,jar2.jar,jar3.jar
--packages	用此命令时将处理所有传递依赖项
--class	指定想运行的 Scala 或 Java 类。这应该是带有包名的完全限定名，例如 org.apache.spark.examples.SparkPi

2）用于 PySpark（Python）程序的参数

当想要 spark-submit 一个 PySpark 应用程序时，需要指定想要运行的 .py 文件，并为依赖库指定 .egg 文件或 .zip 文件。

除了可以使用上面提到的大多数选项和配置外，一些特定于 PySpark 应用程序的选项和配置见表 1-8。

表1-8 用于运行 Python 语言编写的 Spark 应用程序的额外选项

PySpark 专用配置	说 明
--py-files	使用 --py-files 添加 .py、.zip 或 .egg 文件
--config spark.executor.pyspark.memory	PySpark 为每个 executor 进程使用的内存量
--config spark.pyspark.driver.python	用于 PySpark driver 的 Python 二进制可执行文件
--config spark.pyspark.python	用于 PySpark driver 和 executor 的 Python 二进制可执行文件

注意：使用--py-files 指定的文件在集群运行应用程序之前被上传到集群。还可以提前上传这些文件，并在 PySpark 应用程序中引用它们。

下面是提交 PySpark 应用程序的命令示例：

```
$ ./bin/spark-submit \
  --master yarn \
  --deploy-mode cluster \
  wordcount.py
```

下面的示例使用其他 Python 文件作为依赖项，命令如下：

```
$ ./bin/spark-submit \
  --master yarn \
  --deploy-mode cluster \
  --py-files file1.ipynb,file2.ipynb,file3.zip
  wordcount.py
```

1.8.2　提交 pi.ipynb 程序，计算圆周率π值

Spark 安装包中自带了一个使用蒙特卡洛方法求圆周率 π 值的程序。下面使用 spark-submit 将其提交到 PySpark 集群上以 standalone 模式运行，以掌握 spark-submit 提交 PySpark 程序的方法。

建议按以下步骤操作。

（1）打开终端窗口。

（2）确保已经以 standalone 模式启动了 Spark 集群（启动方式见 1.7 节）。

（3）进入 Spark 主目录下，执行的命令如下：

```
$ cd ~/bigdata/spark-3.1.2
$ ./bin/spark-submit --master spark://xueai8:7077 examples/src/main/python/pi.py
```

说明：

（1）--master 参数用于指定要连接的集群管理器，这里是 standalone 模式。

（2）最后一个参数是所提交的 Python 程序。

执行过程如图 1-32 所示。

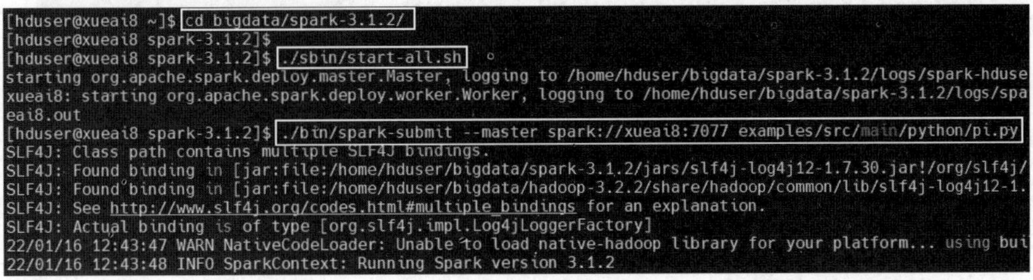

图 1-32　将 PySpark 应用程序提交到集群上运行

运行结果如图1-33所示。

```
22/01/16 12:30:46 INFO DAGScheduler: Job 0 is finished. Cancelling potenti
22/01/16 12:30:46 INFO TaskSchedulerImpl: Killing all running tasks in sta
22/01/16 12:30:46 INFO DAGScheduler: Job 0 finished: reduce at /home/hduse
Pi is roughly 3.148880
22/01/16 12:30:46 INFO SparkUI: Stopped Spark web UI at http://xueai8:4040
22/01/16 12:30:46 INFO MapOutputTrackerMasterEndpoint: MapOutputTrackerMas
22/01/16 12:30:46 INFO MemoryStore: MemoryStore cleared
```

图 1-33 PI 值计算结果

1.8.3 将 PySpark 程序提交到 YARN 集群上执行

也可以将 PySpark 程序运行在 YARN 集群上，由 YARN 来管理集群资源。下面使用 spark-submit 将 pi.py 程序提交到 Spark 集群上以 YARN 模式运行。

建议按以下步骤执行。

（1）打开终端窗口。

（2）不需要启动 Spark 集群。启动 Hadoop/YARN 集群，命令如下：

```
$ start-dfs.sh
$ start-yarn.sh
```

执行过程如图 1-34 所示。

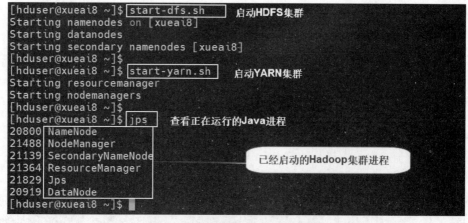

图 1-34 启动并查看 Hadoop 集群

（3）进入 Spark 主目录下，提交 pi.py 程序，命令如下：

```
$ cd ~/bigdata/spark-3.1.2
$ ./bin/spark-submit --master yarn examples/src/main/python/pi.py
```

执行过程如图 1-35 所示。

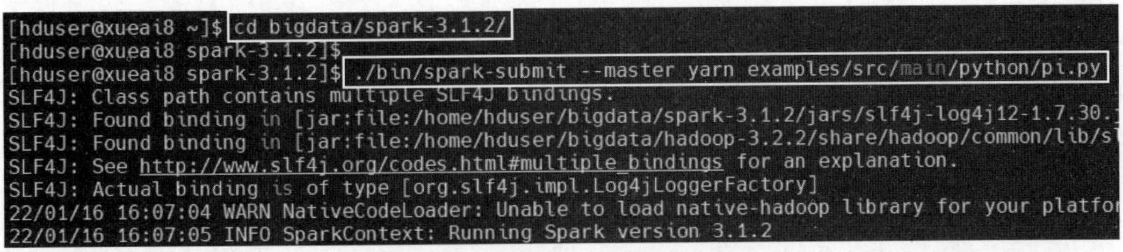

```
[hduser@xueai8 ~]$ cd bigdata/spark-3.1.2/
[hduser@xueai8 spark-3.1.2]$
[hduser@xueai8 spark-3.1.2]$ ./bin/spark-submit --master yarn examples/src/main/python/pi.py
SLF4J: Class path contains multiple SLF4J bindings.
SLF4J: Found binding in [jar:file:/home/hduser/bigdata/spark-3.1.2/jars/slf4j-log4j12-1.7.30.
SLF4J: Found binding in [jar:file:/home/hduser/bigdata/hadoop-3.2.2/share/hadoop/common/lib/s
SLF4J: See http://www.slf4j.org/codes.html#multiple_bindings for an explanation.
SLF4J: Actual binding is of type [org.slf4j.impl.Log4jLoggerFactory]
22/01/16 16:07:04 WARN NativeCodeLoader: Unable to load native-hadoop library for your platfo
22/01/16 16:07:05 INFO SparkContext: Running Spark version 3.1.2
```

图 1-35　在 Spark 集群上执行 PI 程序

执行结果如图 1-36 所示。

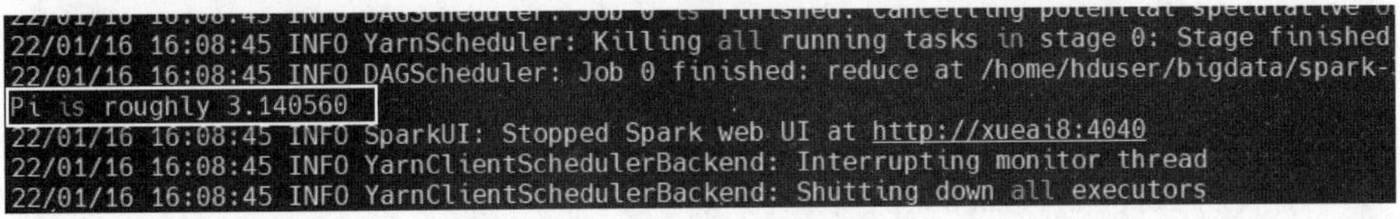

```
22/01/16 16:08:45 INFO DAGScheduler: Job 0 is finished: cancelling potential speculative
22/01/16 16:08:45 INFO YarnScheduler: Killing all running tasks in stage 0: Stage finished
22/01/16 16:08:45 INFO DAGScheduler: Job 0 finished: reduce at /home/hduser/bigdata/spark-
Pi is roughly 3.140560
22/01/16 16:08:45 INFO SparkUI: Stopped Spark web UI at http://xueai8:4040
22/01/16 16:08:45 INFO YarnClientSchedulerBackend: Interrupting monitor thread
22/01/16 16:08:45 INFO YarnClientSchedulerBackend: Shutting down all executors
```

图 1-36　计算 PI 值结果

第 2 章 开发和部署 PySpark 应用程序
CHAPTER 2

要开发 PySpark 应用程序，通常可以采用以下几种开发方式和开发环境：
（1）使用 PySpark Shell，交互式执行。
（2）使用 PyCharm IDE 集成开发环境，先开发测试，然后部署执行。
（3）使用 Jupyter Notebook，交互式开发。
（4）使用 Zeppelin Notebook，交互式开发。

在第 1 章已经了解了如何使用 PySpark Shell 以交互式方式执行 PySpark 程序代码，但是 PySpark Shell 并不适合在生产（工作）环境下使用。在生产（工作）环境中，用户可以根据自己的需求选择后面 3 种开发和执行方式。

在本配置之前，读者需要先在 Linux/CentOS 上安装好 Python 3。笔者是通过 Anaconda 安装的，所以这里的配置中引用的 Python 为 Anaconda 所带的 Python。如果读者安装的 Python 3 与本书不一致，可自行修改。

2.1 使用 PyCharm 开发 PySpark 应用程序

本节向大家介绍如何使用 PyCharm 这个 IDE 来开发 PySpark 应用程序。

本书使用 PyCharm Community Edition 作为 IDE。在本节中，将了解如何使用 PyCharm 设置 PySpark，以及如何将代码部署到集群中。

2.1.1 准备数据文件

下面将构建一个简单的 PySpark 应用程序，用来对莎士比亚文集（shakespeare.txt）执行单词计数任务。需要在两个地方使用这个 shakespeare.txt 数据集。一个在项目开发中用于本地程序测试，另一个在 HDFS（Hadoop 分布式文件系统）中用于集群测试。

先将/home/hduser/data/sparkshakespeare.txt 文件上传到 HDFS，步骤如下：
（1）确保已经启动了 HDFS 集群。
（2）在终端窗口中，执行命令，将文件上传到 HDFS 上，命令如下：

```
$ cd /home/hduser/data/spark
$ hdfs dfs -put shakespeare.txt /data/spark/
```

2.1.2 安装 PyCharm

首先在 PyCharm 官网下载 PyCharm 安装包，根据自己计算机的操作系统选择合适的安装版本。PyCharm 分为收费的企业版和免费的社区版。开发 PySpark 应用程序，使用社区版即可。例如，对于 Windows 系统选择的安装包如图 2-1 所示。

图 2-1　建议下载社区版的 PyCharm 安装包

双击下载的安装包，进行安装。一路单击 Next 按钮即可。安装完成后会在计算机桌面生成启动图标，双击它可以启动 PyCharm。

下面是 PyCharm 中常用的一些快捷键。

（1）Ctrl + Enter：在下方新建行但不移动光标。

（2）Shift + Enter：在下方新建行并移到新行行首。

（3）Ctrl + /：注释(取消注释)选择的行。

（4）Ctrl+d：对光标所在行的代码进行复制。

2.1.3 创建一个新的 PyCharm 项目

要创建一个新的项目，首先需要启动 PyCharm，选择 File→New Project，选择 Pure Python，并在右侧指定项目代码所在的位置，将项目命名为 HelloSpark。单击 Create 按钮创建此项目，如图 2-2 所示。

这样就创建了一个空的项目 HelloSpark。

将数据文件 shakespeare.txt 复制到项目的根目录下，例如~/PythonProjects/HelloSpark，如图 2-3 所示。

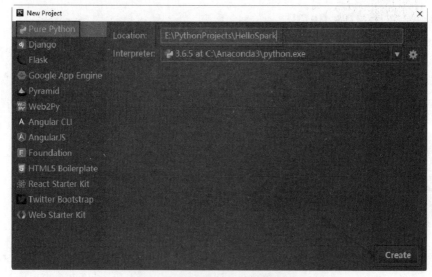

图 2-2 创建一个 PyCharm 项目

图 2-3 将数据文件 shakespeare.txt 复制到项目的根目录下

2.1.4 安装 PySpark 包

要安装 PySpark 包,需要在 PyCharm 中导航到 File→Settings,打开项目设置面板,如图 2-4 所示。

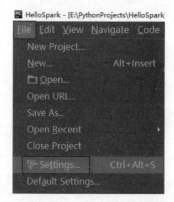

图 2-4 打开 PyCharm 的项目设置面板

在项目设置面板中,选择左侧的 Project:HelloSpark→Project Interpreter,然后单击右边栏的"＋"按钮,如图 2-5 所示。

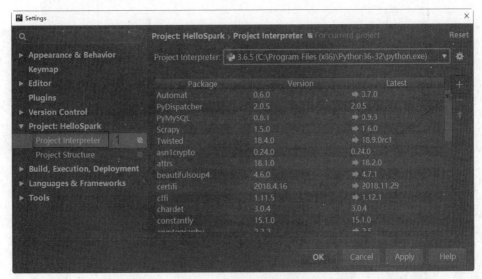

图 2-5　设置项目解释器

在上方的搜索框内输入 pyspark,搜索并选择 pyspark,然后的面板右侧选择指定的版本(说明:虽然截图中显示的版本是 2.3.2,实际上本书所有示例使用的 Spark 版本均基于 3.1.2),最后单击 Install Package 按钮安装,如图 2-6 所示。

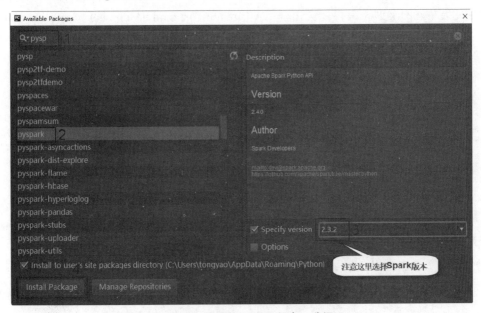

图 2-6　选择安装的 PySpark 版本,选择 3.1.2

安装完毕，关闭此面板，回到 settings 界面，可以看到 PySpark 已经安装成功。单击 OK 按钮即可，如图 2-7 所示。

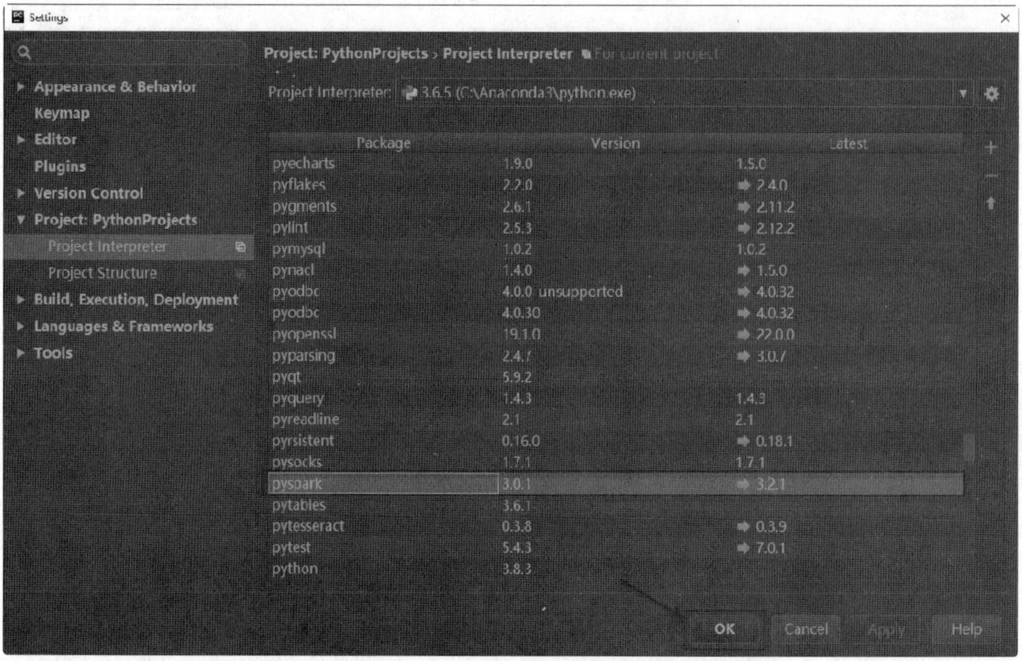

图 2-7　安装完毕，可以在列表中看到安装的 PySpark

在安装 PySpark 时，有可能会出现"AttributeError: module 'pip' has no attribute 'main'"的错误信息。这个错误产生的原因是升级了 pip，而从 pip 10.0.0 开始，不再支持 pip.main() 了。解决方法是，找到 PyCharm 安装目录（这里是笔者的安装目录，读者应替换为自己的安装目录）C:\Program Files (x86)\JetBrains\PyCharm 2016.2.3\helpers\packaging_tool.py 这个文件并打开，在其中找到 do_install 和 do_uninstall 函数，将其修改为以下内容：

```
def do_install(pkgs):
    try:
        #import pip
        try:
            from pip._internal import main
        except Exception:
            from pip import main
    except ImportError:
        error_no_pip()
    return main(['install'] + pkgs)

def do_uninstall(pkgs):
    try:
```

```
    #import pip
    try:
        from pip._internal import main
    except Exception:
        from pip import main
except ImportError:
    error_no_pip()
return main(['uninstall', '-y'] + pkgs)
```

2.1.5 创建 PySpark 应用程序

在刚创建的 HelloSpark 项目目录上，右击，选择 New→File，创建一个新的 Python 源文件，并命名为 Main.py，如图 2-8 所示。

图 2-8 创建一个新的 Python 源文件

编辑 Main.py 文件内容如下：

```
#第2章/Main.py

from pyspark.sql import SparkSession

#构建 SparkSession 和 SparkContext 实例
spark = SparkSession.builder \
    .master("local[2]") \
    .appName("MyFirstStandaloneApp") \
    .getOrCreate()

#加载数据文件
text_file = spark.sparkContext.textFile("./shakespeare.txt")

#进行单词计数
counts = text_file.flatMap(lambda line: line.split(" ")) \
        .map(lambda word: (word, 1)) \
        .reduceByKey(lambda a, b: a + b)

print ("元素数量: " + str(counts.count()))

#将统计结果保存到指定文件中
counts.saveAsTextFile("./shakespeareWordCount")
```

要运行程序，应先从 IntelliJ IDEA 的菜单中单击 Run→Run 'Main'，如图 2-9 所示。

图 2-9　在 IntelliJ IDEA 中运行 PySpark 应用程序

程序计算结果保存在 shakespeareWordCount 文件夹中，该文件夹与源代码的文件夹位于相同的目录下，如图 2-10 所示。

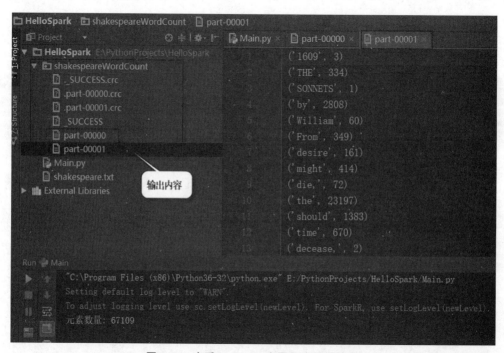

图 2-10　查看 PySpark 应用程序执行结果

2.1.6　部署到集群中运行

23min

接下来将 2.1.5 节开发测试后的 PySpark 作业部署到集群上运行。建议按以下步骤执行。

（1）首先，需要修改源代码，从 HDFS 中读取数据集而不是从本地 PyCharm 项目读取。修改后的代码如下：

```
#第3章/Main.py

from pyspark.sql import SparkSession

#构建 SparkSession 和 SparkContext 实例
```

```python
spark = SparkSession.builder \
    .master("local[2]") \
    .appName("MyFirstStandaloneApp") \
    .getOrCreate()

#加载 HDFS 上的数据文件
text_file = spark.sparkContext.textFile("hdfs://xueai8:8020/data/spark/shakespeare.txt")

#进行单词计数
counts = text_file.flatMap(lambda line: line.split(" ")) \
            .map(lambda word: (word, 1)) \
            .reduceByKey(lambda a, b: a + b)

print ("元素数量: " + str(counts.count()))

#将统计结果保存到指定文件中
counts.saveAsTextFile("hdfs://xueai8:8020/data/spark/shakespeareWordCount")
```

（2）将 Main.py 文件复制到虚拟机上。

（3）使用 spark-submit 将 Main.py 文件提交到 Spark 集群上运行，命令如下：

```
$ spark-submit ./Main.py
```

（4）输出结果被保存到 HDFS 上的/data/spark/shakespeareWordCount 目录中，其内容如下：

```
(u'fawn', 11)
(u'Fame,', 3)
(u'mustachio', 1)
(u'protested,', 1)
(u'sending.', 3)
(u'offendeth', 1)
(u'instant;', 1)
(u'scold', 4)
(u'Sergeant.', 1)
(u'nunnery', 1)
(u'Sergeant,', 2)
...
```

2.2 使用 Zeppelin 进行交互式分析

20min

Apache Zeppelin 是一款基于 Web 的 Notebook，支持交互式数据分析。使用 Zeppelin，可以使用丰富的预构建语言后端（或解释器）制作精美的数据驱动、交互式和协作文档。目前，Apache Zeppelin 支持 Apache Spark、Python、JDBC、Markdown 和 Shell 等多种解释器。

特别是，Apache Zeppelin 提供了内置的 Apache Spark 集成。我们不需要为它构建单独的模块、插件或库。Apache Zeppelin 与 Spark 集成，提供了如下功能：

（1）自动注入 SparkContext 和 SQLContext。
（2）从本地文件系统或 Maven 存储库加载运行时 JAR 依赖项。
（3）取消作业并显示进度。

Apache Zeppelin 专注于企业级应用，Zeppelin Notebook 可以满足企业用户的以下需求：

（1）数据摄取。
（2）数据发现。
（3）数据分析。
（4）数据可视化与协作。

接下来，学习如何安装 Zeppelin 和配置 Zeppelin 解释器，并演示如何使用 Zeppelin Notebook 作为 PySpark 的交互式数据分析工具进行大数据的分析和数据可视化。

2.2.1 下载 Zeppelin 安装包

Apache Zeppelin 的下载网址为 http://zeppelin.apache.org/download.html，选择最新的版本，如图 2-11 所示。

图 2-11 选择下载最新的 Zeppelin 安装包

将下载的安装包复制到 ~/software 目录下。

2.2.2 安装和配置 Zeppelin

建议按以下步骤安装和配置 Zeppelin。

（1）将下载的安装包解压缩到 ~/bigdata 目录下，并改名为 zeppelin-0.9.0，命令如下：

```
$ cd ~/bigdata
$ tar xvf ~/software/zeppelin-0.9.0-bin-netinst.tgz
$ mv zeppelin-0.9.0-bin-netinst zeppelin-0.9.0
```

（2）配置环境变量，命令如下：

```
$ cd
$ sudo nano /etc/profile
```

在文件的最后，添加以下内容：

```
export ZEPPELIN_HOME=/home/hduser/bigdata/zeppelin-0.9.0
export PATH=$PATH:$ZEPPELIN_HOME/bin
```

保存文件并关闭。

（3）执行/etc/profile 文件使配置生效，命令如下：

```
$ source /etc/profile
```

（4）打开 conf/zeppelin-env.sh 文件，命令如下（默认没有，从模板复制一份）：

```
$ cd ~/bigdata/zeppelin-0.9.0/conf
$ cp zeppelin-env.sh.template zeppelin-env.sh
$ nano zeppelin-env.sh
```

在文件的最后添加如下几行内容：

```
export JAVA_HOME=/opt/Java/jdk1.8.0_281
export SPARK_HOME=/home/hduser/bigdata/spark-3.1.2

export HADOOP_HOME=/home/hduser/bigdata/hadoop-3.2.2
export HADOOP_CONF_DIR=/home/hduser/bigdata/hadoop-3.2.2/etc/hadoop

export PYSPARK_PYTHON=/opt/anaconda3/bin/python
export PYSPARK_DRIVER_PYTHON=/opt/anaconda3/bin/python
export PYTHONPATH=$SPARK_HOME/python:$SPARK_HOME/python/lib/py4j-0.10.9-src.zip
```

（5）打开 zeppelin-site.xml 文件，命令如下（默认没有，从模板复制一份）：

```
$ cd ~/bigdata/zeppelin-0.9.0/conf
$ cp zeppelin-site.xml.template zeppelin-site.xml
$ gedit zeppelin-site.xml
```

修改如下两个属性，设置新的端口号，以避免与 Spark Web UI 默认端口发生冲突：

```
<property>
  <name>zeppelin.server.port</name>
  <value>9090</value>
  <description>Server port.</description>
</property>

<property>
  <name>zeppelin.server.ssl.port</name>
  <value>9443</value>
```

```
        <description>Server ssl port. (used when ssl property is set to true)
        </description>
        </property>
```

(6) 启动 Zeppelin 服务。

在终端窗口中启动 Zeppelin 服务，执行的命令如下：

```
$ zeppelin-daemon.sh start
```

执行过程如图 2-12 所示。

```
[hduser@xueai8 ~]$ zeppelin-daemon.sh start
Zeppelin start                                      [ OK ]
[hduser@xueai8 ~]$ jps
2211 Jps
2180 ZeppelinServer
[hduser@xueai8 ~]$
```

图 2-12　启动 ZeppelinServer

(7) 停止 Zeppelin 服务。在终端窗口中，执行命令停止 Zeppelin 服务，命令如下：

```
$ zeppelin-daemon.sh stop
```

2.2.3　配置 Spark/PySpark 解释器

注意：如果使用 Spark local 模式，则此步骤可省略。如果使用 Spark standalone 模式，则需要配置 Spark 解释器。

首先启动浏览器，在浏览器网址栏输入地址 http://xueai8:9090/，打开访问界面，单击右上角的小三角按钮，打开下拉菜单，单击 Interpreter 菜单项，打开解释器配置界面，如图 2-13 所示。

图 2-13　启动 ZeppelinServer

在打开的解释器配置界面中找到 Spark 解释器，添加一个 SPARK_HOME 属性，然后将 master 属性值修改为 spark://xucai8:7077（这实际上是连接到的集群管理器，这里使用的是 spark standalone 模式，相当于启动 PySpark Shell 时指定--master 参数），如图 2-14 所示。

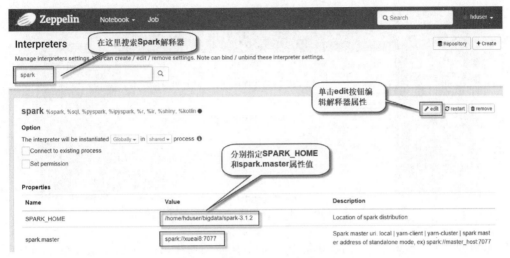

图 2-14　配置 Spark 解释器属性

再新增加两个 PySpark Python 的相关设置，如图 2-15 所示。

图 2-15　新增加两个 PySpark Python 的相关设置

最后单击 Save 按钮保存。

2.2.4　创建和执行 Notebook 文件

回到浏览器 Zeppelin 首页，选择 Create new note，创建一个新的 Notebook 文件，如图 2-16 所示。

图 2-16　创建一个新的 Notebook 文件

在弹出的创建窗口填写相应信息，然后单击 Create 按钮即可，如图 2-17 所示。

图 2-17　填写 Notebook 文件的路径，解释器默认是 Spark

在新打开的 Notebook 界面，执行 Python 代码。需要在第 1 行键入%pyspark，以告诉 Zeppelin 使用 PySpark 解释器，如图 2-18 所示。

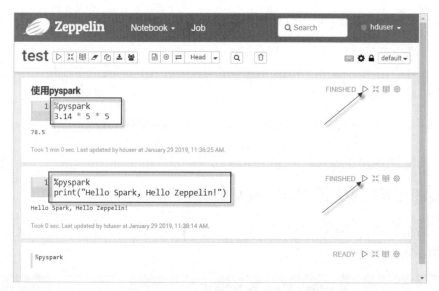

图 2-18　在 Zeppelin Notebook 文件中交互式执行 PySpark 代码

2.3　使用 Jupyter Notebook 进行交互式分析

数据分析师最喜欢的一个交互式分析工具是 Jupyter Notebook，因此也希望在应用 Spark 进行大数据分析时也使用 Jupyter。下面我们就配置 PySpark 与 Jupyter 的组合。

有以下两种方法可以使 PySpark 在 Jupyter Notebook 中可用。

（1）配置 PySpark 驱动程序使用 Jupyter Notebook：运行 PySpark 将自动打开一个 Jupyter Notebook。

（2）加载一个普通的 Jupyter Notebook，并使用 findSpark 包加载 PySpark。

第 1 种方法更快捷，但是特定于 Jupyter 笔记本；第 2 种方法是一种更广泛的方法，可以在自己喜欢的 IDE 中使用 PySpark。

2.3.1 配置 PySpark Driver 使用 Jupyter Notebook

建议按以下步骤配置和启动 Spark 及 Jupyter Notebook。

（1）启动 Spark 集群，命令如下：

```
$ cd ~/bigdata/spark-3.1.2
$ ./sbin/start-all.sh
```

（2）指定驱动程序（driver）使用 Jupyter Notebook。在终端窗口中，执行的命令如下：

```
$ export PYSPARK_DRIVER_PYTHON="Jupyter"
$ export PYSPARK_DRIVER_PYTHON_OPTS="Notebook --no-browser --ip=0.0.0.0"
```

需要注意的是，如果读者是以 root 账户在进行操作，则还需要加上 --allow-root 参数，将上述第 2 行命令修改如下：

```
$ export PYSPARK_DRIVER_PYTHON_OPTS="Notebook --allow-root --no-browser --ip=0.0.0.0"
```

（3）使用 cd 命令进入 Notebook 文档存放的目录位置，然后启动 PySpark Shell。在终端窗口中，执行的命令如下：

```
$ mkdir pyspark_demos
$ cd pyspark_demos

$ pyspark --master spark://xueai8:7077
```

执行过程如图 2-19 所示。

图 2-19 启动 PySpark Shell

不要关闭此窗口,以便使服务一直处于运行状态。

(4) 在 Windows 系统下,通过浏览器访问 Jupyter。到 Windows 系统下,打开浏览器,粘贴上一步复制的 URL,按 Enter 键访问,打开 Notebook 页面,如图 2-20 所示。

图 2-20　在浏览器中打开 Jupyter Notebook

(5) 编写 PySpark 代码。新建一个 Notebook,输入代码执行,如图 2-21 所示。

```
In [1]: from pyspark.sql import SparkSession

In [2]: spark = SparkSession.builder \
            .master("spark://xueai8:7077") \
            .appName("pyspark demo") \
            .getOrCreate()

In [3]: spark.version
Out[3]: '3.1.2'

In [4]: numbers = [1,2,3,4,5,6,7,8,9,0]
        rdd = spark.sparkContext.parallelize(numbers)

In [5]: rdd.collect()
Out[5]: [1, 2, 3, 4, 5, 6, 7, 8, 9, 0]

In [6]: rdd.map(lambda n: n*n).collect()
Out[6]: [1, 4, 9, 16, 25, 36, 49, 64, 81, 0]
```

图 2-21　在 Jupyter Notebook 中交互执行 PySpark 代码

(6) 查看 Spark Web UI。另打开一个浏览器窗口,访问 http://xueai8:8080,打开 Spark Web UI 如图 2-22 所示。

(7) 查看正在执行的 Spark 作业的 Web UI 信息。另打开一个浏览器窗口,访问地址 http://xueai8:4040,可以看到正在执行和已经执行完成的作业情况,如图 2-23 所示。

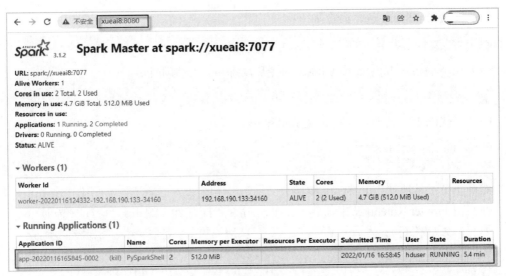

图 2-22　查看 Spark Web UI

图 2-23　查看正在执行的 Spark 作业的 Web UI

2.3.2　使用 findSpark 包

在 Jupyter 笔记本中使用 PySpark 还有另一种方法，即使用 findSpark 包使 Spark 上下文在代码中可用。这个 findSpark 包不是特定于 Jupyter 笔记本的，读者也可以在自己喜欢的 IDE 中使用这个技巧。

安装和使用 findSpark 包的步骤如下。

(1) 安装 findSpark 包，命令如下：

```
$ sudo /opt/anaconda3/bin/pip install findspark
```

(2) 在终端输入 Jupyter Notebook 命令启动即可，命令如下：

```
$ Jupyter Notebook --no-browser --ip=0.0.0.0
```

(3) 在代码中要先导入并初始化 findSpark，代码如下：

```
import findspark
findspark.init()
```

然后便可正常地使用 PySpark 编码。

(4) 在 Jupyter Notebook 中新建一个 Python 3 程序文件，编辑并执行的代码如下：

```
from pyspark.sql import SparkSession

#创建 SparkSession 实例，这是 PySpark 程序入口
spark = SparkSession.builder \
    .master("spark://xueai8:7077") \
    .appName("pyspark_demo") \
    .getOrCreate()

spark.version          #查看 Spark 版本号

#构造一个 RDD
rdd = spark.sparkContext.parallelize(range(10))
rdd.collect()

rdd.getNumPartitions()      #查看 RDD 的分区数
```

第 3 章 PySpark 核心编程

CHAPTER 3

Spark Core 模块包含 Spark 的基本功能，包括任务调度组件、内存管理、故障恢复、与存储系统交互等。在 Spark Core 模块中，核心的数据抽象被称为"弹性分布式数据集（RDD）"。RDD 是 Spark Core 的用户级 API，要真正理解 Spark 的工作原理，就必须理解 RDD 的本质。

Spark 为 Scala、Java、R 和 Python 编程语言提供了编程 API。Spark 本身是用 Scala 编写的，但 Spark 通过 PySpark 支持 Python。PySpark 构建在 Spark 的 Java API 之上（使用 Py4J）。通过 Spark（PySpark）上的交互式 Shell，可以对大数据进行交互式数据分析。数据科学界大多选择 Scala 或 Python 进行 Spark 程序开发和数据分析。

3.1 理解数据抽象 RDD

8min

在 Spark 的编程接口中，每个数据集都被表示为一个对象，称为 RDD。RDD 是一个只读的（不可变的）、分区的（分布式的）、容错的、延迟计算的、类型推断的和可缓存的记录集合。

RDD（Resilient Distributed Dataset，弹性分布式数据集）具有以下特性。

（1）Resilient：不可变的、容错的。

（2）Distributed：数据分散在不同节点（机器，进程）。

（3）Dataset：一个由多个分区组成的数据集。

Spark/PySpark RDD 是对跨集群分布的各个分区的引用的集合，如图 3-1 所示。

RDD 是分布式内存的一个抽象概念，提供了一种高度受限的共享内存模型。通常 RDD 很大，会被分成很多个分区，分别保存在不同的节点上。RDD 是不可变的、容错的、并行的数据结构，允许用户显式地将中间结果持久化到内存中，控制分区以优化数据放置，并使用一组丰富的操作符来操作它们。

RDD 被设计成不可变的，这意味着用户不能具体地修改数据集中由 RDD 表示的特定行。如果调用一个 RDD 操作来操纵 RDD 中的行，则该操作将返回一个新的 RDD。原 RDD 保持不变，新的 RDD 将以希望的方式包含数据。RDD 的不变性本质上要求 RDD 携带"血统（lineage）"信息，Spark 利用这些信息有效地提供容错能力。

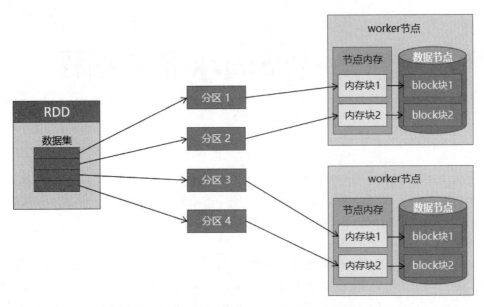

图 3-1 Spark RDD 是对跨集群分布的各个分区的引用的集合

RDD 提供了一组丰富的常用数据处理操作。它们包括执行数据转换、过滤、分组、连接、聚合、排序和计数的能力。关于这些操作需要注意的一点是，它们在粗粒度级别上进行操作，这意味着相同的操作应用于许多行，而不是任何特定的行。

1. RDD 结构

综上所述，RDD 只是一个逻辑概念，它可能并不对应磁盘或内存中的物理数据。根据 Spark 官方描述，RDD 由以下五部分组成：

（1）一组 partition（分区），即组成整个数据集的块。
（2）每个 partition（分区）的计算函数（用于计算数据集中所有行的函数）。
（3）所依赖的 RDD 列表(父 RDD 列表)。
（4）（可选的）对于 key-value 类型的 RDD，则包含一个分区器（分区程序）。
（5）（可选的）每个分区数据驻留在集群中的位置；如果数据存放在 HDFS 上，则它就是块所在的位置。

Spark 运行时使用这 5 条信息来调度和执行通过 RDD 操作表示的用户数据处理逻辑。前三段信息组成"血统"信息，Spark 将其用于两个目的。第 1 个是确定 RDD 的执行顺序，第 2 个是用于故障恢复。

2. 容错

Spark/PySpark 通过"血统"信息重建失败的部分将自动地代表其用户处理故障。每个 RDD 或 RDD 分区都知道如何在出现故障时重新创建自己。它有转换的日志，或者血统，可依据此从稳定存储器或另一个 RDD 中重新创建自己,因此,任何使用 Spark 的程序都可以确保内置的容错能力,而不考虑底层数据源和 RDD 类型。

3. RDD 特性

作为 Spark/PySpark 中最核心的数据抽象,RDD 具有以下特征。

(1) In-Memory:RDD 会优先使用内存。
(2) Immutable(Read-Only):一旦创建不可修改。
(3) Lazy Evaluated:惰性执行。
(4) Cacheable:可缓存,可复用。
(5) Parallel:可并行处理。
(6) Typed:强类型,单一类型数据。
(7) Partitioned:分区的。
(8) Location-Stickiness:可指定分区优先使用的节点。

3.2 RDD 编程模型

在 Spark/PySpark 中,使用 RDD 对数据进行处理,通常遵循如下的模型:
(1) 首先,将待处理的数据构造为 RDD。
(2) 对 RDD 进行一系列操作,包括 Transformation 和 Action 两种类型操作。
(3) 最后,输出或保存计算结果。
这个处理流程如图 3-2 所示。

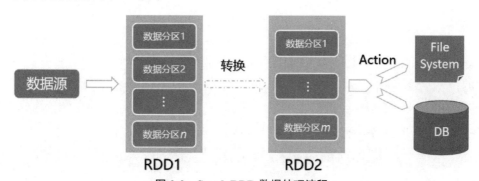

图 3-2 Spark RDD 数据处理流程

接下来通过一个具体的示例来掌握 RDD 编程的一般流程。

3.2.1 单词计数应用程序

下面使用 Spark RDD 实现经典的单词计数应用程序。

【例3-1】 使用 Spark RDD 实现单词计数。这里使用 Jupyter Notebook 作为开发工具，读者可以根据自己的喜好选择任意其他工具。

建议按以下步骤操作。

（1）启动 HDFS 集群和 Spark 集群，命令如下：

```
$ start-dfs.sh

$ cd ~/bigdata/spark-3.1.2
$ ./sbin/start-all.sh
```

（2）准备一个文本文件 word.txt，内容如下：

```
good good study
day day up
```

（3）将该文本文件上传到 HDFS 的 /data/spark/ 目录下，命令如下：

```
$ hdfs dfs -put word.txt /data/spark/
```

（4）在 Jupyter 中新建一个 Notebook。在 Notebook 的单元格中，执行代码。

首先创建 SparkSession 和 SparkContext 的实例，代码如下：

```python
from pyspark.sql import SparkSession

#构建 SparkSession 和 SparkContext 实例
spark = SparkSession.builder \
    .master("spark://xueai8:7077") \
    .appName("pyspark demo") \
    .getOrCreate()

sc = spark.sparkContext
```

读取数据源文件，构造一个 RDD，代码如下：

```python
source = "/data/spark/word.txt"
textFile = sc.textFile(source)
```

将每行数据按空格拆分成单词，使用 flatMap() 转换，代码如下：

```python
words = textFile.flatMap(lambda line: line.split(" "))
```

将各个单词加上计数值 1，使用 map() 转换，代码如下：

```python
wordPairs = words.map(lambda word: (word,1))
```

对所有相同的单词进行聚合相加求各单词的总数，使用 reduceByKey() 转换，代码如下：

```python
wordCounts = wordPairs.reduceByKey(lambda a,b: a + b)
```

将结果返给 Driver 程序,这一步才触发 RDD 开始实际的计算,代码如下:

```
wordCounts.collect()
```

可以看到输出结果如下:

```
[('good', 2), ('study', 1), ('day', 2), ('up', 1)]
```

或者,也可以将计算结果保存到文件中,代码如下:

```
dataSink = "/data/spark/word-result"
wordCounts.saveAsTextFile(dataSink)
```

(5)在 Jupyter Notebook 中交互式数据处理的过程如图 3-3 所示。

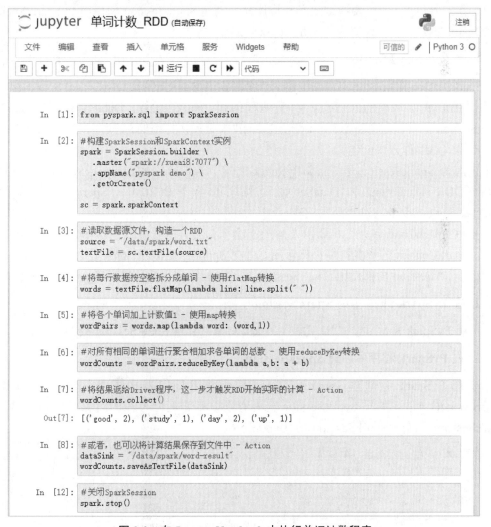

图 3-3　在 Jupyter Notebook 中执行单词计数程序

(6) 以上代码也可以进行精简，代码如下：

```
source = "/data/spark/word.txt"
sc.textFile(source) \
  .flatMap(lambda line: line.split(" ")) \
  .map(lambda word: (word,1)) \
  .reduceByKey(lambda a,b: a + b) \
  .collect()
```

3.2.2 理解 SparkSession

从 Spark 2.0 开始，SparkSession 已经成为 Spark/PySpark 与 RDD、DataFrame 和 DataSet 一起工作的入口点。它主要通过定义和描述如何创建 SparkSession 和使用 PySpark Shell 默认的 SparkSession 变量来解释什么是 SparkSession。

1. 什么是 SparkSession

SparkSession 在版本 Spark 2.0 中引入，全限定名称为 org.apache.spark.sql.SparkSession。它是 Spark 底层功能的入口点，用于编程创建 Spark RDD、DataFrame 和 DataSet。SparkSession 的实例对象 spark 在 PySpark Shell 中是默认可用的，它可以通过 SparkSession 构建器模式以编程方式创建。

正因为 SparkSession 是 Spark/PySpark 的一个入口点，创建 SparkSession 实例将是使用 RDD、DataFrame 和 DataSet 编写程序的第 1 个语句。SparkSession 将使用 SparkSession.builder()构建器模式创建。

虽然 SparkContext 是 2.0 版本之前的一个入口点，但它并没有被 SparkSession 完全取代，SparkContext 的许多特性仍然可用，并在 Spark 2.0 和以后的版本中使用。我们还应该知道 SparkSession 内部使用 SparkSession 提供的配置创建 SparkConfig 和 SparkContext。

2. 在 PySpark Shell 中的 SparkSession

调用 PySpark Shell 时，默认提供了 spark 对象，它是 SparkSession 类的一个实例。

3. 在 PySpark 程序中创建 SparkSession

要在 PySpark 中创建 SparkSession，需要使用构建器模式方法 builder 并调用 getOrCreate()方法。如果 SparkSession 已经存在，则它返回存在的对象，否则创建新的 SparkSession。创建 SparkSession 的代码如下：

```
from pyspark.sql import SparkSession

spark = SparkSession.builder \
  .master("local[*]") \
  .appName("pyspark demo") \
  .getOrCreate()
```

3.2.3 理解 SparkContext

SparkContext 是从 Spark 1.x 引入的（对于 Java API 来讲是 JavaSparkContext），在 2.0 版本中引入 SparkSession 之前，用来作为 Spark 和 PySpark 的入口点。使用 RDD 编程和连接到 Spark Cluster 的第 1 步就是创建 SparkContext。SparkContext 是在 org.apache.spark 包中定义的，它用于在集群中通过编程方式创建 Spark RDD、累加器和广播变量。

注意：每个 JVM 只能创建一个 SparkContext。

在任何给定时间，每个 JVM 应该只有一个 SparkContext 实例是活动的。如果想创建另一个新的 SparkContext，则应该在创建一个新的 SparkContext 之前停止现有的 SparkContext（使用 stop()方法）。

1. 在 PySpark Shell 中的 SparkContext

在 PySpark Shell 中默认提供了一个名为 sc 的对象，该对象是 SparkContext 类的一个实例。用户可以在需要时直接使用该对象。

2. 在 PySpark 程序中创建 SparkContext

当使用 Scala、PySpark 或 Java 编程时，首先需要创建一个 SparkConf 实例，并分配应用名称和设置 master（分别使用 SparkConf 的静态方法 setAppName()和 setMaster()），然后将 SparkConf 对象作为参数传递给 SparkContext 构造器来创建 SparkContext，代码如下：

```
from pyspark.conf import SparkConf
from pyspark.context import SparkContext

#创建Spark 配置对象
sparkConf = SparkConf()
sparkConf.setMaster("local[*]").setAppName("spark examples")

#构造SparkContext 实例
sc = SparkContext(conf = sparkConf)
```

SparkContext 构造函数在 2.0 版本中已经弃用，因此建议使用静态方法 getOrCreate() 来创建 SparkContext。该函数用于获取或实例化 SparkContext，并将其注册为一个单例对象，代码如下：

```
sc = SparkContext.getOrCreate(sparkConf)
```

一旦创建了 SparkContext 对象，就可以使用它来创建 Spark RDD。

3. 在 Spark 2.x 中创建 SparkContext

自从 Spark 2.0 以来，主要使用 SparkSession，SparkContext 中的大多数方法也存在于 SparkSession 中，并且 SparkSession 内部创建了 SparkContext 并公开了 sparkContext 变量以

供使用。在 Spark 2.0 之后,创建获取对 SparkContext 引用的代码如下:

```
from pyspark.sql import SparkSession

#构建 SparkSession 和 SparkContext 实例
spark = SparkSession.builder \
    .master("spark://xueai8:7077") \
    .appName("pyspark demo") \
    .getOrCreate()

#SparkSession 的 sparkContext 变量
sc = spark.sparkContext
```

3.3 创建 RDD

在对数据进行任何 Transformation 或 Action 操作之前,必须先将这些数据构造为一个 RDD。PySpark 提供了创建 RDD 的 3 种方法,分别如下:

(1)第 1 种方法是将现有的集合并行化。
(2)第 2 种方法是加载外部存储系统中的数据集,例如文件系统。
(3)第 3 种方法是在现有 RDD 上进行转换,以此得到新的 RDD。

3.3.1 将现有的集合并行化以创建 RDD

创建 RDD 的第 1 种方法是将对象集合并行化,这意味着将其转换为可以并行操作的分布式数据集。这种方法最简单,是开始学习 PySpark 的好方法,因为它不需要任何数据文件。这种方法通常用于快速尝试一个特性或在 PySpark 中做一些试验。对象集合的并行化是通过调用 SparkContext 类的 parallelize()方法实现的。

例如,对 Python 的数组或列表调用 parallelize()方法构造 RDD,代码如下:

```
#第 3 章/rdd_demo01.ipynb

from pyspark.sql import SparkSession

spark = SparkSession.builder \
    .master("spark://xueai8:7077") \
    .appName("pyspark demo") \
    .getOrCreate()

sc = spark.sparkContext

#通过并行集合(数组)创建 RDD
array1 = [1,2,3,4,5,6,7,8,9,10]
```

```
rdd1 = sc.parallelize(array1)
#rdd1 = sc.parallelize([1,2,3,4,5,6,7,8,9,10])

rdd1.collect()

#或者
list2 = list(range(11))    #range(start=0,end,step=0)
rdd2 = sc.parallelize(list2)

rdd2.collect()

#通过并行集合（数组）创建RDD
strList = ["明月几时有","把酒问青天","不知天上宫阙","今夕是何年"]
strRDD = sc.parallelize(strList)
strRDD.collect()
```

执行过程及输出结果如图3-4所示。

```
In [1]: from pyspark.sql import SparkSession

        spark = SparkSession.builder \
           .master("spark://xueai8:7077") \
           .appName("pyspark demo") \
           .getOrCreate()

        sc = spark.sparkContext
```

```
In [2]: # 通过并行集合（数组）创建RDD
        array1 = [1,2,3,4,5,6,7,8,9,10]
        rdd1 = sc.parallelize(array1)
        # rdd1 = sc.parallelize([1,2,3,4,5,6,7,8,9,10])

        rdd1.collect()
```
Out[2]: [1, 2, 3, 4, 5, 6, 7, 8, 9, 10]

```
In [3]: # 或者并行化list
        list2 = list(range(11))    # range (start=0,end,step=0)
        rdd2 = sc.parallelize(list2)

        rdd2.collect()
```
Out[3]: [0, 1, 2, 3, 4, 5, 6, 7, 8, 9, 10]

```
In [4]: # 通过并行集合（数组）创建RDD
        strList = ["明月几时有","把酒问青天","不知天上宫阙","今夕是何年"]
        strRDD = sc.parallelize(strList)

        strRDD.collect()
```
Out[4]: ['明月几时有', '把酒问青天', '不知天上宫阙', '今夕是何年']

图 3-4　在 Jupyter Notebook 中执行单词计数程序

3.3.2 从存储系统读取数据集以创建 RDD

创建 RDD 的第 2 种方法是从存储系统读取数据集，存储系统可以是本地计算机文件系统、HDFS、Cassandra、Amazon S3 等。Spark 可以从 Hadoop 支持的任何数据源创建 RDD，包括其本地文件系统、HDFS、Cassandra、HBase、Amazon S3 等。PySpark 支持 Hadoop InputFormat 支持的任何格式。

例如，要从文件系统中加载一个文本文件中的内容，代码如下：

```
#从文件系统中加载数据创建 RDD
file = "/data/spark/rdd/wc.txt"          #HDFS
rdd3 = sc.textFile(file)                 #sc 是 SparkContext 的实例

print(type(rdd3))                        #查看 rdd3 的类型

rdd3.collect()
```

SparkContext 类的 textFile()方法假设每个文件是一个文本文件，并且每行由一个新行分隔。这个 textFile()方法返回一个 RDD，它表示所有文件中的所有行。需要注意的重要一点是，textFile()方法是延迟计算的，这意味着如果指定了错误的文件或路径，或者错误地拼写了目录名，则在采取其中一项 Action 操作之前，这个问题不会出现（因此也不会被发现）。

3.3.3 从已有的 RDD 转换得到新的 RDD

创建 RDD 的第 3 种方法是调用现有 RDD 上的一个转换操作。例如，通过对 rdd4 的转换得到一个新的 RDD rdd5，代码如下：

```
#将字符转换为大写，得到一个新的 RDD
rdd4 = rdd3.map(lambda line: line.upper())
rdd4.collect()
```

注意：关于 map 函数，在稍后部分讲解。

3.3.4 创建 RDD 时指定分区数量

PySpark 在集群的每个分区上运行一个任务，因此必须谨慎地决定优化计算工作。尽管 PySpark 会根据集群自动设置分区数量，但用户可以将其作为第 2 个参数传递给并行化函数。

例如，使用 parallelize()方法创建一个 RDD，并将分区数指定为 3，代码如下：

```
sc.parallelize(data,3)          #3 个分区
```

为了进一步理解 RDD 的分区概念，参看下面这个示例。假设现在要创建一个 RDD，包含 14 条记录（或元组），并将该 RDD 的分区数指定为 3，即这 14 条记录分布在 3 个节点上，则该 RDD 的数据分区和对应的执行任务分布，如图 3-5 所示。

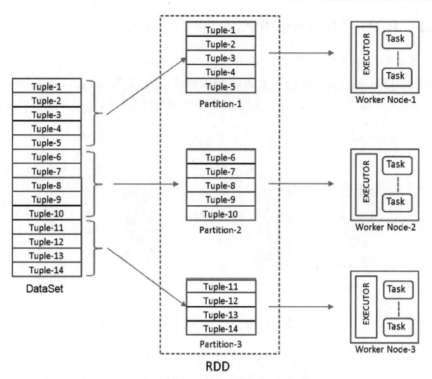

图 3-5　创建一个 RDD，并将分区数指定为 3

3.4　操作 RDD

创建了 RDD 之后，就可以编写 PySpark 程序对 RDD 进行操作了。RDD 操作分为两种类型：转换（Transformation）和动作（Action）。转换是用来创建 RDD 的方法，而动作是使用 RDD 的方法，如图 3-6 所示。

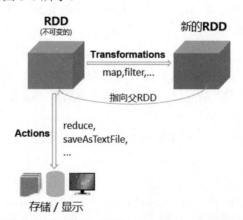

图 3-6　RDD 操作分为两种类型：转换（Transformation）和动作（Action）

3.4.1 RDD 上的 Transformation 和 Action

RDD 支持两种类型的操作：Transformation 和 Action。

Transformation 是定义如何构建 RDD 的延迟操作。大多数 Transformation 转换接受单个函数作为输入参数。每当在 RDD 上执行 Transformation 转换时，都会生成一个新的 RDD，如图 3-7 所示。

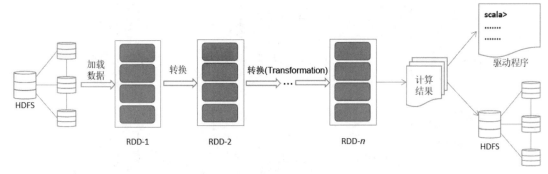

图 3-7　Spark RDD 上的 Transformation 转换操作

RDD 操作作用在粗粒度级别上。数据集中的每行都表示为 Java 对象，这个 Java 对象的结构对于 PySpark 来讲是不透明的。RDD 的用户可以完全控制如何操作这个 Java 对象。

因为 RDD 是不可变的（只读的）数据结构，因此任何转换都会产生新的 RDD。转换操作被延迟计算，称为"惰性转换"，这意味着 PySpark 将延迟对被调用的操作的执行，直到采取 Action 操作。换句话说，Transformation 转换操作仅仅记录指定的转换逻辑，并没有立即应用这些转换，而是在稍后才应用它们。调用 Action 操作将触发对它之前的所有转换的求值，然后向驱动程序返回一些结果，或者将数据写入存储系统，如 HDFS 或本地文件系统。基于延迟计算概念的一个重要优化技术是在执行期间将类似的转换折叠或组合为单个操作，即优化转换步骤。例如，如果 Action 操作要求返回第 1 行，则 PySpark 就只计算单个分区，然后跳过其余部分。

简而言之，RDD 是不可变的，RDD 的 Transformation 转换是延迟计算的，RDD 的 Action 操作是即时计算的，并触发数据处理逻辑的计算，而在 RDD 的内部实现机制中，底层接口则是基于迭代器的，从而使数据访问变得更高效，也避免了大量中间结果对内存的消耗。

通过应用程序操作 RDD 与操作数据的本地集合类似。以一个简单应用为例，代码如下：

```
lines = sc.textFile("hdfs://path/to/the/file")
filteredLines = lines.filter(lambda line: line.contains("spark")).cache()
result = filteredLines.count()
```

上面这段代码的意思是,从 HDFS 上加载指定的日志文件,找出包含单词 spark 的行数。其在内存中的计算和转换过程如下。

(1)一个 300MB 的日志文件,分布式存储在 HDFS 上,如图 3-8 所示。

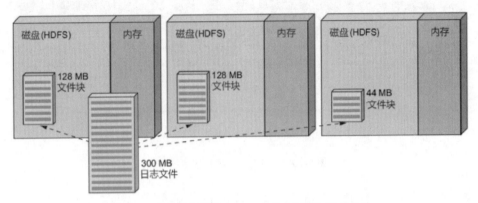

图 3-8　分布式存储在 HDFS 上的 300MB 日志文件

(2)将该日志文件加载到分布式的内存中,代码如下:

```
lines = sc.textFile("hdfs://path/to/the/file")
```

如图 3-9 所示。

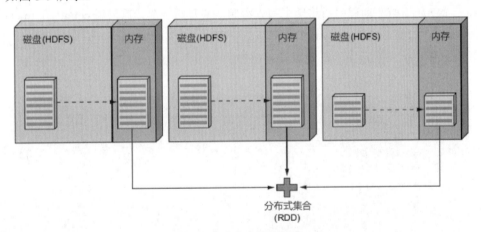

图 3-9　分布式内存数据集合 RDD

(3)过滤满足条件的行(只包含单词 spark 的行),这是原始数据集的一个子集,并将这个中间结果缓存到内存中,代码如下:

```
filteredLines = lines.filter(lambda line: line.contains("spark")).cache()
```

结果如图 3-10 所示。

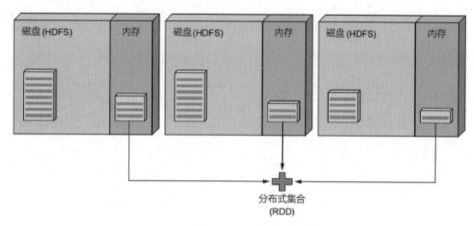

图 3-10　在 RDD 上执行过滤操作

（4）统计过滤后的行数，返给驱动程序 Driver，代码如下：

```
result = filteredLines.count()
```

3.4.2　RDD Transformation 操作

21min

Transformation 用于操作 RDD 并返回一个新的 RDD，如 map()和 filter()方法，而 Action 用于将一个结果返回给驱动程序或将结果写入存储，并开始一个计算，如 count()和 first()。

PySpark 对于 Transformation RDD 是延迟计算的，只在遇到 Action 时才真正进行计算。许多转换只作用于元素范围内，也就是一次作用于一个元素。

现在假设有一个 RDD，包含的元素为{1, 2, 3, 3}。首先，构造出这个 RDD，代码如下：

```
from pyspark.sql import SparkSession

#构建 SparkSession 和 SparkContext 实例
spark = SparkSession.builder \
    .master("spark://xueai8:7077") \
    .appName("pyspark demo") \
    .getOrCreate()

sc = spark.sparkContext

#构造一个 RDD
rdd = sc.parallelize([1,2,3,3])
```

接下来，学习普通 RDD 上的一些常见的 Transformation 转换操作方法。

1）map(func)

算子 map()使用 func 函数转换每个 RDD 元素并返回一个新的 RDD，如图 3-11 所示。

应用 map() 转换，代码如下：

```
#map 转换
rdd1 = rdd.map(lambda x: x + 1)   #transformation
rdd1.collect()      #action
```

执行以上代码，输出结果如下：

```
[2,3,4,4]
```

2）flatMap(func)

这个转换操作使用 func 函数来转换 RDD 中的每个元素，并将一个或多个元素返到新的 RDD 中，如图 3-12 所示。

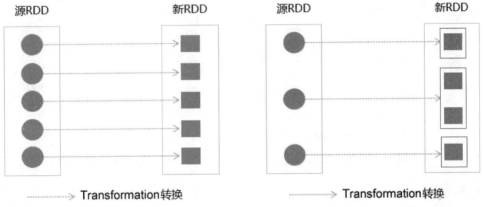

图 3-11　在 RDD 上执行 map 转换操作　　图 3-12　在 RDD 上执行 flatMap 转换操作

应用 flatMap() 转换，代码如下：

```
rdd2 = rdd.flatMap(lambda x: range(x,4))
rdd2.collect()
```

执行以上代码，输出结果如下：

```
[1, 2, 3, 2, 3, 3, 3]
```

3）filter(func)

这个转换操作使用 func 函数来过滤 RDD 中的每个元素，当 func 返回值为 true 时，被判断的元素添加到新的 RDD 中；当 func 返回值为 false 时，则丢弃被判断的元素，因此经过 filter() 函数转换之后的新 RDD 总是源 RDD 的一个子集，如图 3-13 所示。

应用 filter() 转换，代码如下：

```
#filter 转换
rdd3 = rdd.filter(lambda x: x!=1)
rdd3.collect()
```

执行以上代码,输出结果如下:

```
[2, 3, 3]
```

4)distinct(numPartitions: Optional[int] = None)

这个转换操作用于对 RDD 中的元素去重,源 RDD 中重复的元素在新 RDD 中只保留唯一的一个。在应用这个转换操作时,可以通过参数指定分区数。该转换的原理如图 3-14 所示。

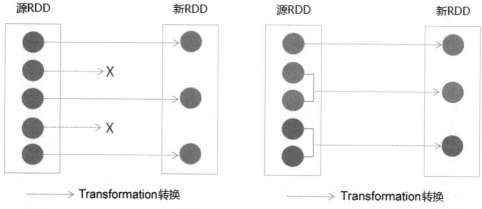

图 3-13 在 RDD 上执行 filter() 转换操作   图 3-14 在 RDD 上执行 distinct() 转换操作

应用 distinct()转换,代码如下:

```
rdd4 = rdd.distinct()
rdd4.collect
```

执行以上代码,输出结果如下:

```
[2, 1, 3]
```

5)sample(withReplacement: bool, fraction: float, seed: Optional[int] = None)

这是一个抽样方法,返回这个 RDD 的一个抽样子集,其中各参数的含义如下。

(1)withReplacement:是否可以对元素进行多次采样(采样后替换)。

(2)fraction:抽样因子。对于 without replacement,每个元素被选中的概率,fraction 值必须在[0,1]区间上;对于 with replacement,每个元素被选择的期望次数,fraction 值必须大于或等于 0。

(3)seed:用于随机数生成器的种子。

例如,抽取 50%的样本子集到新的 RDD 中,代码如下:

```
rdd5 = rdd.sample(False,0.5)
rdd5.collect()
```

执行以上代码,输出结果如下:

```
[2, 3]
```

注意:这个操作并不能保证精确地提供给定 RDD 的计数的比例。

6) keyBy(f: Callable[[T], K])

当在类型为 T 的 RDD 上调用时,返回一个(K, T)元组对作为元素的新 RDD。通过应用 func 函数创建这个新 RDD 中元素的元组。例如,对于由单词组成的一个 RDD,应用 keyBy()函数转换将 RDD 的元素转换为一个二元组,单词的首字母为 key,单词本身为 value,代码如下:

```
x = sc.parallelize(["John", "Fred", "Anna", "James"])
y = x.keyBy(lambda w: w[0])
y.collect()
```

以上代码指定将 RDD 中的元素(单词)转换为键-值对形式的元组,其中 k 来自单词的首字母。执行上面的代码,输出结果如下:

```
[('J', 'John'), ('F', 'Fred'), ('A', 'Anna'), ('J', 'James')]
```

7) groupBy(f: Callable[[T], K])

当在类型为 T 的 RDD 数据集上调用 groupBy()操作时,返回一个由 Tuple[K, Iterable[T]] 元素组成的新 RDD。每个组由一个 key 和一系列映射到该 key 的元素组成。每个组内元素的顺序不能得到保证,甚至在每次计算结果 RDD 时可能会有所不同。这种方法有可能会引起数据 shuffle。例如,构造一个简单的 RDD,并应用 groupBy()转换,代码如下:

```
#创建单词的 RDD
x = sc.parallelize(["Joseph", "Jimmy", "Tina", "Thomas", "James", "Cory",
"Christine", "Jackeline", "Juan"], 4)

#在 x 上应用 groupBy 操作
y = x.groupBy(lambda word: word[0])

for t in y.collect():
    print((t[0],[i for i in t[1]]))
```

在上面的代码中,在 x 这个 RDD 上调用 groupBy 转换操作,它会按所有单词元素的首字母进行分组,把首字母相同的单词分到同一个组中,以首字母为 key,组成新的键-值对 RDD。输出结果如下:

```
('J', ['Joseph', 'Jimmy', 'James', 'Jackeline', 'Juan'])
('T', ['Tina', 'Thomas'])
('C', ['Cory', 'Christine'])
```

8）sortBy(keyfunc: Callable[[T], S])

这个操作对源 RDD 中的元素按给定的 func 函数进行排序，并可以指定正序或倒序（默认为正序），返回一个新的 RDD，新的 RDD 中包含排好序的元素。例如，构造一个简单的 RDD，并使用 sortBy()转换进行排序和倒序操作，代码如下：

```
rdd = sc.parallelize([3,1,90,3,5,12])
rdd.collect()         #[3, 1, 90, 3, 5, 12]

//默认升序
rdd.sortBy(lambda x: x).collect()                    #[1, 3, 3, 5, 12, 90]

//降序
rdd.sortBy(lambda x: x, ascending=False).collect()   #[90, 12, 5, 3, 3, 1]
```

也可以在排序的同时改变分区数量，代码如下：

```
#默认不改变分区数
result1 = rdd.sortBy(lambda x: x, ascending=False)
print("分区数: ", result1.getNumPartitions())  #分区数: 2

#将分区数变为1
result2 = rdd.sortBy(lambda x: x, ascending=False, numPartitions=1)
print("分区数: ", result2.getNumPartitions())  #分区数: 1
```

在上面的示例中，对 rdd 中的元素进行排序，并对排序后的 RDD 分区个数进行了修改，其中 result1 就是排序后的 RDD，默认的分区个数是 2，result2 是修改分区以后的 RDD，分区数变为了 1。

9）randomSplit(weights, seed)

这个转换操作使用提供的权重随机分割源 RDD，以数组形式返回拆分后的子 RDD（拆分后的 RDD 组成的数组并返回），其中各参数的含义如下。

（1）weights：分割的权重，如果它们的和不等于 1，则将被标准化。

（2）seed：随机种子。

例如，构造一个简单的 RDD，并按 80/20 的比例将其分割为两个子 RDD，代码如下：

```
#构造一个 RDD
rdd1 = sc.parallelize([1,2,3,4,5,6,7,8,9,10])

#按 80/20 分割数据集
splitedRDD = rdd1.randomSplit([0.8, 0.2])

#查看
splitedRDD[0].collect()       #[1, 2, 3, 6, 7, 8, 9]
splitedRDD[1].collect()       #[4, 5, 10]
```

PySpark 也支持对两个 RDD 执行集合运算。为了便于演示，首先创建两个 RDD，分别包含元素{1,2,3,3}和{3,4,5}，代码如下：

```
#构造这两个RDD
rdd1 = sc.parallelize([1,2,3,3])
rdd2 = sc.parallelize([3,4,5])

for i in rdd1.collect():
   print(i, end=" ")

print()

rdd2.collect()              #[3,4,5]
```

对两个 RDD 可执行如下集合操作。

10) union(other: pyspark.rdd.RDD[U])

返回此 RDD 与另一个 RDD 的并集。返回的新 RDD 包含两个 RDD 中的全部元素（相当于 SQL 语言中的 union all）。例如，将 rdd1 和 rdd2 合并在一起，代码如下：

```
#union 转换
rdd3 = rdd1.union(rdd2)
rdd3.collect()              #[1, 2, 3, 3, 3, 4, 5]
```

11) intersection(other: pyspark.rdd.RDD[T])

返回此 RDD 与另一个 RDD 的交集。输出将不包含任何重复元素，即使输入 RDD 中有重复元素也一样。例如，要统计 rdd1 和 rdd2 的交集，代码如下：

```
#intersection 转换
rdd4 = rdd1.intersection(rdd2)
rdd4.collect()            #[3]
```

12) subtract(other: pyspark.rdd.RDD[T])

返回一个 RDD，其中包含其他 RDD 中不存在的元素，这相当于求两个 RDD 的差集。例如，找出 rdd1 中存在而 rdd2 中不存在的元素，代码如下：

```
#subtract 转换
rdd5 = rdd1.subtract(rdd2)
rdd5.collect()              #[1, 2]
```

13) cartesian(other: pyspark.rdd.RDD[U])

返回这个 RDD 和另一个 RDD 的笛卡儿积，即返回的新 RDD 包含所有元素对(a, b)，其中 a 来自当前 RDD，b 来自另一个 RDD。例如，对 rdd1 和 rdd2 执行笛卡儿连接，代码如下：

```
#cartesian 转换(笛卡儿积)
rdd6 = rdd1.cartesian(rdd2)
rdd6.collect()
```

执行上面代码，输出结果如下：

```
[(1, 3),
 (2, 3),
 (1, 4),
 (2, 4),
 (1, 5),
 (2, 5),
 (3, 3),
 (3, 3),
 (3, 4),
 (3, 4),
 (3, 5),
 (3, 5)]
```

14）zip(other: pyspark.rdd.RDD[U])

将这个 RDD 与另一个 RDD 执行"拉链"操作，返回的新 RDD 中包含的元素为键-值对，每个键-值对中的第 1 个元素来自第 1 个 RDD，第 2 个元素来自另一个 RDD。两个 RDD 必须有相同数量的元素。例如，将 rdd1 和 rdd2 组合在一起，创建一个新的 rdd3[(key,value)]，其中 key 来自 rdd1，value 来自 rdd2，代码如下：

```
rdd1 = sc.parallelize(["aa","bb","cc"])
rdd2 = sc.parallelize([1,2,3])

rdd3 = rdd1.zip(rdd2)
rdd3.collect()          #[('aa', 1), ('bb', 2), ('cc', 3)]
```

3.4.3 RDD Action 操作

Action 操作指的是将一个结果返给驱动程序或将结果写入存储的操作，并开始执行一个计算，如 count()和 first()。

一旦创建了 RDD，就只有在执行了 Action 时才会执行各种 Transformation 转换。可以将一个 Action 的执行结果写回存储系统，或者返回驱动程序，以便在本地进行进一步的计算。常用的 action 操作函数见表 3-1。

表 3-1　常用的 action 操作函数

action 操作函数	描　　述
reduce(func)	使用函数 func 对 RDD 中的元素进行聚合计算
collect()	将 RDD 操作的所有结果返给驱动程序。这通常对产生足够小的 RDD 的操作很有用
count()	返回 RDD 中的元素数量
first()	返回 RDD 的第 1 个元素。它的工作原理类似于 take(1)函数

续表

action 操作函数	描述
take(n)	返回 RDD 的前 n 个元素。它首先扫描一个分区，然后使用该分区的结果来估计满足该限制所需的其他分区的数量。这种方法应该只在预期得到的数组很小的情况下使用，因为所有的数据都被加载到驱动程序的内存中
top(n)	按照指定的隐式排序从这个 RDD 中取出最大的 n 个元素，并维护排序。这与 takeOrdered 相反。这种方法应该只在预期得到的数组很小的情况下使用，因为所有的数据都被加载到驱动程序的内存中
takeSample	返回一个数组，其中包含来自 RDD 的元素的抽样子集
takeOrdered(n)	返回 RDD 的前 n 个（最小的）元素，并维护排序。这和 top 是相反的。这种方法应该只在预期得到的数组很小的情况下使用，因为所有的数据都被加载到驱动程序的内存中
saveAsTextFile(path)	将 RDD 的元素作为文本文件(或文本文件集)写入本地文件系统、HDFS 或任何其他 Hadoop 支持的文件系统的给定目录中。Spark 将对每个元素调用 toString，将其转换为文件中的一行文本
foreach(func)	在 RDD 的每个元素上运行函数 func
aggregate	类似于 reduce，执行聚合运算，但它可以返回具有与输入元素数据类型不同的结果。这个函数聚合每个分区的元素，然后使用给定的 combine 组合函数和一个中性的"零值"，对所有分区的结果进行聚合

下面通过几个示例来掌握几个常用的 action 函数的用法。

【例 3-2】假设有一个包含元素{1, 2, 3, 3}的 RDD，计算该 RDD 中所有元素的和。

实现代码如下：

```
#第3章/rdd_demo02.ipynb

#构造RDD
rdd = sc.parallelize([1,2,3,3])

#reduce(func)
result = rdd.reduce(lambda x,y: x + y)

#输出结果
print(result)
```

执行上面的代码，输出结果如下：

```
9
```

【例 3-3】假设有一个包含整数元素[1,2,3,4]的 RDD，分区数为 2，使用 Spark RDD 的 aggregate()函数计算该 RDD 中所有元素的平均值。

这个aggregate()函数的签名如下：

```
aggregate(zeroValue: U, seqOp: Callable[[U, T], U], combOp: Callable[[U, U],
U]) → U
```

这个函数使用给定的组合函数和中立的"零值"聚合每个分区的元素，然后聚合所有分区的结果，其中各参数的含义分别如下。

（1）zeroValue：当执行 seqOp 运算时每个分区用来累积结果的初始值，当执行 combOp 运算时不同分区合并结果的初始值。它通常是中性元素（例如求和计算时初始值 0 或求积计算时初始值 1）。

（2）seqOp：用于在分区内累积结果的运算函数。

（3）combOp：用于合并来自不同分区结果的合并运算函数。

要计算 RDD 中元素的平均值，需要计算出两个值，一个是 RDD 的各元素的累加值，另一个是元素计数。对于加法计算，要初始化为(0, 0)。计算过程如图 3-15 所示。

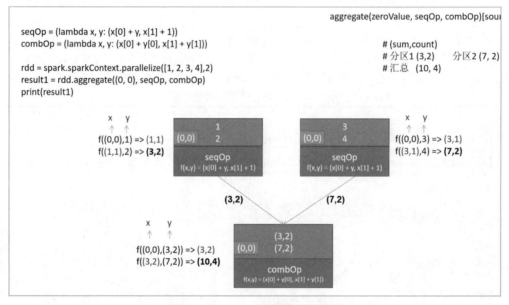

图 3-15　aggregate()函数执行过程

实现代码如下：

```
#第3章/rdd_demo03.ipynb

from pyspark.sql import SparkSession

#构建 SparkSession 和 SparkContext 实例
spark = SparkSession.builder \
    .master("spark://xueai8:7077") \
```

```python
    .appName("pyspark demo") \
    .getOrCreate()

sc = spark.sparkContext

#构造 RDD
rdd = sc.parallelize([1,2,3,4,5,3,2])

#定义分区计算函数
seqOp = (lambda accu, v: (accu[0] + v, accu[1] + 1))

#定义分区结果合并函数
combOp = (lambda accu1, accu2: (accu1[0] + accu2[0], accu1[1] + accu2[1]))

#执行 aggregate 聚合计算
result = rdd.aggregate((0, 0), seqOp, combOp)

#输出
print(result)          #(20, 7)

#计算平均值
avg = result[0] / result[1]
print("RDD 元素的平均值为", avg)       #2.857142857142857

#停止
sc.stop()
spark.stop()
```

执行上面的代码，输出结果如下：

```
(20, 7)
RDD 元素的平均值为 2.857142857142857
```

3.4.4　RDD 上的描述性统计操作

PySpark 在包含数值数据的 RDD 上提供了许多描述性统计操作。描述性统计都是在数据的单次传递中计算的，代码如下：

```
#构造一个 RDD
rdd1 = sc.parallelize(range(1,21,2))

rdd1.collect()       #[1, 3, 5, 7, 9, 11, 13, 15, 17, 19]

#描述性统计方法
rdd1.sum()           #100
rdd1.max()           #19
rdd1.min()           #1
```

```
rdd1.count()          #10
rdd1.mean()           #10.0
rdd1.variance()       #33.0
rdd1.stdev()          #5.744562646538029
```

也可以用直方图统计数据分布，代码如下：

```
rdd1.histogram([1.0, 8.0, 20.9])   #([1.0, 8.0, 20.9], [4, 6])
```

上面代码的意思是，统计 rdd1 中[1.0, 8.0)和[8.0, 20.9) 范围内元素的数量，此列中前者有 4 个，后者有 6 个。也可以指定平均划分的区域数量，代码如下：

```
rdd1.histogram(3)            #([1, 7, 13, 19], [3, 3, 4])
```

结果的意思是，[1.0, 7.0)范围内有 3 个元素，[7.0, 13.0)范围内有 3 个元素，[13.0, 19.0)范围内有 4 个元素。

如果需要多次调用描述性统计方法，则可以使用 StatCounter 对象。可以通过调用 stats() 方法返回一个 StatCounter 对象，代码如下：

```
#通过调用stats()方法返回一个StatCounter对象
status = rdd1.stats()

print(status.count())
print(status.mean())
print(status.stdev())
print(status.max())
print(status.min())
print(status.sum())
print(status.variance())
```

执行以上代码，输出结果如下：

```
10
10.0
5.744562646538029
19.0
1.0
100.0
33.0
```

3.5 Key-Value Pair RDD

有一类特殊的 RDD，其元素是以<key,value>对的形式出现，称为 Pair RDD。针对 Key-Value Pair RDD，PySpark 专门提供了一些操作，这些操作只在键-值对的 Pair RDD 上可用。

3.5.1 创建 Pair RDD

PySpark 在包含键-值对的 Pair RDD 上提供了专门的 Transformation API，包括 reduceByKey()、groupByKey()、sortByKey()和 join()等。Pair RDD 能够在 key 上并行操作，或者跨网络重新组织数据。Pair RDD 常被用于执行聚合操作，以及常被用来完成初始的 ETL（Extract-Transform-Load，抽取-转换-加载）以获取 key-value 格式数据。

注意：除了 count()操作之外，大多数操作通常涉及 shuffle，因为与 key 相关的数据可能驻留在不同的分区上。

创建 Pair RDD 的方式有多种。假设在 HDFS 的/data/spark/目录下存在一个 word.txt 文件，其内容如下：

```
good good study
day day up
```

可以从 word.txt 文件中加载数据以创建键-值对 RDD，代码如下：

```
file = "/data/spark/rdd/wc.txt"
lines = sc.textFile(file)

#通过转换，生成 Pair RDD
pairRDD = lines
    .flatMap(lambda line: line.split(" "))
    .map(lambda word: (word,1))

pairRDD.collect()
```

执行以上代码，输出内容如下：

```
[('good',1), ('good',1), ('study',1), ('day',1), ('day',1), ('up',1)]
```

可以看到，转换之后的 RDD 中元素是二元组，其中第 1 个元组元素称为 key，第 2 个元组元素称为 value。

当然，也可以通过并行化内存集合创建键-值对 RDD，或通过对已有 RDD 的转换来创建一个新的键-值对 RDD，代码如下：

```
rdd = sc.parallelize(["Hadoop","Spark","Hive","Spark"])
pairRDD = rdd.map(lambda word: (word,1))
pairRDD.collect()
```

执行上面的代码，输出结果如下：

```
[('Hadoop', 1), ('Spark', 1), ('Hive', 1), ('Spark', 1)]
```

也可以使用 keyBy 转换来构建一个 Pair RDD，代码如下：

```
a = sc.parallelize(["black", "blue", "white", "green", "grey"], 2)

#通过应用指定的函数来创建该RDD中元素的元组(参数函数生成对应的key)，返回一个Pair RDD
b = a.keyBy(lambda x:len(x))

for t in b.collect():
    print(t)
```

执行以上代码，输出结果如下：

```
(5, 'black')
(4, 'blue')
(5, 'white')
(5, 'green')
(4, 'grey')
```

3.5.2　操作 Pair RDD

在键-值对 RDD 上除了可执行普通 RDD 上一样的 Transformation 转换操作外，PySpark 还专门为这一类 RDD 提供了专用的 Transformation 转换操作和 Action 操作(这些专用转换操作只能作用于键-值对 RDD，不能在普通 RDD 上调用)。

为了演示这些专用于键-值对 RDD 上的转换操作，首先构造一个键-值对 RDD，代码如下：

```
#构造 Pair RDD
pairRDD = sc.parallelize([(1,2),(3,4),(3,6)])
pairRDD.collect()
```

执行以上代码，输出结果如下：

```
[(1, 2), (3, 4), (3, 6)]
```

基于上面构造的键-值对 RDD，下面介绍一些专用的 Transformation 操作。

1) keys

这个转换操作会返回一个包含所有 key 的新 RDD，代码如下：

```
#keys
p1 = pairRDD.keys()
p1.collect()
```

执行以上代码，输出结果如下：

```
[1, 3, 3]
```

2）values

这个转换操作会返回一个包含所有 value 的新 RDD，代码如下：

```
#values
p2 = pairRDD.values()
p2.collect()
```

执行以上代码，输出结果如下：

```
[2, 4, 6]
```

3）mapValues(func)

将 func 函数应用到键-值对 RDD 中的每个元素上，只对每个键-值对中的 value 执行 map 转换，不对 key 做任何改变，代码如下：

```
p3 = pairRDD.mapValues(lambda x: x*x)
p3.collect()
```

执行以上代码，输出结果如下：

```
[(1, 4), (3, 16), (3, 36)]
```

4）flatMapValues(func)

将 func 函数应用到键-值对 RDD 中的每个元素上，只对每个键-值对中的 value 执行 flatMap 转换，不对 key 做任何改变。这也保留了原始的 RDD 分区，代码如下：

```
p4 = pairRDD.flatMapValues(lambda x: range(x,6))
p4.collect()
```

执行以上代码，输出结果如下：

```
[(1, 2), (1, 3), (1, 4), (1, 5), (3, 4), (3, 5)]
```

5）sortByKey([ascending], [numPartitions])

对键-值对 RDD 中的元素按照 key 进行排序，默认为升序排序。可通过布尔类型的 ascending 参数指定排序顺序，代码如下：

```
#p5 = pairRDD.sortByKey()                    #升序
p6 = pairRDD.sortByKey(ascending=False)      #降序
p6.collect()
```

执行以上代码，输出结果如下：

```
[(3, 4), (3, 6), (1, 2)]
```

6）groupByKey([numPartitions])

将 RDD 中的元素按 key 进行分组，具有相同 key 的元素值被分到一个序列中。例如，当对(K,V)类型的 RDD 调用此转换方法时，返回一个(K, Iterable<V>)类型的新 RDD，代码如下：

```
p7 = pairRDD.groupByKey()
p7.map(lambda x : (x[0], list(x[1]))).collect()
```

执行以上代码，输出结果如下：

```
[(1, [2]), (3, [4, 6])]
```

注意：groupByKey 是一种宽依赖的操作，会导致对 RDD 执行哈希分区，从多个分区 shuffle 数据，并且不使用分区本地的 combiner 来减少数据传输，因此开销很大。当需要对分组数据进行进一步聚合时，不建议使用此操作。

7）subtractByKey(other)

这个转换操作对两个 RDD 按 key 求差集，返回的新 RDD 中只包含那些 key 仅在当前 RDD 中有而在另一个 RDD 中没有的那些键-值对，代码如下：

```
#构造另一个 RDD
other = sc.parallelize([(3,5),(4,6)])

#求两个 RDD 的 key 差集
rdd7 = rdd5.subtractByKey(other)

#查看
rdd7.collect()
```

执行以上代码，输出结果如下：

```
[(1,2)]
```

8）join(other)

对两个 RDD 按 key 进行连接。例如，当对类型(K,V)和(K,W)的 RDD 调用此操作时，会返回(K,(V,W))类型的 RDD，其中包含每个 key 的所有元素对，代码如下：

```
#构造两个键-值对 RDD
pairRDD1 = sc.parallelize([(1,2),(3,4),(3,6),(5,8)])
pairRDD2 = sc.parallelize([(1,3),(3,7),(4,9)])

#执行 join 连接
joinedRDD = pairRDD1.join(pairRDD2)

#查看
joinedRDD.collect()
```

执行以上代码，输出结果如下：

```
[(1,(2,3)), (3,(4,7)), (3,(6,7))]
```

9）leftOuterJoin(other)

对两个 RDD 按 key 执行左外连接，返回的新 RDD 中包含左侧 RDD 中的所有 key 和右侧 RDD 中满足连接条件的 key，代码如下：

```
#构造两个键-值对 RDD
pairRDD1 = sc.parallelize([(1,2),(3,4),(3,6),(5,8)])
pairRDD2 = sc.parallelize([(1,3),(3,7),(4,9)])

#执行左外连接
joinedRDD = pairRDD1.leftOuterJoin(pairRDD2)

#查看
joinedRDD.collect()
```

执行以上代码，输出结果如下：

```
[(1, (2, 3)), (5, (8, None)), (3, (4, 7)), (3, (6, 7))]
```

10）rightOuterJoin(other)

对两个 RDD 按 key 执行右外连接，返回的新 RDD 中包含右侧 RDD 中的所有 key 和左侧 RDD 中满足连接条件的 key，代码如下：

```
#构造两个键-值对 RDD
pairRDD1 = sc.parallelize([(1,2),(3,4),(3,6),(5,8)])
pairRDD2 = sc.parallelize([(1,3),(3,7),(4,9)])

#执行右外连接
joinedRDD = pairRDD1.rightOuterJoin(pairRDD2)

#查看
joinedRDD.collect()
```

执行以上代码，输出结果如下：

```
[(4, (None, 9)), (1, (2, 3)), (3, (4, 7)), (3, (6, 7))]
```

11）fullOuterJoin(other)

对两个 RDD 按 key 执行全外连接，返回的新 RDD 中包含两侧 RDD 中的所有 key，代码如下：

```
#构造两个键-值对 RDD
pairRDD1 = sc.parallelize([(1,2),(3,4),(3,6),(5,8)])
pairRDD2 = sc.parallelize([(1,3),(3,7),(4,9)])

#执行全外连接
joinedRDD = pairRDD1.fullOuterJoin(pairRDD2)

#查看
```

```
for line in joinedRDD.collect():
    print(line)
```

执行以上代码，输出结果如下：

```
(4, (None, 9))
(1, (2, 3))
(5, (8, None))
(3, (4, 7))
(3, (6, 7))
```

12) cogroup(other)

对来自两个 RDD 的数据按 key 分组，代码如下：

```
r5 = pairRDD1.cogroup(pairRDD2)

for t in r5.collect():
    print(t[0],end=": ")
    for v in t[1]:
        for i in v:
            print(i, end=" ")
    print()
```

执行以上代码，输出结果如下：

```
1: 2
3: 4 6 9
```

以下是 Pair RDD 上的 Action 操作。

1) countByKey()

这是一个 Action 操作，用于计算每个 key 的元素数量，将结果收集到一个本地字典中，代码如下：

```
stus = [("计算机系","张三"),("数学系","李四"),("计算机系","王老五"),("数学系","赵老六")]
rdd1 = sc.parallelize(stus)

kvRDD1 = rdd1.keyBy(lambda x:x[0])         #转换为 Pair RDD
kvRDD1.countByKey()
```

执行以上代码，输出结果如下：

```
defaultdict(<class 'int'>, {'计算机系': 2, '数学系': 2})
```

2) collectAsMap()

这是一个 Action 操作，将这个 RDD 中的键-值对作为字典返给 master。这种方法只应该在结果数据很小的情况下使用，因为所有的数据都被加载到驱动程序的内存中，代码如下：

```
stus = [("计算机系","张三"),("数学系","李四"),("计算机系","王老五"),("数学系",
"赵老六")]
rdd1 = sc.parallelize(stus)

kvRDD1 = rdd1.keyBy(lambda x:x[0])      #转换为 Pair RDD
kvRDD1.collectAsMap()
```

执行以上代码，输出结果如下：

```
{'计算机系': ('计算机系', '王老五'), '数学系': ('数学系', '赵老六')}
```

除了以上 Pair RDD 上的 Action 操作，还有较为复杂的 reduceByKey()、aggregateByKey()、combineByKey()等，后面会详细讲解。

3.5.3 关于 reduceByKey() 操作

在 3.5.2 节讲过，对 Pair RDD 执行 groupByKey()转换操作时会导致数据 shuffling，并且不会进行分区内优化，因此当需要对数据进行聚合操作时，不建议使用该操作。对于聚合操作，PySpark 提供了另外几个带有优化措施的方法，分别是 reduceByKey()、aggregateByKey()和 combinerByKey()，分别适用于不同的场景。首先来了解 reduceByKey()转换操作。

Pair RDD 的 reduceByKey()操作使用 reduce()函数合并每个 key 的值，这也是一个宽依赖的操作，因此也有可能发生跨分区的数据 shuffling，但是与 groupByKey()转换操作不同的是，当 reduceByKey()操作重复地应用于具有多个分区的同一组 RDD 数据时，它首先使用 reduce()函数在各个分区本地执行合并，然后跨分区发送记录以准备最终结果。也就是说，在跨分区发送数据之前，它还使用相同的 reduce()函数在本地合并数据，以减少数据 shuffling，优化传输性能。

【例 3-4】有如下一组数据，代码如下：

```
data = [("a", 1), ("b", 1), ("a", 1), ("a", 1), ("b", 1), \
        ("b", 1), ("a", 1), ("b", 1), ("a", 1), ("b", 1)]
```

将其构造为具有 3 个分区的 RDD，代码如下：

```
rdd = spark.sparkContext.parallelize(data, 3)
```

现在需要对该 RDD 中的数据求和，那么可以应用 reduceByKey() 操作，其工作过程如图 3-16 所示。

可以看到，这个 RDD 有 3 个分区。数据首先在各个分区执行 reduceByKey()操作，然后各个分区计算结果再通过 shuffling，相同的 key 会被分配到同一子分区中，然后在子分区上执行 reduceByKey()操作。由于在 shuffling 之前各个分区先进行了本地合并，所以会极大地减少 shuffling 的数据量。

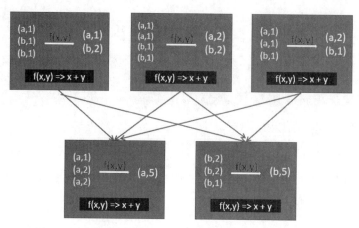

图 3-16　在键-值对 RDD 上执行 reduceByKey() 转换操作

以上过程的完整实现代码如下：

```
#假设，现在有这么一组数据
data = [("a", 1), ("b", 1), ("a", 1), ("a", 1), ("b", 1),
    ("b", 1), ("a", 1), ("b", 1), ("a", 1), ("b", 1)]

#创建 PairRDD x
x = spark.sparkContext.parallelize(data, 3)

#在 x 上应用 reduceByKey 操作
y = x.reduceByKey(lambda accum, n: accum + n)
print(y.collect())
```

执行上面的代码，输出结果如下：

```
[('b', 5), ('a', 5)]
```

当计算逻辑比较复杂时，也可以单独定义 reduce 函数，代码如下：

```
#单独定义关联函数
def sumFunc(accum, n):
    return accum + n

y = x.reduceByKey(sumFunc)
print(y.collect())
```

执行上面的代码，输出结果如下：

```
[('b', 5), ('a', 3)]
```

当应用 reduceByKey() 操作执行聚合计算时，要求计算结果的数据类型必须与输入元素的数据类型保持一致，也就是说不能改变 RDD 中元素的数据类型。例如上例中，父 RDD 的元素数据类型为 Tuple(String, Int)，那么子 RDD 中的计算结果的数据类型也必须是

Tuple(String, Int)。那么,如果希望在计算过程中改变 RDD 元素的数据类型(也就是子 RDD 的元素类型和父 RDD 的元素类型不同),就不能使用 reduceByKey()了,这时可以使用 PySpark 提供的另一个聚合转换操作 aggregateByKey()。

3.5.4　关于 aggregateByKey() 操作

PySpark 的 aggregateByKey()转换操作用于聚合每个 key 的值,使用给定的聚合函数和一个中性的"零值",并为该 key 返回不同类型的值。

这个 aggregateByKey 函数总共接受以下 3 个参数。

(1)zeroValue:它是累加值或累加器的初值。如果聚合类型是对所有的值求和,则它可以是 0。如果聚合目标是找出最小值,这个值则可以是 Double.MaxValue。如果聚合目标是找出最大值,这个值则可以使用 Double.MinValue。或者,如果只是想要一个各自的集合作为每个 key 的输出,则可以使用一个空的数组或字典对象。

(2)seqOp:是聚合单个分区的所有值的操作。它将一种类型 V 的数据转换/合并为另一种类型 U 的序列操作函数。

(3)combOp:类似于 seqOp,进一步聚合来自不同分区的所有聚合值。它将多个转换后的类型 U 合并为一个单一类型 U 的组合操作函数。

【例 3-5】有这么一组代表学生课目和成绩的数据,代码如下:

```
student_data = [
        ("Joseph", "Maths", 83), ("Joseph", "Physics", 74),
        ("Joseph", "Chemistry", 91), ("Joseph", "Biology", 82),
        ("Jimmy", "Maths", 69), ("Jimmy", "Physics", 62),
        ("Jimmy", "Chemistry", 97), ("Jimmy", "Biology", 80),
        ("Tina", "Maths", 78), ("Tina", "Physics", 73),
        ("Tina", "Chemistry", 68), ("Tina", "Biology", 87),
        ("Thomas", "Maths", 87), ("Thomas", "Physics", 93),
        ("Thomas", "Chemistry", 91), ("Thomas", "Biology", 74),
        ("Cory", "Maths", 56), ("Cory", "Physics", 65),
        ("Cory", "Chemistry", 71), ("Cory", "Biology", 68),
        ("Jackeline", "Maths", 86), ("Jackeline", "Physics", 62),
        ("Jackeline", "Chemistry", 75), ("Jackeline", "Biology", 83),
        ("Juan", "Maths", 63), ("Juan", "Physics", 69),
        ("Juan", "Chemistry", 64), ("Juan", "Biology", 60)], 2)
    ]
```

其中元素三元组分别代表一个学生的姓名、课目、成绩。每个学生都有四门课的成绩,分别是 Maths、Physics、Chemistry 和 Biology。现在要求找出每个学生的最好成绩,以(姓名,成绩)的形式输出,该怎么实现呢?

从需求可以得知,最终输出 RDD 的元素类型为 Tuple(String, Double),而输入 RDD

的元素类型为 Tuple3(String, String, Double)，因此不适用 reduceByKey()操作，这时可以使用 aggregateByKey()方法。实现过程如图 3-17 所示（RDD 元素是随机分布到各个分区的）。

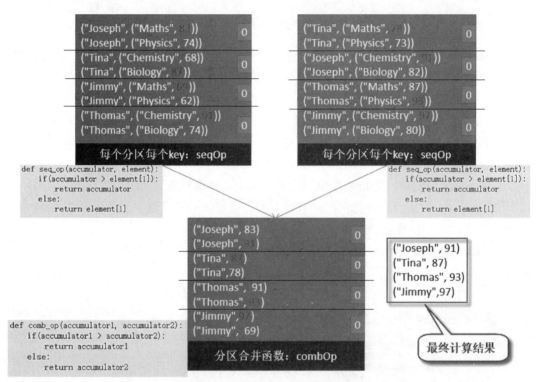

图 3-17　在键-值对 RDD 上执行 aggregateByKey() 转换操作

以上过程的实现代码如下：

```
#使用键-值对创建 PairRDD studentRDD
student_data = [
        ("Joseph", "Maths", 83), ("Joseph", "Physics", 74),
        ("Joseph", "Chemistry", 91), ("Joseph", "Biology", 82),
        ("Jimmy", "Maths", 69), ("Jimmy", "Physics", 62),
        ("Jimmy", "Chemistry", 97), ("Jimmy", "Biology", 80),
        ("Tina", "Maths", 78), ("Tina", "Physics", 73),
        ("Tina", "Chemistry", 68), ("Tina", "Biology", 87),
        ("Thomas", "Maths", 87), ("Thomas", "Physics", 93),
        ("Thomas", "Chemistry", 91), ("Thomas", "Biology", 74),
        ("Cory", "Maths", 56), ("Cory", "Physics", 65),
        ("Cory", "Chemistry", 71), ("Cory", "Biology", 68),
        ("Jackeline", "Maths", 86), ("Jackeline", "Physics", 62),
        ("Jackeline", "Chemistry", 75), ("Jackeline", "Biology", 83),
```

```
            ("Juan", "Maths", 63), ("Juan", "Physics", 69),
            ("Juan", "Chemistry", 64), ("Juan", "Biology", 60)], 2)
]
student_rdd = spark.sparkContext.parallelize(student_data, 2)

#定义Sequence Operation 和 Combiner Operations
#Sequence operation : 从单个分区查找最高成绩
def seq_op(accumulator, element):
    if(accumulator > element[1]):
        return accumulator
    else:
        return element[1]

#Combiner Operation : 从所有分区累加器中找出最高成绩
def comb_op(accumulator1, accumulator2):
    if(accumulator1 > accumulator2):
        return accumulator1
    else:
        return accumulator2

#在此情况下，零值将是0，因为要寻找最高成绩
zero_val = 0
aggr_rdd = student_rdd \
    .map(lambda t: (t[0], (t[1], t[2]))) \
    .aggregateByKey(zero_val, seq_op, comb_op)

#查看输出
for tpl in aggr_rdd.collect():
    print(tpl)
```

执行以上代码，输出结果如下：

```
('Jimmy', 97)
('Tina', 87)
('Thomas', 93)
('Joseph', 91)
('Cory', 71)
('Jackeline', 86)
('Juan', 69)
```

在上例的基础上，修改代码，要求同时输出每个学生的最高成绩及该成绩所属的课程。实现代码如下：

```
#定义Sequence Operation 和 Combiner Operations
def seq_op(accumulator, element):
    if(accumulator[1] > element[1]):
        return accumulator
```

```
        else:
            return element

#Combiner Operation：从所有分区累加器中找出最高成绩
def comb_op(accumulator1, accumulator2):
    if(accumulator1[1] > accumulator2[1]):
        return accumulator1
    else:
        return accumulator2

#在此情况下，零值将是 0，因为要寻找最高成绩
zero_val = ('', 0)
aggr_rdd = student_rdd \
    .map(lambda t: (t[0], (t[1], t[2]))) \
    .aggregateByKey(zero_val, seq_op, comb_op)

#查看输出
for tpl in aggr_rdd.collect():
    print(tpl)
```

执行以上代码，输出结果如下：

```
('Jimmy', ('Chemistry', 97))
('Tina', ('Biology', 87))
('Thomas', ('Physics', 93))
('Joseph', ('Chemistry', 91))
('Cory', ('Chemistry', 71))
('Jackeline', ('Maths', 86))
('Juan', ('Physics', 69))
```

在上例的基础上，修改代码，要求计算所有学生的平均成绩。要计算平均成绩，意味着要计算出每个学生的总成绩及课目数，两者相除，结果就是该学生的平均成绩。

实现代码如下：

```
#定义 Sequencial Operation 和 Combiner Operations
def seq_op(accumulator, element):
    return (accumulator[0] + element[1], accumulator[1] + 1)

#Combiner Operation：从所有分区累加器中找出最高成绩
def comb_op(accumulator1, accumulator2):
    return (accumulator1[0] + accumulator2[0], accumulator1[1] + accumulator2[1])

#Zero Value：在此情况下，零值将是 0，因为需要计算各科成绩之和
zero_val = (0, 0)
aggr_rdd = student_rdd.map(lambda t: (t[0], (t[1], t[2])))\
                .aggregateByKey(zero_val, seq_op, comb_op)\
```

```
                .map(lambda t: (t[0], t[1][0]/t[1][1]*1.0))

#查看输出
for tpl in aggr_rdd.collect():
    print(tpl)
```

执行以上代码，输出结果如下：

```
('Jimmy', 77.0)
('Tina', 76.5)
('Thomas', 86.25)
('Joseph', 82.5)
('Cory', 65.0)
('Jackeline', 76.5)
('Juan', 64.0)
```

注意：aggregateByKey()是一种宽依赖的操作，会导致数据跨分区的 shuffling，但它是一个对性能进行优化了的 Transformation 操作，在 shuffling 之前会先在各个分区上执行聚合，再 shuffling 聚合之后的数据。与 reduceByKey()相比，当聚合要求输入和输出 RDD 类型不同时，应该使用 aggregateByKey()；当聚合要求输入和输出 RDD 类型相同时，应该使用 reduceByKey()。

3.5.5 关于 combineByKey() 操作

除了前面讲过的 reduceByKey()和 aggregateByKey()这两个转换操作之外，PySpark 还为 Pair RDD 提供了一个更有用的 combineByKey()操作。它使用一组自定义的聚合函数组合每个 key 的元素，在其内部 combineByKey()操作会按分区合并元素。

Pair RDD 的 combineByKey()转换与 Hadoop MapReduce 编程中的 combiner 非常相似。它也是一个宽依赖的操作，在最后阶段需要 shuffle 数据。

combineByKey()操作使用如下 3 个函数作为参数。

（1）createCombiner()：在第 1 次遇到一个 Key 时创建组合器函数，将 RDD 数据集中的 V 类型 value 值转换 C 类型值（V => C）。

（2）mergeValue()：mergeValue 是合并值函数，当再次遇到相同的 Key 时，将 createCombiner 的 C 类型值与这次传入的 V 类型值合并成一个 C 类型值（C,V)=>C。

（3）mergeCombiners()：合并组合器函数，将 C 类型值两两合并成一个 C 类型值。

这个转换操作是 PySpark RDD 中最复杂最难理解的一个操作，下面通过一个应用示例来理解它的用法。

【例 3-6】假设有一组销售数据，数据采用键-值对的形式（公司，收入），代码如下：

```
val data = Array(
    ("company-1",92),("company-1",85),("company-1",82),
```

```
    ("company-1",93),("company-1",86),("company-1",83),
    ("company-2",78),("company-2",96),("company-2",85),
    ("company-3",88),("company-3",94),("company-3",80)
)
```

现在要求使用 combineByKey() 统计出每个公司的总收入和平均收入。实现原理和过程如图 3-18 所示（考虑到 RDD 中的元素是随机分布在各个分区上的）。

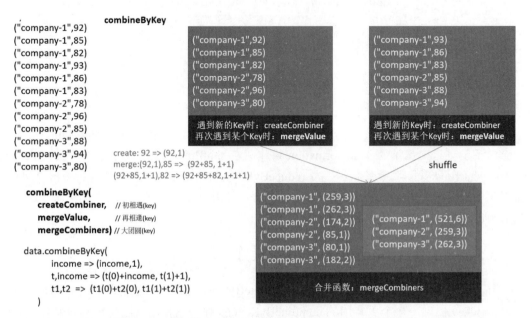

图 3-18 在键-值对 RDD 上执行 combineByKey() 转换操作

基于图 3-18 中的转换步骤，进一步理解 combineByKey() 操作各个函数参数的含义。

1）createCombiner() 函数

这个函数是 combineByKey() 操作的第 1 个参数，它是每个 key 的第 1 个聚合步骤。当在一个分区中发现任何新的 key 时，该函数将被执行。这个函数的执行在每个单独的值上对一个节点的分区都是局部的，它类似于 aggregateByKey() 函数的第 1 个参数（zeroVal 零值）。

2）mergeValue() 函数

这个函数是 combineByKey() 操作的第 2 个参数，它在将同一 key 的下一个值赋给 combiner 时执行。它还在节点的每个分区上本地执行，并合并同一个 key 的所有值。这个函数的参数是一个累加器和一个新的输入值，它在现有累加器中合并输入的新值。这个函数类似于 aggregateByKey() 转换操作的第 2 个参数（seqOp）。

3）mergeCombiners() 函数

这个函数是 combineByKey() 操作的第 3 个参数，用于组合如何跨分区合并单个 key 的两个累加器（combiner）以生成最终的预期结果。它的参数是两个累加器（combiner），用于合并来自不同分区的单个 key 的结果。这个函数类似于 aggregateByKey()函数的第 3 个参数（combOp）。

基于上面的理解，使用 combineByKey 实现的代码如下：

```
#各个公司的销售数据
data = [
    ("company-1",92),("company-1",85),("company-1",82),
    ("company-1",93),("company-1",86),("company-1",83),
    ("company-2",78),("company-2",96),("company-2",85),
    ("company-3",88),("company-3",94),("company-3",80)
]


#构造一个RDD，并指定两个分区
rdd = sc.parallelize(data, 2)

#执行聚合转换
cbk = rdd.combineByKey(
        lambda income: return (income,1),
        lambda acc, income: return (acc[0]+income, acc[1]+1),
        lambda acc1, acc2: return (acc1[0]+acc2[0], acc1[1]+acc2[1])
        )

#聚合结果
for v in cbk.collect():
    print(v)
```

执行上面的代码，输出结果如下：

```
(company-2,(259,3))
(company-3,(262,3))
(company-1,(521,6))
```

上面输出的数据形式是"(公司,(总销售额,月份数量))"，而要求是统计出每个公司的总收入和平均收入，因此需要以上面的计算结果进一步进行转换，代码如下：

```
#提取每个元素值，并计算出平均收入
res = cbk.map(lambda t: (t, v[0], v[0]/v[1]))

#输出结果
for r in res.collect():
    print(r)
```

执行上面的代码，最终计算结果如下：

```
(company-2,259,86.333336)
```

```
(company-3,262,87.333336)
(company-1,521,86.833336)
```

注意：combineByKey()是一种宽依赖的操作，它在聚合的最后阶段会shuffle数据并创建另一个RDD。它是一个通用的转换，groupByKey()、reduceByKey()和aggregateByKey()转换的内部实现都使用了combineByKey()。combineByKey()转换操作可灵活地执行map端或reduce端combine合并操作，因此combineByKey()转换的使用更加复杂。要实现combineByKey()，总需要实现3个函数：createCombiner()、mergeValue()、mergeCombiner()。

3.6 持久化RDD

PySpark中最重要的功能之一是跨操作在内存中持久化（或缓存）数据集。当持久化一个RDD时，每个节点在内存中存储它计算的任何分区，并在该数据集（或从该数据集派生的数据集）上的其他操作中重用它们。这使后续的操作要快得多（通常超过10倍）。缓存是迭代算法和快速交互使用的关键工具。

3.6.1 缓存RDD

在PySpark中，RDD采用惰性求值的机制，每次遇到Action操作时，PySpark都会从头重新计算RDD及其所有的依赖。这对于迭代计算而言，代价是很大的，因为迭代计算经常需要多次重复使用同一组数据。例如，多次计算同一个RDD，代码如下：

```
list = ["Hadoop","Spark","Hive"]
input = sc.parallelize(list)    #构造一个RDD

result = input.map(lambda x: x.upper())    #Transformation

result.count()                  #Action操作，触发一次真正从头到尾的计算
",".join(result.collect())      #Action操作，又触发一次真正从头到尾的计算
```

在上面的代码中，有两次Action操作：count()和collect()。每遇到一次Action操作，都重新构造RDD、执行map转换。可以通过PySpark中的持久化（缓存）机制避免这种重复计算的开销。

缓存机制是PySpark提供的一种将数据缓存到内存（或磁盘）的机制，主要用途是使中间计算结果可以被重用。要缓存RDD，常用到两个函数：cache()和persist()。

可以使用persist()方法将一个RDD标记为持久化。之所以说"标记为持久化"，是因为出现persist()语句的地方，并不会马上计算生成RDD并把它持久化，而是要等到遇到第1个Action操作触发真正计算以后，才会把计算结果进行持久化。持久化后的RDD分区将会被保留在计算节点的内存中，可被后面的Action操作重复使用。

如果一个 RDD 数据集被要求参与几个 Action，则持久化该 RDD 数据集会节省大量的时间、CPU 周期、磁盘输入/输出和网络带宽。容错机制也适用于缓存分区。当由于节点故障而丢失任何分区时，它将使用血统图重新计算。

PySpark 还会在随机操作（如 reduceByKey()）中自动保存一些中间数据，甚至不需要用户调用 persist()。这样做是为了避免节点在 shuffling 期间失败时重新计算整个输入。

重写前面的示例，加入对 RDD 进行缓存的代码，代码如下：

```
list = ["Hadoop","Spark","Hive"]
input = sc.parallelize(list)
result = input.map(lambda x: x.upper())        #Transformation

#缓存RDD，但是，语句执行到这里，并不会缓存RDD，这时RDD还没有被计算生成
#使用cache()方法时，会调用persist(MEMORY_ONLY)，即
result.persist(StorageLevel.MEMORY_ONLY)
result.cache()

#判断RDD是否缓存
result.is_cached

#清除RDD的缓存
result.unpersist()

#查看RDD是否缓存
result.is_cached

#第1次Action操作，触发一次真正从头到尾的计算，这时才会执行上面的rdd.cache()，
把这个RDD放到缓存中
result.count()

#第2次Action操作，不需要触发从头到尾的计算，只需重复使用上面缓存中的RDD
",".join(result.collect())

#把持久化的RDD从缓存中移除
result.unpersist()
```

PySpark RDD 的 cache() 方法默认将 RDD 计算保存到存储级别 MEMORY_ONLY，这意味着它将把数据作为未序列化的对象存储在 JVM 堆中。

PySpark RDD 类中的 cache() 方法在内部实际上调用了 persist() 方法，persist() 方法使用 sparkSession.sharedState.cacheManager.cacheQuery 来缓存 RDD 的结果集。

如果可用内存不足，PySpark 则会将持久的分区溢写到磁盘上。开发人员可以使用 unpersist() 方法删除不需要的 RDD。PySpark 会自动监控缓存，并使用 LRU（Least Recently Used，最近最少使用）算法删除旧分区。PySpark 的缓存不仅能将数据缓存到内存，也能

使用磁盘，甚至同时使用内存和磁盘，这种缓存的不同存储方式，称作 StorageLevel（存储级别）。使用 persist() 方法指定存储级别，代码如下：

```
import pyspark

rddPersist = rdd.persist(pyspark.StorageLevel.MEMORY_ONLY)
```

PySpark persist() 方法有两个签名，第 1 个签名不带任何参数，默认保存为 MEMORY_ONLY 存储级别，第 2 个签名以 StorageLevel 作为参数，将以不同的存储级别进行存储。

在 Python 中，存储的对象将始终使用 Pickle 库进行序列化，所以是否选择序列化级别并不重要。Python 中可用的存储级别包括 MEMORY_ONLY、MEMORY_AND_DISK、MEMORY_ONLY_SER、MEMORY_AND_DISK_SER、DISK_ONLY、MEMORY_ONLY_2、MEMORY_AND_DISK_2。

PySpark 会自动监控每次 persist() 和 cache() 调用，并检查每个节点上的使用情况，如果未使用或使用 LRU 算法，则删除持久化数据。也可以使用 unpersist() 方法手动删除。unpersist() 将 RDD 标记为非持久化，并从内存和磁盘中删除它的所有块，代码如下：

```
rddUnPersist = rddPersist.unpersist()
```

3.6.2 RDD 缓存策略

PySpark 支持的所有不同存储级别都可以在 org.apache.spark.storage.StorageLevel 类中找到。存储级别用于定义如何及在哪里存储 RDD。RDD 块可以在多个存储（内存、磁盘、堆外）中以序列化或非序列化格式缓存。

（1）MEMORY_ONLY：这是 RDD cache() 方法的默认行为，并将 RDD 作为反序列化对象存储到 JVM 内存中。当没有足够的可用内存时，它将不会保存一些分区的 RDD，并在需要时重新计算这些分区。这需要更多的存储空间，但运行速度更快，因为从内存中读取所需的 CPU 周期很少。

（2）MEMORY_ONLY_SER：这与 MEMORY_ONLY 相同，但不同的是它将 RDD 作为序列化对象存储到 JVM 内存中。它比 MEMORY_ONLY 占用更少的内存（节省空间），因为它以序列化的方式保存对象，并且为了反序列化而多占用一些 CPU 周期。

（3）MEMORY_ONLY_2：与 MEMORY_ONLY 存储级别相同，但将每个分区复制到两个集群节点。

（4）MEMORY_ONLY_SER_2：与 MEMORY_ONLY_SER 存储级别相同，但将每个分区复制到两个集群节点。

（5）MEMORY_AND_DISK：在这个存储级别中，RDD 将作为反序列化对象存储在 JVM 内存中。当所需的存储空间大于可用内存时，它将一些多余的分区存储到磁盘中，并在需要时从磁盘读取数据。由于涉及 I/O，所以速度较慢。当重新计算很昂贵且内存资源稀缺时，建议使用这种操作模式。

（6）MEMORY_AND_DISK_SER：这与 MEMORY_AND_DISK 存储级别的差异是一样的，当空间不可用时，它将 RDD 对象序列化在内存和磁盘上。

（7）MEMORY_AND_DISK_2：与 MEMORY_AND_DISK 存储级别相同，但将每个分区复制到两个集群节点。

（8）MEMORY_AND_DISK_SER_2：与 MEMORY_AND_DISK_SER 存储级别相同，但将每个分区复制到两个集群节点。

（9）DISK_ONLY：在这种存储级别中，数据仅以序列化格式缓存在磁盘上。由于涉及 I/O，所以 CPU 计算时间较长。

（10）DISK_ONLY_2：与 DISK_ONLY 存储级别相同，但必须将每个分区复制到两个集群节点。

（11）OFF_HEAP：与 MEMORY_ONLY_SER 类似，但必须将数据存储在堆外内存中，例如，在 Alluxio 上。这需要启用堆外内存。

上面的缓存策略还可以使用序列化来以序列化格式存储数据。序列化增加了处理成本，但减少了大型数据集的内存占用。将 _SER 后缀附加到上述策略名上表示以使用序列化，例如，MEMORY_ONLY_SER、MEMORY_AND_DISK_SER、DISK_ONLY 和 OFF_HEAP 始终以序列化格式写入数据。

数据也可以通过在 StorageLevel 上添加 _2 后缀来表示将缓存复制到另一个节点，例如 MEMORY_ONLY_2、MEMORY_AND_DISK_SER_2。当集群的一个节点（或执行器）出现故障时，复制有助于加速恢复。

那么如何选择合适的缓存策略呢？

Spark 的不同存储级别意味着在内存使用和 CPU 效率之间可以进行权衡。建议通过以下步骤来选择一个：

（1）如果 RDD 适合默认存储级别（MEMORY_ONLY），就保留它们。这是 CPU 效率最高的选项，允许 RDD 上的操作尽可能快地运行。

（2）如果没有，则可以尝试使用 MEMORY_ONLY_SER 并选择一个快速序列化库，以使对象更节省空间，但访问速度仍然相当快（Java 和 Scala）。

（3）不要溢出到磁盘，除非计算数据集的函数开销很大，或者它们过滤了大量数据。否则重新计算分区的速度可能与从磁盘读取分区的速度一样快。

（4）如果希望快速恢复故障（例如，如果使用 Spark 为来自 Web 应用程序的请求提供服务），则可使用复制的存储级别。通过重新计算丢失的数据，所有存储级别都提供了完全的容错能力，但是复制的存储级别允许在 RDD 上继续运行任务，而无须等待重新计算丢失的分区。

3.6.3　检查点 RDD

通过链接任意数量的 Transformation 转换操作，RDD lineage（血统）可以任意增长。PySpark 提供了一种方法，可以将整个 RDD 持久化到稳定的存储器中，存储的数据包括

RDD 计算后的数据和分区器，然后，在发生节点故障时，PySpark 不需要从头开始计算丢失的 RDD 碎片，而是从存储的快照那里开始计算 lineage 中其余的部分。这个特性称为"检查点（checkpoint）"。

简单来讲，检查点是一种截断 RDD 依赖链并把 RDD 数据持久化到存储系统（通常是 HDFS 或本地）的过程。它的主要作用是截断 RDD 的依赖关系，防止任意增长的 lineage 导致的堆栈溢出。在检查点之后，RDD 的依赖项，以及它的父 RDD 的信息会被擦除，因为重新计算不再需要它们了。

必须先调用 SparkContext 的 setCheckpointDir() 方法设置保存数据的目录，然后通过调用 checkpoint 操作来对 RDD 设置检查点，这时该 RDD 将被保存到检查点目录中的一个文件中，所有对其父 RDD 的引用将被删除。

RDD 检查点的调用必须在这个 RDD 上执行任何作业之前。当在 RDD 上调用 checkpoint() 方法时，仅是将此 RDD 标记为检查点。必须在 RDD 上调用了 Action 操作才能完成检查点。

将 RDD 标记为检查点，该 RDD 将被保存到用 SparkContext.setCheckpointDir() 设置的检查点目录下的一个文件中，所有对其父 RDD 的引用将被删除。该函数必须在该 RDD 上执行任何作业之前调用。强烈建议将该 RDD 持久化到内存中，否则将其保存到文件中将需要重新计算。

例如，设置 RDD checkpoint，代码如下：

```
sc.setCheckpointDir("hdfs://localhost:8020/ck/rdd")  #设置 RDD 的检查点目录

data = sc.textFile("hdfs://localhost:8020/input")    #加载数据源
rdd = data.map(...).reduceByKey(...)

rdd.checkpoint()  #对 RDD 标记 checkpoint，不会真正执行，直到遇到第 1 个 Action 算子
rdd.count()       #第 1 个 Action 算子，触发之前的代码执行
```

在使用检查点时，需要注意它与缓存的区别。缓存采用临时保存，Executor 挂掉会导致数据丢失，但是数据可以重新计算，而检查点则用于截断依赖链，可靠方式下 Executor 挂掉不会丢失数据，但数据一旦丢失则不可恢复。

尽管 PySpark 会自动管理（包括创建和回收）cache() 和 persist() 持久化的数据，但是 checkpoint 持久化的数据需由用户自己管理。检查点会清除 RDD 的血统信息，避免血统过长导致序列化开销增大，而 cache() 和 persist() 不会清除 RDD 的血统。

3.7 数据分区

34min

数据分区（partition）是 PySpark 中的重要概念，是 PySpark 在集群中的多个节点之间划分数据的机制。分区是 RDD 的最小单元，RDD 是由分布在各个节点上的分区组成的。

PySpark 使用分区来管理数据，分区的数量决定了任务（task）的数量，每个任务对应着一个数据分区。这些分区有助于并行化分布式数据处理。

3.7.1 获取和指定 RDD 分区数

默认情况下，为每个 HDFS block 创建一个分区，该分区默认为 128MB（Spark 2.x）。例如，当从本地文件系统将一个文本文件加载到 PySpark 时，文件的内容会被分成几个分区，这些分区均匀地分布在集群中的节点上。在同一个节点上可能会出现不止一个分区。所有这些分区的总和形成了 RDD。这就是"弹性分布数据集"中"分布"一词的来源。

例如，加载一个数据集，并指定该数据集被分割为 3 个分区，分别存储在集群的 3 个节点上，如图 3-19 所示。

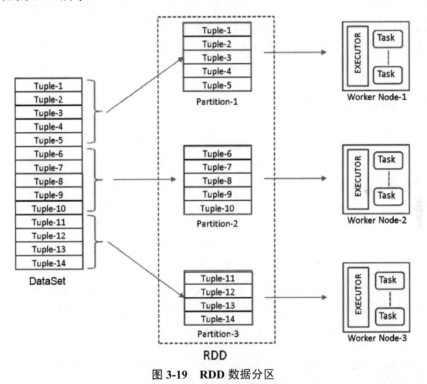

图 3-19 RDD 数据分区

每个 RDD 维护一个分区的列表和一个可选的首选位置列表，用于计算分区。分区的首选位置是一个分区的数据所驻留的主机名或 executors 的列表，这样计算就可以更靠近数据了（如果 PySpark 获得了首选位置的列表，则 PySpark 调度程序会试着在数据实际存在的执行器上运行任务，这样就不需要进行数据传输了。这对性能有很大的影响）。可以从 RDD 的 partitions 字段获得 RDD 分区的列表。它是一个数组（Array），所以可以通过读取 RDD 的 partitions.size 字段来获得该 RDD 分区的数量。

例如，查看 RDD 的分区数量，代码如下：

```
#构造一个 Pair RDD
pairs = sc.parallelize( [(1, 1), (2, 2), (3, 3)] )

#查看分区数量
pairs.getNumPartitions()
```

也可以在创建 RDD 时指定分区的数量。例如，在调用 textFile()和 parallelize()方法创建 RDD 时，可以手动指定分区个数。这两种方法的签名如下：

```
SparkContext.textFile(name: str, minPartitions: Optional[int] = None,
use_unicode: bool = True) → pyspark.rdd.RDD[str]

SparkContext.parallelize(c: Iterable[T], numSlices: Optional[int] = None) →
pyspark.rdd.RDD[T]
```

如果在创建 RDD 时未指定 RDD 的分区数量，则 PySpark 将使用默认分区数，默认值为 spark.default.parallelism 配置的参数值。

3.7.2 调整 RDD 分区数

从数据源生成 RDD 时，数据通常被随机分配到不同的分区或者保持数据源的分区。RDD 分区数的多少，会对 PySpark 程序的执行产生一定的影响。因为除了影响整个集群中的数据分布之外，它还直接决定了将要运行 RDD 转换的任务的数量。

如果分区数量太少，则直接的影响是集群计算资源不能被充分利用。例如分配 8 个核，但分区数量为 4，则将有一半的核没有利用到。此外，因为数据集可能会变得很大，当无法装入 executor 的内存中时可能会导致内存问题。

如果分区数量太多，虽然计算资源能够充分利用，但会导致 task 数量过多，而 task 数量过多会影响执行效率，主要是 task 在序列化和网络传输过程中会带来较大的时间开销。

根据 Spark RDD Programming Guide 上的建议，集群节点的每个核分配 2~4 个分区比较合理，也就是说，建议将分区数设置为集群中 CPU 核数的 3~4 倍，如图 3-20 所示。

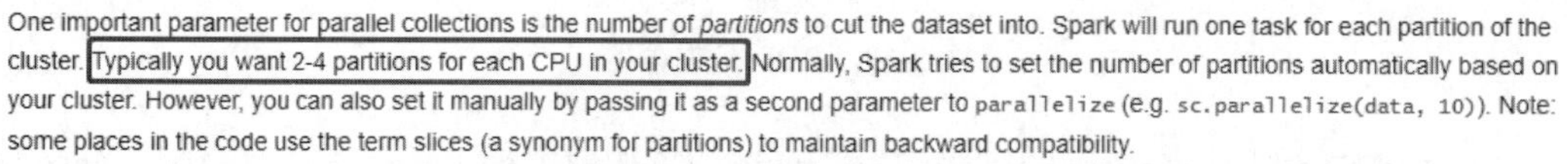

One important parameter for parallel collections is the number of *partitions* to cut the dataset into. Spark will run one task for each partition of the cluster. Typically you want 2-4 partitions for each CPU in your cluster. Normally, Spark tries to set the number of partitions automatically based on your cluster. However, you can also set it manually by passing it as a second parameter to `parallelize` (e.g. `sc.parallelize(data, 10)`). Note: some places in the code use the term slices (a synonym for partitions) to maintain backward compatibility.

图 3-20　Spark RDD Programming Guide 建议的分区数

在某些情况下，为了更有效地分配工作负载或避免内存问题，需要显式地重新划分 RDD 的分区。例如，从 HDFS 上加载压缩文件时，因为压缩文件不能被分片，所以只能有一个 RDD 分区，即使在 sc.textFile("xxx.gz", 100)中指定了分区数也不会真的进行分区。在这种情况下，就需要对 RDD 调用 repartition(N)方法进行重分区。

最主要的两种调整数据分区的方法是 coalesce()和 repartition()函数。函数 coalesce()用于减少或增加分区的数量。完整的方法签名如下：

```
RDD.coalesce(numPartitions: int, shuffle: bool = False) →
pyspark.rdd.RDD[T]
```

第 2 个（可选）布尔参数 shuffle 用于指定是否应该执行 shuffle（默认值为 False）。也就是说，coalesce()是否触发 RDD 的 shuffling，这取决于 shuffle 标志（默认禁用，即 False）。在减少分区时，coalesce()并没有对所有数据进行移动，仅仅是在原来分区的基础之上进行了合并而已，这样的操作可以减少数据的移动，所以效率较高。

函数 repartition()也可以增加或减少分区数量。完整的方法签名如下：

```
RDD.repartition(numPartitions: int) → pyspark.rdd.RDD[T]
```

如果查看 repartition()函数的源码，则可以看到 repartition()直接调用了 coalesce(numPartitions, shuffle=True)，repartition(N)方法相当于 coalesce(N, true)方法。不同的是，调用 repartition()函数时，还会产生 shuffle 操作（该操作与 HiveQL 的 DISTRIBUTE BY 操作类似），而 coalesce()函数可以控制是否 shuffling，但当 shuffle 为 False 时，只能减小分区数，而无法增大分区数，因此，如果要减少 RDD 的分区数，则建议使用 coalesce()而不是 repartition()函数，这样可以避免数据跨分区间的 shuffling。

例如，对一个 RDD 重新进行分区，并查看分区数，代码如下：

```
#在创建时，指定分区个数
rdd1 = sc.parallelize([1,2,3,4,5,6,7,8], 4)
print(rdd1.collect())              #[1, 2, 3, 4, 5, 6, 7, 8]
print(rdd1.getNumPartitions())            #rdd1 的分区数量，目前为 4

#对于通过转换得到的新 RDD，直接调用 repartition 方法重新分区
rdd2 = rdd1.map(lambda x: x*x)     #转换得到 rdd2
print(rdd2.collect())              #[1, 4, 9, 16, 25, 36, 49, 64]

rdd3 = rdd2.repartition(8)         #重新分区，得到 rdd3
print(rdd3.getNumPartitions())     #rdd3 的分区数量，这时为 8
```

执行上面的代码，输出内容如下：

```
4
8
```

3.7.3　内置数据分区器

当需要对 Pair RDD 进行重分区时，RDD 的分区由 portable_hash()函数执行，该分区程序将一个分区索引赋给每个 RDD 元素（在每个 key 和分区 ID 间建立起映射，分区 ID 的

值从 0 到 numPartitions - 1）。portable_hash()函数是 PySpark 的默认分区程序，这个函数为内置类型返回一致的哈希码，特别是对于 None 和具有 None 的元组，其定义如下：

```
def portable_hash(x):
    """
    This function returns consistent hash code for builtin types, especially
    for None and tuple with None.
    The algorithm is similar to that one used by CPython 2.7

    Examples
    --------
    >>> portable_hash(None)
    0
    >>> portable_hash((None, 1)) & 0xffffffff
    219750521
    """

    if "PYTHONHASHSEED" not in os.environ:
        raise RuntimeError("Randomness of hash of string should be disabled via PYTHONHASHSEED")

    if x is None:
        return 0
    if isinstance(x, tuple):
        h = 0x345678
        for i in x:
            h ^= portable_hash(i)
            h *= 1000003
            h &= sys.maxsize
        h ^= len(x)
        if h == -1:
            h = -2
        return int(h)
    return hash(x)
```

分区索引是准随机的，因此，分区很可能不会拥有完全相同的大小，然而，在具有相对较少分区的大型数据集中，该算法可能会在其中均匀地分布数据。

当使用 portable_hash()时，数据分区的默认数量是由配置参数 spark.default.parallelism 决定的。如果用户没有指定该参数，则它将被设置为集群中的核的数量。

可以使用 partitionBy()方法对 Pair RDD 进行重分区，该方法的定义如下：

```
RDD.partitionBy(numPartitions: Optional[int], partitionFunc: Callable[[K], int] = <function portable_hash>) → pyspark.rdd.RDD[Tuple[K, V]][source]
```

当在 Pair RDD 上调用 partitionBy()方法进行重分区时，需要向它传递一个参数，即期望的分区数量。如果分区程序与之前使用的分区程序相同，则保留分区，RDD 保持不变。

否则就会安排一次 shuffle，并创建一个新的 RDD。

【例 3-7】调用 partitionBy()方法对指定的 RDD 进行重分区，并指定分区数。
代码如下：

```
pairs = sc.parallelize([(1, 1), (2, 2), (3, 3)])
print(pairs.partitioner)            #查看所使用的分区器        None
print(pairs.getNumPartitions())     #初始分区数量              2

repairs = pairs.repartition(4)      #重新分区
print(repairs.getNumPartitions())   #重分区之后的数量          4

partitionedRDD = pairs.partitionBy(2)   #使用指定的分区数
partitionedRDD.persist()            #持久化，以便后续操作重复使用 partitionedRDD

print(partitionedRDD.partitioner)   #查看所使用的分区器
print(partitionedRDD.getNumPartitions())    #查看分区数量       2
```

在上面的代码中，传给 partitionBy()的参数值为 4，代表重分区的数量，它将控制将来在该 RDD 上执行操作时的并行任务数。一般来讲，这个值至少与集群中核的数量一样多。

需要注意的是，只有当一个 RDD 数据集在面向 key 的操作中被重用多次的情况下（例如 join()操作），通过 partitionBy()控制分区才有意义。例如，不带 partitionBy()的 Pair RDD join 过程如图 3-21 所示。

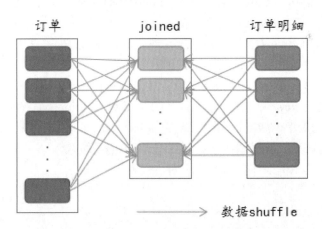

图 3-21　不带 partitionBy() 的 Pair RDD join 过程

从图 3-21 可以看出，因为 Pair RDD 中的 key 默认为随机分布的，所以同一个 key 可分散分布在一个 Pair RDD 的各个分区中，如果不事先用 partitionBy() 进行重分区，则在执行 join 连接时，就会产生大量的数据跨分区 shuffle，严重影响性能。

在这种情况下，可先在 Pair RDD 上调用 partitionBy()控制重分区，让相同的 key 事先分布到同一个分区上，然后执行 join 连接，过程如图 3-22 所示。

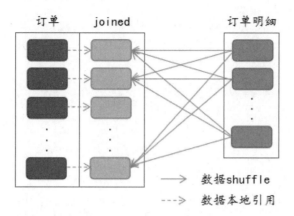

图 3-22 带 partitionBy() 的 Pair RDD join 过程

从图 3-22 可以看出，由于订单数据集先按 key 进行了分区，相同 key 的数据位于同一个分区上，因此在执行 join 连接时，这些数据在本地引用不会产生数据 shuffle。

另外，partitionBy()也经常用于在存储 RDD 时控制输出的结果文件数量。因为默认情况下 RDD 的每个分区对应一个输出文件，当分区数量很大时，将数据写到文件中会产生大量小文件，所以可以将数据写到文件之前通过 partitionBy()来调整分区数，从而控制输出结果文件的数量。

从前面的内容可知，repartition()方法和 partitionBy()方法都可以调整 RDD 的分区数，但这两种方法是有区别的。repartition()不能指定分区函数，它主要用于根据内核数量和数据量指定分区的数量。repartition()实际上是通过在现有值上添加随机键来在内部使用 Pair RDD，因此它不能对输出数据的分布提供强有力的保证。

partitionBy()可以接受一个分区函数，或者默认使用 portable_hash()，所以在 partitionBy()函数中，所有相同的 key 应该在同一个分区中。使用 partitionBy()方法的主要目的是用于提高 shuffle 函数的效率，如 reduceByKey()、join()、cogroup()等。它只有在 RDD 被多次使用的情况下才有好处，所以通常它后面跟着 persist()以缓存。

通过一个简单的示例来体会两者的区别，代码如下：

```
pairs = sc
   .parallelize([1, 2, 3, 4, 2, 4, 1, 5, 6, 7, 7, 5, 5, 6, 4])
   .map(lambda x: (x, x))

pairs.partitionBy(4).glom().collect()

pairs.repartition(4).glom().collect()
```

执行以上代码，输出结果如下：

```
[[],
 [(5, 5), (6, 6), (7, 7), (7, 7), (5, 5), (5, 5), (6, 6), (4, 4)],
```

```
[],
[(1, 1), (2, 2), (3, 3), (4, 4), (2, 2), (4, 4), (1, 1)]]
```

实际上，PySpark 中的任何重分区都是使用 Pair RDD 来处理的。如果需要，PySpark 只需添加虚拟键或虚拟值使其工作。

3.7.4 自定义数据分区器

当需要精确地在分区中放置数据时，也可以自定义分区程序。自定义分区程序只可以在 Pair RDD 上使用。

如果 Pair RDD 转换没有指定一个分区程序，则所使用的分区数量将是父 RDD（转换为这个分区的 RDD）的最大分区数。如果父 RDD 中没有定义一个分区程序，则将使用 portable_hash()，分区数由 spark.default.parallelism 参数指定。

例如，在下面这个示例中，有一个整数列表[10,20,30,40,50,10,20,35]。现在想将它分为两个分区，例如 P1 和 P2。P1 包含所有小于 30 的元素，P2 包含所有大于 30 的元素。

【例 3-8】使用自定义分区程序对 Pair RDD 进行重分区，将 Pair RDD 中的所有偶数写到一个输出文件，并将所有奇数写到另一个输出文件。

代码如下：

```
#自定义分区程序
def myPartitioner(k):
    if k%2==0:
        return 0
    else:
        return 1

#构造一个 RDD
rdd1 = sc.parallelize(range(1,100001))
print("rdd1 的全部分区数: {}".format(rdd1.getNumPartitions()))

#使用自定义的分区器进行重分区，并指定分区数量
rdd2 = rdd1.map(lambda x: (x,None)).partitionBy(2, myPartitioner)
print("rdd2 的全部分区数: {}".format(rdd1.getNumPartitions()))

#将结果输出到文件中存储
rdd2.map(lambda x: x[0])
    .saveAsTextFile("/data/spark/output")
```

也可以在终端窗口中使用 Linux 命令来查看生成的结果文件，命令如下：

```
$ ls /home/hduser/data/spark/files-output2
$ head -10 /home/hduser/data/spark/files-output2/part-00000
$ head -10 /home/hduser/data/spark/files-output2/part-00001
```

因为在自定义分区器中指定了两个分区，所以存储文件相应地也有两个，如图 3-23 所示。

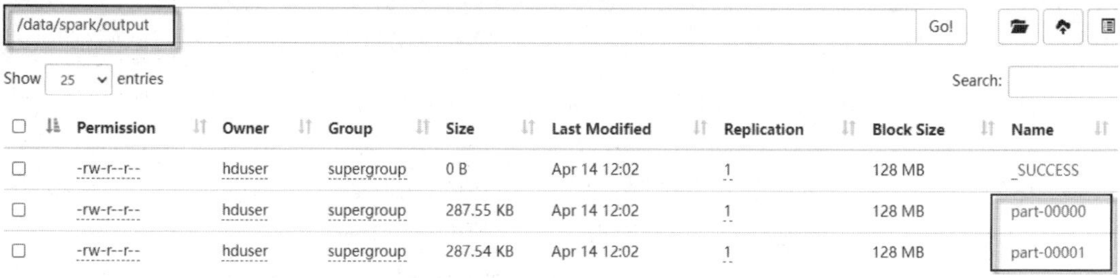
/data/spark/output
Go!
Show 25 entries
Search:
Permission
Owner
Group
Size
Last Modified
Replication
Block Size
Name
-rw-r--r--
hduser
supergroup
0 B
Apr 14 12:02
1
128 MB
_SUCCESS
-rw-r--r--
hduser
supergroup
287.55 KB
Apr 14 12:02
1
128 MB
part-00000
-rw-r--r--
hduser
supergroup
287.54 KB
Apr 14 12:02
1
128 MB
part-00001

图 3-23　一个分区对应一个存储文件

加载上面存储的结果文件 part-00000，会发现都是偶数，代码如下：

```
sc.textFile("/data/spark/output/part-00000") \
  .map(lambda x: int(x)) \
  .takeOrdered(10)
```

执行上面的代码，输出内容如下：

```
[2, 4, 6, 8, 10, 12, 14, 16, 18, 20]
```

加载上面存储的结果文件 part-00001，会发现都是奇数，代码如下：

```
sc.textFile("/data/spark/output/part-00001") \
  .map(lambda x: int(x)) \
  .takeOrdered(10)
```

执行上面的代码，输出内容如下：

```
[1, 3, 5, 7, 9, 11, 13, 15, 17, 19]
```

3.7.5　避免不必要的 shuffling

PySpark RDD 的 shuffling 是一种重新分配或重新分区数据的机制，以便数据在不同的分区上分组。根据数据大小，可能需要使用 spark.sql.shuffle.partitions 配置或通过代码来减少或增加 RDD 的分区数量。

当需要将来自多个分区的数据组合起来以构建新的分区时，就会发生数据 shuffle。例如，当按 key 对元素进行分组时，PySpark 需要检查 RDD 的所有分区，找到具有相同 key 的元素，然后对它们进行物理分组，从而形成新的分区。

PySpark RDD 中的某些操作会触发 shuffle 事件，例如 groupByKey()、reducebyKey()、join()、union()、groupBy()、aggregateByKey()等转换操作。在 shuffle 之前和之后的任务分

别被称为 map 和 reduce 任务。map 任务的结果被写到中间文件（通常只针对操作系统的文件系统缓存），并通过 reduce 任务读取。Spark shuffling 是一项昂贵的操作，因为它在执行器之间甚至在集群的工作节点之间移动数据，涉及磁盘 I/O、数据序列化和反序列化及网络 I/O，所以在 PySpark 作业中尽量减少 shuffling 的次数是很重要的。当在 PySpark 作业上遇到性能问题时，应该查看涉及转换的 PySpark 转换。

例如，在创建 RDD 时，PySpark 并不需要将所有 key 的数据存储在一个分区中，因为在创建 RDD 时，无法为 RDD 设置 key，因此，当运行 reduceByKey() 操作来聚合 key 上的数据时，PySpark 会执行以下操作：

（1）PySpark 首先在所有分区上运行 map 任务，按照每个 key 将所有值分组。
（2）将 map 任务的结果保存在内存中。
（3）当结果与内存不匹配时，PySpark 会将数据存储到磁盘中。
（4）PySpark 将 map 后的数据跨分区进行 shuffling，有时还会将 shuffling 后的数据存储到磁盘中，以便在需要重新计算时重用。
（5）运行垃圾回收。
（6）最后基于 key 在每个分区上运行 reduce 任务。

虽然 reduceByKey() 会触发数据 shuffle，但不会改变分区数，因为子 RDD 从父 RDD 继承了分区大小。

尽管大多数 RDD 转换不需要进行 shuffle，但在特定条件下其中一些转换会引起数据的跨分区 shuffling，因此，为了最小化 shuffle 出现的次数，需要了解引起数据 shuffling 的这些条件。

（1）在明确改变分区时会进行 shuffling。

当使用一个自定义分区器时，总会引发 shuffling。当在转换过程中使用一个与之前的 HashPartitioner 不同的 HashPartitioner 时（分区数量不同），也会引起 shuffling。例如，下面的代码总会引起 shuffling：

```
rdd.aggregateByKey((zeroValue, 100), seqFunc, comboFunc).collect()
```

（2）由删除分区引起的 shuffling。

有时，转换会导致 shuffling，尽管使用的是默认的分区器。例如，在下面的代码中，第 2 行不会引起 shuffling，但是第 3 行会：

```
rdd = sc.parallelize(range(1, 10000))

#不会引起 shuffle
rdd.map(lambda x: (x, x*x)).map(lambda x: (x[1],x[0])).count()

#会引起 shuffle
rdd.map(lambda x: (x, x*x)).reduceByKey(lambda v1, v2: v1+v2).count()
```

在 PySpark 中有两个 shuffle 实现，分别是基于 sort 排序的实现和基于 hash 的实现。从 Spark 1.2 开始默认使用的是基于 sort 排序的 shuffle，因为它的内存更高效，文件更少。可以通过设置 spark.shuffle.manager 参数的值为 hash 或 sort 定义要使用哪个 shuffle 实现。

可通过 spark.shuffle.consolidateFiles 参数指定在一个 shuffle 期间是否要合并中间文件。出于性能上的考虑，如果使用 EXT4 或 XFS 文件系统，建议将这个参数设为 true（默认值为 false）。

一般来讲，shuffling 可能需要大量的内存用于聚合和 join 分组。设置 spark.shuffle.spill 参数指定用于这些任务的内存数量是否应该被限制（默认值为 true）。在这种情况下，任何多余的数据都会溢写到磁盘上。对于内存的限制由参数 spark.shuffle.memoryFraction 指定（默认为 0.2）。此外，spark.shuffle.spill.compress 参数告诉 Spark 是否为溢写的数据启用压缩（默认情况下也是 true）。

溢写阈值不应该设得太高，否则会导致内存溢出异常，但是如果溢写阈值设得太低了，溢写就会频繁发生，所以找到一个好的平衡点是很重要的。在大多数情况下，保持默认值应该很合适。

此外，以下几个参数也很有用。

（1）spark.shuffle.compress：指定是否压缩中间文件（默认为 true）。

（2）spark.shuffle.spill.batchSize：指定当溢写到磁盘时将被序列化或反序列化的对象的数量。默认值为 10 000。

（3）spark.shuffle.service.port：如果启用了外部 shuffle 服务，则需指定服务器将侦听的端口。

3.7.6 基于数据分区的操作

截至目前，学习到的 RDD Transformation 操作都是应用于 RDD 中的每个元素上的，如 map()、flatMap()、filter()等。PySpark 另外还提供了一些操作，可以将这些操作应用于每个分区上，而不是每个元素上。这样的操作有 mapPartitions()、mapPartitionsWithIndex()、glom()、foreachPartition()等。这些基于分区的操作函数见表 3-2。

表 3-2 基于分区的操作函数

函 数 名	传 入 参 数	返 回 内 容	函 数 签 名
mapPartitions	该分区内元素的迭代器	返回元素的迭代器	f: (Iterator[T]) -> Iterator[U]
mapPartitionsWithIndex	集成分区号和该分区内元素的迭代器	返回元素的迭代器	f: (Int, Iterator[T]) -> Iterator[U]
foreachPartition	元素的迭代器	无	f: (Iterator[T]) -> Unit
glom			

1. mapPartitions()

PySpark 的 mapPartitions() 是一个转换操作，应用于 RDD 中的每个分区。PySpark 中的 RDD 存储在分区中，而 mapPartitions() 用于在 PySpark 架构中的 RDD 分区上应用一个函数。它可以作为 map() 和 foreach() 的替代方法。

简单来讲，map() 转换作用于 RDD 上的每个元素，对 RDD 中的元素进行一对一的转换，而 mapPartitions() 作用于 RDD 上的每个分区，对每个分区进行一对一转换。查阅文档，该方法的签名如下：

```
RDD.mapPartitions(f: Callable[[Iterable[T]], Iterable[U]],
preservesPartitioning: bool = False) → pyspark.rdd.RDD[U]
```

可以看到，mapPartitions() 转换将函数 f 应用在每个分区上，每个分区中的数据行以迭代器的形式传入函数 f，返回一个新的 Iterator。mapPartitions() 将结果保存在内存中，直到所有的行都在分区中处理完毕。另外 mapPartitions() 还接受一个可选参数 preservePartitioning，默认为 False。如果它被设置为 True，新的 RDD 将保留父 RDD 的分区。如果它被设置为 False，则分区器将被移除。

使用 mapPartitions() 可以帮助更有效地解决一些问题：需要一次性调用每个分区的数据模型的大量初始化，可以通过 mapPartitions() 来完成。例如，在连接到数据库时，因为打开数据库连接是开销很大的操作，所以需要在每个分区上应用一次 mapPartitions()，这样数据库连接便基于数据分区，而不是基于每个元素。应用 mapPartitions() 的模板代码如下：

```
def f(partitionData):
  #执行繁重的初始化，如数据库连接
  for element in partitionData:
    #对分区中的元素执行操作
  #返回更新后的数据

#应用 mapPartitions 函数
df.rdd.mapPartitions(f)
```

在下面的示例中，应用了 mapPartitions() 转换，代码如下：

```
#构造一个 RDD
rdd1 = spark.sparkContext.parallelize([1,2,3,4,5,6,7,8,9,10], 2)

#自定义函数，传入每个分区的迭代器，返回每个分区的元素和
def f(it):
    for e in it:
        yield [e * e]

#在 mapPartitions 转换中应用自定义函数 f
rdd2 = rdd1.mapPartitions(f)
```

```
#输出结果
print(rdd2.collect())
```

执行以上代码,输出内容如下:

```
[[1], [4], [9], [16], [25], [36], [49], [64], [81], [100]]
```

也可以不使用列表推导式,修改后的代码如下:

```
#自定义函数,传入每个分区的迭代器,返回每个分区的元素和
def f(it):
    s = []
    for e in it:
        s.append(e * e)
    return iter(s)

#在mapPartitions转换中应用自定义函数f
rdd3 = rdd1.mapPartitions(f)

#输出结果
print(rdd3.collect())
```

执行以上代码,输出结果如下:

```
[1, 4, 9, 16, 25, 36, 49, 64, 81, 100]
```

2. mapPartitionsWithIndex()

与 mapPartitions()类似,但又有所不同,mapPartitionsWithIndex()的 map()函数另外还接受传入的分区索引:(Int, Iterator T)=Iterator U,然后,分区的索引就可以在map()函数中使用,用于跟踪原始分区。

这个转换都接受一个额外的可选参数 preservePartitioning(),默认为 False。如果它被设置为 True,新的 RDD 将保留父 RDD 的分区。如果它被设置为 False,则分区器将被移除。

参看下面的应用示例:

```
#构造一个RDD,将分区数指定为2
x = spark.sparkContext.parallelize([1,2,3,4,5,6,7,8,9,10], 2)

#定义函数 f,它有两个传入参数:分区索引和分区元素迭代器
def f(partitionIndex, it):
    for e in it:
        yield [(partitionIndex, e * e)]

#执行转换
y = x.mapPartitionsWithIndex(f)

#查看
```

```
for t in y.collect():
    print(t)
```

执行以上代码,输出结果如下:

```
[(0, 1)]
[(0, 4)]
[(0, 9)]
[(0, 16)]
[(0, 25)]
[(1, 36)]
[(1, 49)]
[(1, 64)]
[(1, 81)]
[(1, 100)]
```

3. glom()

这种方法用于合并每个分区内的所有元素,并返回由这些合并后的元素创建的新 RDD。新 RDD 中元素的数量等于其分区的数量。在这个过程中,分区器会被删除。

简单使用 glom()方法示例,代码如下:

```
#构造 RDD
x = spark.sparkContext.parallelize([1,2,3,4,5,6,7,8,9,10], 2)

x.glom().collect()
```

执行上面的代码,输出结果如下:

```
[[1, 2, 3, 4, 5], [6, 7, 8, 9, 10]]
```

在下面的示例代码中,创建一个具有 30 个分区的 RDD 并对其执行 glom()转换。新 RDD 中的数组对象的计数包含来自每个分区的数据,也是 30,代码如下:

```
import random

#用随机生成的 500 个 100 以内的整数构造一个 list
list=[]
#随机生成 500 个点
for i in range(500):
    list.append(random.randint(0, 100))

#构造一个 RDD,并用 glom()方法收集分区数据
rdd = spark.sparkContext.parallelize(list, 30).glom()

#返回 RDD 数据
#rdd.collect()
```

```
#统计RDD中元素的数量
print(rdd.count())
```

输出结果如下：

```
30
```

可以看出，glom()方法将每个分区中的元素聚集到一个数组中，并以这些数组作为新RDD的元素。因为有30个分区，所以新RDD中的元素个数是30个。glom()可以作为一种快速的方法将所有RDD的元素放到一个数组中。

4. foreachPartition()

这是一个Action操作，用于执行分区迭代。其方法签名如下：

```
RDD.foreachPartition(f: Callable[[Iterable[T]], None]) → None
```

它将函数f作用于RDD的每个分区上，因此传入的是一个包含分区元素的迭代器。它是一个Action操作，因此不返回值，而是在每个分区上执行输入函数。

该方法类似于foreach()方法,但是提供了性能上的优化。在PySpark中，foreachPartition()用于初始化繁重的操作（如数据库连接），并希望在每个分区初始化一次，而foreach()用于在RDD分区的每个元素上应用一个函数。

使用foreachPartition的模板代码如下：

```python
#构造一个RDD
rdd = spark.sparkContext.parallelize([1,2,3,4,5,6,7,8,9])

#定义一种方法
def f(iterator):
#初始化任何数据库连接
    ...
    for x in iterator:
        #应用函数，例如数据落地
        ...

#调用foreachPartition()方法
rdd.foreachPartition(f)
```

3.8 使用共享变量

除了RDD，PySpark中还提供了另一个数据抽象"共享变量"。共享变量可以在并行操作中使用。

默认情况下，当传递给PySpark操作（如map()或reduce()）的函数在远程集群节点上执行时，它会将函数中使用的每个变量的副本发送给每个任务，这些变量被复制到每台机器,而对远程机器上的变量的更新不会传播回驱动程序,但是有时,用户需要在任务之间共享同一个变量,或者在任务和驱动程序之间共享同一个变量。

PySpark 支持两种类型的共享变量:广播变量(Broadcast Variable)和累加器(accumulator),广播变量可用于在所有节点的内存中缓存一个值,累加器是只"添加"到其中的变量,例如计数器和 sum 求和。广播变量和累加器能够维护一个全局状态,或者在 PySpark 程序中的任务和分区之间共享数据。

3.8.1 广播变量

默认情况下,在驱动程序中创建的变量,被序列化并由 executor 所需的任务一起发送,但是一个驱动程序可以在几个作业中重用相同的变量,并且一些任务可能会被发送到同一个 executor,作为同一作业的一部分,因此,一个大变量可能会被串行化并在网络上传输超过必要的次数。在这些情况下,最好使用广播变量。

当运行一个定义和使用了广播变量的 PySpark RDD 作业时,PySpark 按以下过程执行:
(1) PySpark 将作业分解为具有分布式 shuffling 的 stages,并在 stage 中执行操作。
(2) 后续的 stage 也被分解为多个任务(tasks)。
(3) PySpark 自动广播每个阶段中任务所需的公共数据(可重用)。
(4) 被广播的数据会以序列化的格式进行缓存,在执行每个任务之前反序列化。

这意味着,只有当跨多个阶段(stage)的任务需要相同的数据或者以反序列化的形式缓存数据非常重要时,显式地创建广播变量才有用。

使用广播变量,PySpark 在每台机器上保持一个缓存的只读变量,而不是将其副本与任务一起发送。PySpark 会尝试使用高效的广播算法来分发广播变量,而且只传输一次。

广播变量是用 SparkContext.broadcast(value)方法创建的,它返回一个 Broadcast 类型的对象。值可以是任何可序列化的对象,然后,executors 可以使用 Broadcast.value 方法进行读取。广播变量可以从整个集群中共享和访问,但它们不能被 executors 修改。

例如,创建了一个广播变量并在 map()转换操作中读取该广播变量的值用于计算,代码如下:

```
#第3章/rdd_demo08.ipynb

from pyspark.sql import SparkSession

#构建 SparkSession 和 SparkContext 实例
spark = SparkSession.builder \
    .master("spark://xueai8:7077") \
    .appName("pyspark demo") \
    .getOrCreate()

sc = spark.sparkContext
```

```
broads = sc.broadcast(3)                    #创建广播变量,变量可以是任意类型

listRDD = sc.parallelize([1,2,3,4,5])  #构造一个rdd

#map 转换,读取广播变量
results = listRDD.map(lambda x: x * broads.value)

print("结果是: ")
for line  in results.collect():
    print(line)
```

执行上面的代码,输出内容如下:

```
结果是:
3
6
9
12
15
```

广播变量的值可以是任意可序列化的数据。例如,将一个数组作为广播变量,代码如下:

```
from pyspark.sql import SparkSession

#构建 SparkSession 和 SparkContext 实例
spark = SparkSession.builder \
   .master("spark://xueai8:7077") \
   .appName("pyspark demo") \
   .getOrCreate()
sc = spark.sparkContext

broadcastVar = sc.broadcast([1, 2, 3])   #创建广播变量

listRDD = sc.parallelize([1,2,3,4,5])    #构造一个RDD

#flatMap 转换,读取广播变量,参与计算
results = listRDD.flatMap(lambda x: [v*x for v in broadcastVar.value])

print("结果是: ")
for line  in results.collect():
    print(line)
```

执行以上代码,输出内容如下:

```
结果是:
1
2
3
```

```
2
4
6
3
6
9
4
8
12
5
10
15
```

需要注意,广播变量不会通过 sc.broadcast(var)调用发送给执行程序,而是在它们第 1 次被使用时被发送给执行程序。

在一个 PySpark 程序中,可以有多个广播变量,代码如下:

```
#第3章/rdd_demo09.ipynb

from pyspark.sql import SparkSession

#构建 SparkSession 和 SparkContext 实例
spark = SparkSession.builder \
    .master("spark://xueai8:7077") \
    .appName("pyspark demo") \
    .getOrCreate()
sc = spark.sparkContext

states = {"豫":"河南省","粤":"广东省","鲁":"山东省"}
countries = {"CHINA":"中国","USA":"美国"}

#创建广播变量
broadcastStates = sc.broadcast(states)
broadcastCountries = sc.broadcast(countries)

#构造 RDD
data = [("张三","CHINA","豫"),\
        ("李四","CHINA","粤"), \
        ("王老五","CHINA","豫"), \
        ("赵小六","CHINA","鲁")]
rdd = sc.parallelize(data)

#使用广播变量
def func(t):
    country = t[1]
    state = t[2]
```

```
        fullCountry = broadcastCountries.value[country]
        fullState = broadcastStates.value[state]
        return (t[0], fullCountry, fullState)

rdd2 = rdd.map(func)

#输出结果
for row in rdd2.collect():
    print(row)
```

执行以上代码,输出结果如下:

```
(张三,中国,河南省)
(李四,中国,广东省)
(王老五,中国,河南省)
(赵小六,中国,山东省)
```

当不再需要广播变量时,可以调用 destroy()方法销毁它。所有关于它的信息都将被删除(从 executors 和驱动程序),并且该变量将不可用。如果试图在调用 destroy()之后访问它,则将抛出一个异常。

另一种方法是调用 unpersist()方法,它只从 executors 的缓存中删除变量值。如果尝试在 unpersist()之后使用它,则它将再次被发送到 executors。

广播哈希连接(Broadcast Hash Join)指的是,当有两个 RDD 执行 join 连接时,先将其中一个较小的 RDD 广播到每个工作节点,然后,它与较大的 RDD 的每个分区进行 map 端连接。例如,将一个较大的 rddA 的每个分区和一个较小的 rddB 中相关的值进行 join 连接,如图 3-24 所示。

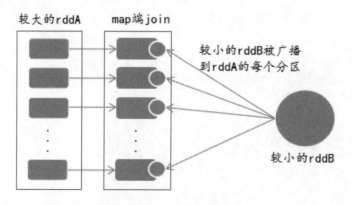

图 3-24 Broadcast Hash Join

如果一个 RDD 可以装入内存,或者经过处理后可以使其装入内存,则进行 Broadcast Hash Join 总是有益的,因为它不需要 shuffle。

PySpark 并没有提供 Broadcast Hash Join 的直接实现，不过可以手动实现 Broadcast Hash Join 的一个版本，方法是将较小的 RDD 收集到驱动程序中（通过调用 collect()方法）作为一个 map，然后广播结果，并使用 mapPartitions()组合元素。

在下面的示例中手工实现了 Broadcast Hash Join，用于连接较大和较小的 RDD。在这个示例中不考虑那些 key 不在两个 RDD 中出现的元素，代码如下：

```python
#第3章/rdd_demo10.ipynb
from pyspark.sql import SparkSession

#构建 SparkSession 和 SparkContext 实例
spark = SparkSession.builder \
    .master("spark://xueai8:7077") \
    .appName("pyspark demo") \
    .getOrCreate()

sc = spark.sparkContext

#构造一个较大的 RDD
bigData = [("a",1),("b",2),("c",3),("a",4),("b",5),("c",6)]
bigRDD = sc.parallelize(bigData, 2)

#构造一个较小的 RDD
smallData = [("a",11),("b",12)]
smallRDD = sc.parallelize(smallData, 2)

#将较小的 RDD 回收到 driver 端
smallRDDLocal = smallRDD.collectAsMap()

#创建广播变量
smallBroadcast = sc.broadcast(smallRDDLocal)

#执行 mapPartitions 转换
def func(it):
    s = []
    for k,v1 in it:
        #如果在小的 RDD 中找到匹配的 key
        if k in smallBroadcast.value:
            s.append((k,v1,smallBroadcast.value[k]))
    return s

joinedRDD = bigRDD.mapPartitions(func,preservesPartitioning = True)

#输出结果
for row in joinedRDD.collect():
    print(row)
```

执行上面的代码，输出结果如下：

```
(a,(1,11))
(b,(2,12))
(a,(4,11))
(b,(5,12))
```

有时并不是所有较小的 RDD 都能装入内存，但是有些 key 在大 RDD 中有较大比例，所以希望只广播最常见的 key。在这种情况下，可以在大型 RDD 上使用 countByKeyApprox 来大致了解哪些 key 适宜广播，然后，只针对这些 key 过滤出较小的 RDD，并在 HashMap 中本地收集结果。使用 sc.broadcast() 可以广播 HashMap，以便每个工作节点只有一个副本，并对 HashMap 手动执行连接，然后，使用相同的 HashMap，可以筛选大型 RDD，使其不包含大量重复的键，并执行标准连接，将其与手工连接的结果联合起来。这种方法非常复杂，但可以处理用其他方法无法处理的高度倾斜的数据。

3.8.2 累加器

累加器（Accumulators）用来把 executor 端变量信息聚合到 driver 端。在 driver 程序中定义的变量，在 executor 端的每个任务都会得到这个变量的一份新的副本，每个任务更新这些副本的值后，传回 driver 端进行合并。

累加器是跨 executors 之间共享的分布式只写共享变量，每个 executor 之间是不可见的，只有 driver 端可以对其进行读操作。可以使用它们实现 PySpark 作业中的全局求和与计数。

可以通过调用 SparkContext.accumulator() 方法来创建数值累加器，以累积值，然后，可以在任务中调用累加器的 add() 方法来修改值，但是，无法在任务中读取它的值。只有 driver 端驱动程序可以通过调用累加器的 value 方法读取累加器的值。

例如，在下面的示例中，定义了一个累加器 acc，初始值为 0，然后在 executors 上执行时通过调用 acc.add() 方法累加值，代码如下：

```
#第3章/rdd_demo11.ipynb

from pyspark.sql import SparkSession

#构建 SparkSession 和 SparkContext 实例
spark = SparkSession.builder \
    .master("spark://xueai8:7077") \
    .appName("pyspark demo") \
    .getOrCreate()

sc = spark.sparkContext

acc = sc.accumulator(0)
```

```
print("累加器初始值: ", acc.value)

list1 = sc.parallelize([1, 2, 3, 4])

#在executors上执行
list1.foreach(lambda x: acc.add(x))

#在driver端读累加器的值
print("现在累加器的值: ", acc.value)
```

执行以上代码，输出结果如下：

```
累加器初始值: 0
现在累加器的值: 10
```

在 PySpark 的执行模型中，只有当计算被触发（例如，执行一个 action）时，PySpark 才会添加累加器。

在下面这个示例中，要求删除指定数组中既能被 2 整除也能被 3 整除的数，并统计删除了多少个数。这里使用累加器来统计删除了多少个数，代码如下：

```
#第3章/rdd_demo12.ipynb

from pyspark.sql import SparkSession

#构建SparkSession和SparkContext实例
spark = SparkSession.builder \
    .master("spark://xueai8:7077") \
    .appName("pyspark demo") \
    .getOrCreate()

sc = spark.sparkContext

#定义计数器
acc = sc.accumulator(0)

#构造RDD
rdd1 = sc.parallelize(range(1,13))

#filter method
def filterFun(number):
    if (number%2)==0 and (number%3)==0:
        acc.add(1)
        return False
    else:
        return True

#filter
```

```
rdd2 = rdd1.filter(filterFun).cache()
print(rdd2.collect())

#在driver端才能读取累加器的值
print("删除了符合条件的元素有{}个".format(acc.value))
```

执行以上代码，输出结果如下：

```
[1, 2, 3, 4, 5, 7, 8, 9, 10, 11]
删除了符合条件的元素有两个
```

3.9 PySpark RDD 可视化

3min

PySpark RDD 还没有任何绘图功能。如果想绘制一些内容，则可以将数据从 SparkContext 中取出并放入"本地"Python 会话中，在那里可以使用 Python 的任意一个绘图库来处理它。对于 RDD，调用 collect()方法，将数据返回 driver 再绘制。

在下面这个示例中，创建一个数据正态分布的 RDD，然后调用 collect()将处理过的数据返回 driver 端，并使用 Python 的绘图库绘制直方图，代码如下：

```
#第3章/rdd_demo13.ipynb

from pyspark.sql import SparkSession
import random
import matplotlib
import matplotlib.ipynbplot as plt

#构建 SparkSession 和 SparkContext 实例
spark = SparkSession.builder \
    .master("spark://xueai8:7077") \
    .appName("pyspark rdd demo") \
    .getOrCreate()

sc = spark.sparkContext

#创建一个包含100个随机数的RDD
x = [random.normalvariate(0,1) for i in range(100)]
rdd = sc.parallelize(x)

#绘制 RDD 中的数据
num_bins = 50
n, bins, patches = plt.hist(rdd.collect(), num_bins, density=1,
facecolor='green', alpha=0.5)
```

执行以上代码，执行结果如图 3-25 所示。

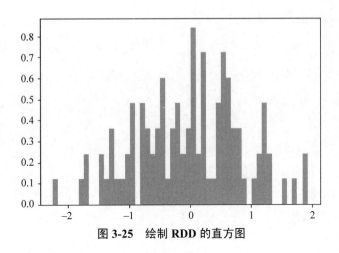

图 3-25　绘制 RDD 的直方图

3.10　PySpark RDD 编程案例

本节通过几个不同类型的应用场景的编程案例，进一步深入学习 PySpark RDD 及应用 RDD 解决一些业务问题。

3.10.1　合并小文件

10min

在使用 Hadoop 时，经常会遇到小文件问题。当系统中有大量小文件时，读写这些小文件会遇到严重的性能问题。

在 PySpark 中，使用 SparkContext 的 wholeTextFiles()方法和 colleasc()方法，可以实现对小文件的合并，其中 wholeTextFile()的方法签名如下：

```
SparkContext.wholeTextFiles(path: str, minPartitions: Optional[int] = None,
use_unicode: bool = True) → pyspark.rdd.RDD[Tuple[str, str]]
```

该方法的第 1 个参数 path 代表输入文件的目录，它可以是以逗号分隔的输入文件列表；第 2 个参数是建议的最小分区数。返回的 RDD 包含以文件路径和对应的文件内容组成的元组元素。

该方法从 HDFS、本地文件系统（所有的节点都可访问）或任何 Hadoop 支持的文件系统 URI 读取一个文本文件目录。每单个读取为单行记录，以键-值对的形式返回，其中 key 是每个文件的路径，value 是每个文件的内容。这些文本文件必须以 UTF-8 编码。

例如，如果在 HDFS 上有以下这些文件：

```
hdfs://a-hdfs-path/part-00000
hdfs://a-hdfs-path/part-00001
...
hdfs://a-hdfs-path/part-nnnnn
```

将这些文件内容加载到 RDD 中,代码如下:

```
rdd = sparkContext.wholeTextFile("hdfs://a-hdfs-path")
```

返回的 RDD 将包含以下内容:

```
(a-hdfs-path/part-00000, its content)
(a-hdfs-path/part-00001, its content)
...
(a-hdfs-path/part-nnnnn, its content)
```

下面通过一个示例来掌握此用法。

【例 3-9】应用 PySpark RDD 技术,合并 Hadoop HDFS 上的小文件。

首先在本地文件系统准备 3 个文本文件。

file1.txt:

```
good good study
```

file2.txt:

```
day day up
```

file3.txt:

```
to be or not to be,this is a question.
```

然后将上述 3 个文本文件上传到 HDFS 的/data/spark/inputs/目录下(如果 HDFS 上没有该目录,则应先自行创建),命令如下:

```
$ hdfs dfs -put file1.txt /data/spark/inputs/
$ hdfs dfs -put file2.txt /data/spark/inputs/
$ hdfs dfs -put file3.txt /data/spark/inputs/
```

编写代码,合并这 3 个小文件,代码如下:

```
#第3章/rdd_demo14.ipynb

from pyspark.sql import SparkSession

#构建 SparkSession 实例
spark = SparkSession.builder \
    .master("spark://xueai8:7077") \
    .appName("pyspark demo") \
    .getOrCreate()

#加载数据源,构造 RDD
textFiles = spark.sparkContext.wholeTextFiles("/data/spark/inputs/*")

#查看内容
```

```
for item in textFiles.collect():
    print(item)
```

执行以上代码，输出内容如下：

```
(hdfs://xueai8:8020/data/spark/inputs/file1.txt,good good study)
(hdfs://xueai8:8020/data/spark/inputs/file2.txt,day day up)
(hdfs://xueai8:8020/data/spark/inputs/file3.txt,to be or not to be,this is a question.)
```

取元素二元组中第 2 个字段的值(文件内容)，合并到一个文件中，最后写到 HDFS 上，代码如下：

```
textFiles \
    .map(lambda line: line[1]) \
    .coalesce(1) \
    .saveAsTextFile("/data/spark/outputs")
```

这时去查看 HDFS 的 /data/spark/outputs 目录，可以看到只有一个文件存在。例如，通过 Web UI 查看 HDFS 文件目录，如图 3-26 所示。

图 3-26 将多个小文件合并到一个文件中

使用 HDFS Shell 命令可将原目录删除，命令如下：

```
$ hdfs dfs -rm -r /data/spark/inputs
```

3.10.2 二次排序实现

所谓二次排序，指的是对于 Tuple(key,value) 类型的数据，不但按 key 排序，而且每个 key 对应的 value 也是有序的。

7min

【例 3-10】假设有一个文本文件 data.txt，内容如下：

```
2018,5,22
2019,1,24
2018,2,128
2019,3,56
2019,1,3
```

```
2019,2,-43
2019,4,5
2019,3,46
2018,2,64
2019,1,4
2019,1,21
2019,2,35
2019,2,0
```

其中,每行文本是用逗号分隔的年、月和总数量。现在想要对这些数据排序,期望的输出结果如下:

```
2018-2    64,128
2018-5    22
2019-1    3,4,21,24
2019-2    -43,0,35
2019-3    46,56
2019-4    5
```

PySpark 二次排序解决方案如下:需要将年和月组合起来构成一个 key,将第 3 列作为 value,并使用 groupByKey()函数将同一个 key 的所有 value 全部分组到一起,然后对同一个 key 的所有 value 值列表进行排序即可。

首先,将本地的 data.txt 上传到 HDFS 的指定位置,命令如下:

```
$ hdfs dfs -put data.txt /data/spark/input/
```

然后,编写 PySpark 程序代码,进行二次排序,代码如下:

```
#第3章/rdd_demo15.ipynb

from pyspark.sql import SparkSession

#构建 SparkSession 实例
spark = SparkSession.builder \
    .master("spark://xueai8:7077") \
    .appName("pyspark demo") \
    .getOrCreate()

#加载数据集
inputPath = "/data/spark/input/data.txt"
inputRDD = sc.textFile(inputPath)

#定义 map 函数
def customMapFun(line):
    arr = line.split(",")              #按逗号分隔
    key = arr[0] + "-" + arr[1]        #组合年和月
    value = arr[2]                     #值
```

```
        return (key, int(value))        #返回键-值对
#实现二次排序
sortedRDD = inputRDD \
    .map(customMapFun) \
    .groupByKey() \
    .map(lambda t: (t[0],sorted(list(t[1])))) \
    .sortByKey()

#结果输出
for t in sortedRDD.collect():
    print(t)
```

执行以上代码，输出结果如下：

```
2018-2      64,128
2018-5      22
2019-1      3,4,21,24
2019-2      -43,0,35
2019-3      46,56
2019-4      5
```

3.10.3 Top N 实现

推荐领域有一个著名的开放测试数据集 movielens，其中包含电影评分数据集 ratings.csv 及电影数据集 movies.csv。

电影数据集 movies.csv 包含 3 个字段：movieId、title 和 genres，分别表示电影的 ID、电影名称和影片类型。查看 movies.csv 中的前 10 行数据记录，内容如下：

```
movieId,title,genres
1,Toy Story (1995),Adventure|Animation|Children|Comedy|Fantasy
2,Jumanji (1995),Adventure|Children|Fantasy
3,Grumpier Old Men (1995),Comedy|Romance
4,Waiting to Exhale (1995),Comedy|Drama|Romance
5,Father of the Bride Part II (1995),Comedy
6,Heat (1995),Action|Crime|Thriller
7,Sabrina (1995),Comedy|Romance
8,Tom and Huck (1995),Adventure|Children
9,Sudden Death (1995),Action
```

电影评分数据集 ratings.csv 包含 4 个字段：userId、movieId、rating 和 timestamp，分别表示用户 ID、电影 ID、用户对该电影的评分、时间戳。查看 ratings.csv 中的前 10 行数据记录，内容如下：

```
userId,movieId,rating,timestamp
1,1,4.0,964982703
```

```
1,3,4.0,964981247
1,6,4.0,964982224
1,47,5.0,964983815
1,50,5.0,964982931
1,70,3.0,964982400
1,101,5.0,964980868
1,110,4.0,964982176
1,151,5.0,964984041
```

假设现在有一个需求，要求统计评分最高的 5 部电影，以[电影名, 评分]的形式输出。那么使用 PySpark RDD 该怎么实现呢？

这是一个典型的 Top N 问题，对 ratings.csv 抽取 movieId 和 rating 两个字段，按 rating 评分值字段倒序取前 5 条（也就是评分最高的 5 部电影）。另外，因为输出时需要输出电影名称而不是电影的 ID（一些观众并不知道电影 ID 代表哪部电影），因此需要对两个数据集执行 join 连接操作，以获取评分最高的 5 部电影的名称。

【例 3-11】使用 PySpark RDD 技术，统计 movielens 电影数据集中评分最高的前 5 部电影。

实现过程和代码如下。

（1）创建 SparkSession 和 SparkContext 对象实例，代码如下：

```
#第 3 章/rdd_demo16.ipynb

from pyspark.sql import SparkSession

#构建 SparkSession 和 SparkContext 实例
spark = SparkSession.builder \
    .master("spark://xueai8:7077") \
    .appName("pyspark demo") \
    .getOrCreate()

sc = spark.sparkContext
```

（2）加载电影数据集文件数据，构造 RDD，代码如下：

```
#加载电影数据集，构造 RDD
movies = "file://home/hduser/data/spark/movielens/movies.csv"
moviesRDD = sc.textFile(movies)

print("电影数据集中数据总记录数量:", moviesRDD.count())
moviesRDD.cache()                    #缓存电影数据集

#查看前两条数据
for line in moviesRDD.take(2):
    print(line)
```

执行以上代码,输出内容如下:

```
电影数据集的总数据量: 9743
movieId,title,genres
1,Toy Story (1995),Adventure|Animation|Children|Comedy|Fantasy
```

可以看出,该文件共包含 9743 行记录,其中第 1 行是标题行,实际包含的电影记录数量为 9742 条。

(3)加载评分数据集文件数据,构造 RDD,代码如下:

```
#加载评分数据集,构造 RDD
ratings = "file://home/hduser/data/spark/movielens/ratings.csv"
ratingsRDD = sc.textFile(ratings)

print("评分数据集中数据总记录数量:", ratingsRDD.count())
ratingsRDD.cache()              #缓存评分数据集

#查看前两条数据
for line in ratingsRDD.take(2):
    print(line)
```

执行以上代码,输出内容如下:

```
评分数据集的总数据量: 100837
userId,movieId,rating,timestamp
1,1,4.0,964982703
```

可以看出,该文件共包含 100 837 行记录,其中第 1 行是标题行,实际包含评分记录数量为 100 836 条。

(4)对数据进行初步处理,过滤掉标题行,将 movieId 和 rating 两个字段抽取出来,组成新 RDD 的元素,代码如下:

```
#自定义 map 转换函数
def ratingsMapFun(line):
    fileds = line.split(",")
    return (int(fileds[1]), float(fileds[2]))

#删除标题行,并抽取(movieId,rating)字段
rating = ratingsRDD \
    .filter(lambda line: not line.startswith("userId")) \
    .map(ratingsMapFun)

print("去掉标题行后的数据量: ", rdd1.count())

#查看前 5 条
for row in rating.take(5):
    print(row)
```

执行以上代码,输出结果如下:

```
去掉标题行后的数据量:100836
(1,4.0)
(3,4.0)
(6,4.0)
(47,5.0)
(50,5.0)
```

可以看到,已经去掉了标题行,并在 rating RDD 中只保留了 movieId(电影 ID)和 rating(评分)这两个字段。

(5)因为要求每部电影的平均评分,所以要以 movieId 为 key 进行分组聚合计算,同时统计每部电影的总评分和打分的人数,实现代码如下:

```
#定义计算平均评分的函数
def avgFun(t):
    avg = sum(t[1]) / len(t[1])
    return (t[0], avg)

#获得(movieid,ave_rating)
movieScores = rating.groupByKey().map(avgFun)

#查看前5条数据
for row in movieScores.take(5):
    print(row)
```

执行上面的代码,输出结果如下:

```
(6, 3.946078431372549)
(50, 4.237745098039215)
(70, 3.5090909090909093)
(110, 4.031645569620253)
(216, 3.326530612244898)
```

可以看出,经过这一步处理后,返回的是一个元组,元组元素为(movieId,平均得分),但是这个结果对用户来讲并不友好,因为它只显示了电影的 ID。下一步显示对应的电影名称及其平均得分。

(6)在上一步的计算结果中显示的是电影的 ID,这对用户是非常不友好的。为此,需要结合 moviesRDD 提取电影名称(title)。先对 moviesRDD 进行预处理,提取(movieId, title)的唯一值,代码如下:

```
#自定义 map 转换函数
def movieMapFun(line):
    fileds = line.split(",")
    return (int(fileds[0]), fileds[1])
```

```
#删除标题行，并抽取(movieId, title)
movieskey = moviesRDD \
    .filter(lambda line: not line.startswith("movieId")) \
    .map(movieMapFun)

#查看前 5 条
for row in movieskey.take(5):
    print(row)
```

执行以上代码，输出结果如下：

```
('1', 'Toy Story (1995)')
('2', 'Jumanji (1995)')
('3', 'Grumpier Old Men (1995)')
('4', 'Waiting to Exhale (1995)')
('5', 'Father of the Bride Part II (1995)')
```

（7）接下来对两个 RDD 执行 join 连接操作，获取平均评分前 5 名的电影名称和平均评分，代码如下：

```
result = movieScores \
    .join(movieskey) \
    .filter(lambda f: f[1][0]>4.0) \
    .map(lambda f: (f[0],f[1][1],f[1][0]))

#查看前 5 条
top5 = result.takeOrdered(5, key=lambda x: -x[2])
for row in top5:
    print(row)
```

执行以上代码，输出平均评分最高的 5 部电影如下：

```
(260, 'Star Wars: Episode IV - A New Hope (1977)', 4.231075697211155)
(296, 'Pulp Fiction (1994)', 4.197068403908795)
(356, 'Forrest Gump (1994)', 4.164133738601824)
(608, 'Fargo (1996)', 4.116022099447513)
(1136, 'Monty Python and the Holy Grail (1975)', 4.161764705882353)
```

3.10.4 数据聚合计算

12min

【例 3-12】给定一组销售数据，数据采用键-值对的形式（公司，收入），求出每个公司的总收入和平均收入。

在本示例中，直接用 sc.parallelize()方法在内存中生成数据，在求每个公司总收入时，先分 3 个分区进行求和，然后把 3 个分区合并起来，代码如下：

```
#第 3 章/rdd_demo17.ipynb
```

```python
from pyspark.sql import SparkSession

#构建SparkSession和SparkContext实例
spark = SparkSession.builder \
    .master("spark://xueai8:7077") \
    .appName("pyspark demo") \
    .getOrCreate()

sc = spark.sparkContext

#构造RDD
data = sc.parallelize([("company-1",92),("company-1",85),("company-1",82),
                       ("company-2",78),("company-2",96),("company-2",85),
                       ("company-3",88),("company-3",94),("company-3",80)], 3)

#定义createCombiner、mergeValue 和 mergeCombiner 函数
def createCombiner(income):
    return (income, 1)

def mergeValue(acc, income):
    return (acc[0] + income, acc[1] + 1)

def mergeCombiner(acc1, acc2):
    return (acc1[0] + acc2[0], acc1[1] + acc2[1])

cbk = data \
    .combineByKey(createCombiner, mergeValue, mergeCombiner) \
    .map(lambda t: (t[0], t[1][0], t[1][0]/float(t[1][1])))

#查看输出
for row in cbk.collect():
    print(row)
```

执行以上代码，输出结果如下：

```
('company-1', 259, 86.33333333333333)
('company-3', 262, 87.33333333333333)
('company-2', 259, 86.33333333333333)
```

第 4 章 PySpark SQL（初级）

CHAPTER 4

PySpark SQL 是 PySpark 用于处理结构化和半结构化数据的接口，允许使用关系操作符表示分布式内存计算。结构化数据指的是任何有模式的数据，如 JSON、Hive 表、Parquet。模式意味着每个记录拥有一组已知的字段。半结构化数据是指模式和数据之间没有分离。

通过 PySpark SQL，用户可以使用熟悉的 SQL 或 DataFrame API 查询 PySpark 程序中的结构化数据。DataFrame 和 SQL 提供了一种通用的方式访问各种数据源，包括 Hive、Avro、Parquet、ORC、JSON 和 JDBC。甚至可以跨这些数据源连接数据。

与 PySpark RDD API 不同，PySpark SQL 提供的接口为 PySpark 提供了有关数据结构和正在执行的计算的更多信息。在内部，PySpark SQL 使用这些额外的信息来执行额外的优化。

通过引入 PySpark SQL，将关系处理与 PySpark 的函数式编程 API 集成在一起，因此，PySpark SQL 被设计用来集成关系型处理和函数式编程的功能，这样复杂的逻辑就可以在分布式计算设置中实现、优化和扩展。

4.1 PySpark SQL 数据抽象

尽管 PySpark 提供了一个函数式编程 API 来操作分布式的数据集合，但使用 RDD 进行编码有些复杂和混乱，而且有时很慢。PySpark SQL 提供了使用结构化和半结构化数据的 3 个主要功能：

（1）它提供了由 Python 语言所支持的 DataFrame 抽象，以简化使用结构化数据集的工作。DataFrame 类似于关系数据库中的表。

（2）它可以读写各种结构化格式的数据（如 JSON、Hive 表、Parquet）。

（3）它允许在 PySpark 程序内部使用 SQL 查询数据，以及通过标准数据库连接器（JDBC/ ODBC）连接到 Spark SQL 的外部工具（如 Tableau 等商业智能工具）中使用 SQL 查询数据。

PySpark SQL 提供了一个统一的接口，使用专门的 DataFrameReader 和 DataFrameWriter 对象在分布式存储系统（如 Cassandra 或 HDFS （Hive, Parquet, JSON））中访问数据。PySpark

SQL 允许对 Hadoop HDFS 或与 Hadoop 兼容的文件系统（如 S3）中的大量数据执行类似 SQL 的查询。它可以访问来自不同数据源（文件或表）的数据。

PySpark SQL 的主要数据抽象是 DataFrame，它表示结构化数据（具有已知模式的记录），如图 4-1 所示。

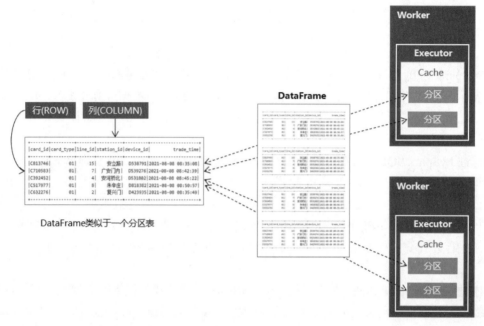

图 4-1 PySpark DataFrame 类似一个分区表

DataFrame 是一个带有 Transformation 和 Action 的结构化查询执行管道的编程接口。在内部，结构化查询（逻辑和物理）是关系运算符和表达式的 Catalyst 树。DataFrame 可以从各种各样的数据源构建，例如结构化数据文件、Hive 中的表、外部数据库或现有的 RDD。

PySpark SQL 的核心是 Catalyst 优化器，它利用 Scala 的高级特性（例如模式匹配）来提供可扩展的查询优化器。开发人员编写的基于 DataFrame 的高级代码被转换为 Catalyst 表达式，然后通过这个执行和优化管道转换为低级 Java 字节码，即使 Catalyst Optimizer 的代码库是用 Scala 编写的，Python 也可以利用 Spark 中的性能优化。基本上，它是一个包含大约 2000 行代码的 Python 包装器，允许 PySpark DataFrame 查询显著加快。

DataFrame 在 RDD 上提供了巨大的性能改进，提供了两个强大的功能。

1. 自定义内存管理（又名 Project Tungsten）

数据以二进制格式存储在堆外内存中（Off-heap Memory），这节省了大量的内存空间。此外，也不涉及垃圾收集开销。通过提前了解数据的模式并有效地以二进制格式存储，还可以避免昂贵的 Java 序列化。

2. 优化的执行计划（又名 Catalyst Optimizer）

使用 Spark Catalyst 优化器创建用于执行的查询计划。在经过一些步骤准备好优化的执行计划之后，最终的执行仅在 RDD 内部进行，但这对用户是完全隐藏的。

4.2 PySpark SQL 编程模型

10min

所有的 PySpark SQL 应用程序都以特定的步骤工作。这些工作步骤如图 4-2 所示。

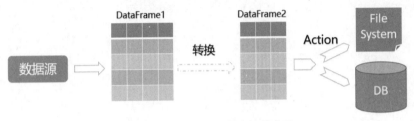

图 4-2　Spark SQL 程序执行步骤

也就是说，每个 PySpark SQL 应用程序都由相同的基本部分组成：
（1）从数据源加载数据，构造 DataFrame。
（2）对 DataFrame 执行转换（Transformation）操作。
（3）将最终的 DataFrame 存储到指定位置。

根据这个步骤，下面实现一个完整的 PySpark SQL 应用程序。

【例 4-1】编写一个 PySpark SQL 应用程序，它从一个 JSON 格式的文件中加载人员信息，并进行统计。

建议按以下步骤执行。

（1）准备源数据文件。

Spark 安装目录中自带了一个 people.json 文件，位于 examples/src/main/resources/"子目录下，其内容如下：

```
{"name":"Michael"}
{"name":"Andy", "age":30}
{"name":"Justin", "age":19}
```

将 people.json 文件复制到 HDFS 的指定目录下。

（2）创建一个 Jupyter Notebook 文件，并编辑代码如下：

```
#第 4 章/dataframe_demo01.ipynb

from pyspark.sql import SparkSession

#(1)构建 SparkSession 实例
```

```
spark = SparkSession.builder \
    .master("spark://xueai8:7077") \
    .appName("pyspark sql demo") \
    .getOrCreate()

#(2)加载数据源,构造 DataFrame
input = "/data/spark/resources/people.json"
df = spark.read.json(input)

#(3)执行转换操作
from pyspark.sql.functions import *

#找出年龄超过21岁的人
resultDF = df.where(col("age") > 21)

#显示 DataFrame 数据
resultDF.show()

#(4)将结果保存到 CSV 文件中
output = "/data/spark/people-output"
resultDF.write.format("csv").save(output)

#(5)停止 SparkSession
spark.stop()
```

(3)执行过程和执行结果如图 4-3 所示。

1. 读数据格式

在 PySpark SQL 中,所有读取数据的 API 都遵循以下调用格式:

```
DataFrameReader \
    .format(...) \
    .option("key", "value") \
    .schema(...) \
    .load()
```

例如,将一个 CSV 文件内容读到 DataFrame 中,代码如下:

```
spark.read \
    .format("csv") \
    .option("mode", "failfast") \            #读取模式
    .option("inferSchema", "True") \         #是否自动推断 schema
    .option("path", "path/to/file(s)") \     #文件路径
    .schema(someSchema) \                    #使用预定义的 schema
    .load()
```

其中读取模式 mode 有 3 种可选项,见表 4-1。

```
In [1]: from pyspark.sql import SparkSession

        #(1) 构建SparkSession 实例
        spark = SparkSession.builder \
              .master("spark://xueai8:7077") \
              .appName("pyspark sql demo") \
              .getOrCreate()

In [2]: #(2) 加载数据源，构造DataFrame
        input = "file:///home/hduser/data/spark/resources/people.json"
        df = spark.read.json(input)

In [5]: #(3) 执行转换操作
        from pyspark.sql.functions import *

        # 找出年龄超过21岁的人
        resultDF = df.where(col("age") > 21)

        # 显示DataFrame数据
        resultDF.show()

        +---+----+
        |age|name|
        +---+----+
        | 30|Andy|
        +---+----+

In [6]: #(4) 将结果保存到/CSV文件中
        output = "/data/spark/people-output"
        resultDF.write.format("csv").save(output)

In [7]: #(5) 停止SparkSession
        spark.stop()
```

图 4-3　PySpark SQL 程序执行过程和执行结果

表 4-1　3 种读取模式

读 取 模 式	描　　述
permissive	当遇到损坏的记录时，将其所有字段设置为 null，并将所有损坏的记录放在名为_corruption t_record 的字符串列中
dropMalformed	删除不正确模式的行
failFast	遇到数据格式不正确时立即失败

2. 写数据格式

在 PySpark SQL 中，所有写数据的 API 都遵循以下调用格式：

```
DataFrameWriter \
   .format(...) \
   .option(...) \
   .partitionBy(...) \
```

```
    .bucketBy(...) \
    .sortBy(...) \
    .save()
```

例如,将一个 DataFrame 中的数据写入一个 CSV 文件,代码如下:

```
dataframe.write \
    .format("csv") \                                    #输出格式
    .option("mode", "overwrite") \                      #写模式
    .option("dateFormat", "yyyy-MM-dd") \               #日期格式
    .option("path", "path/to/file(s)") \                #写入路径
    .save()
```

其中写数据模式 mode 有 4 种可选项,见表 4-2。

表 4-2 4 种写数据模式

写数据模式	描述
error 或 errorifexists	如果给定的路径已经存在文件,则抛出异常。这是默认的模式
append	以追加的方式写入数据
overwrite	以覆盖的方式写入数据
ignore	如果给定的路径已经存在文件,则不做任何操作

4.3 程序入口 SparkSession

在 4.2 节的示例程序中,在代码的开始,首先创建了一个 SparkSession 的实例对象 spark。

在 PySpark 2.0 中,SparkSession 表示在 PySpark 中操作数据的统一入口点。要创建一个基本的 SparkSession,需要使用 SparkSession.builder(),代码如下:

```
from pyspark.sql import SparkSession

#构建 SparkSession 实例
spark = SparkSession.builder \
    .master("spark://localhost:7077") \
    .appName("pyspark sql demo") \
    .getOrCreate()
```

在 PySpark 程序中,使用构建器设计模式实例化 SparkSession 对象,然而,在 REPL 环境中(在 PySpark Shell 会话中),SparkSession 会被自动创建,并通过名为 spark 的实例对象提供给用户使用。也就是说,如果是使用 PySpark Shell 交互式地执行 PySpark 程序语句,则可以直接调用 spark 实例对象,如图 4-4 所示。

SparkSession 对象可以用来配置 PySpark 的运行时配置属性。例如,PySpark 和 YARN 管理的两个主要资源是 CPU 和内存。如果想为 PySpark executor 设置内核数量和堆大小,

```
[hduser@xueai8 spark-3.1.2]$ ./bin/pyspark
Python 3.7.6 (default, Jan  8 2020, 19:59:22)
[GCC 7.3.0] :: Anaconda, Inc. on linux
Type "help", "copyright", "credits" or "license" for more information.
SLF4J: Class path contains multiple SLF4J bindings.
SLF4J: Found binding in [jar:file:/home/hduser/bigdata/spark-3.1.2/jars/slf4j-log4j12
SLF4J: Found binding in [jar:file:/home/hduser/bigdata/hadoop-3.2.2/share/hadoop/comm
SLF4J: See http://www.slf4j.org/codes.html#multiple_bindings for an explanation.
SLF4J: Actual binding is of type [org.slf4j.impl.Log4jLoggerFactory]
22/01/16 13:31:28 WARN NativeCodeLoader: Unable to load native-hadoop library for you
Setting default log level to "WARN".
To adjust logging level use sc.setLogLevel(newLevel). For SparkR, use setLogLevel(new
22/01/16 13:31:31 INFO SharedState: spark.sql.warehouse.dir is not set, but hive.meta
of hive.metastore.warehouse.dir ('/user/hive/warehouse').
22/01/16 13:31:31 INFO SharedState: Warehouse path is '/user/hive/warehouse'.
Welcome to
      ____              __
     / __/__  ___ _____/ /__
    _\ \/ _ \/ _ `/ __/  '_/
   /__ / .__/\_,_/_/ /_/\_\   version 3.1.2
      /_/

Using Python version 3.7.6 (default, Jan  8 2020 19:59:22)
Spark context Web UI available at http://xueai8:4040
Spark context available as 'sc' (master = local[*], app id = local-1642311090515).
SparkSession available as 'spark'.
>>>
```

（PySpark Shell 已经默认创建了两个实例：sc 和 spark）

图 4-4　PySpark Shell 启动时会自动创建 SparkSession 的实例 spark

则可以通过分别设置 spark.executor.cores 和 spark.executor.memory 属性实现这一点。例如，将 PySpark executor 运行时属性的核设置为 2 个，将内存设置为 4GB，代码如下：

```
from pyspark.conf import SparkConf

conf = SparkConf()
conf.set("spark.executor.cores", "2")
conf.set("spark.executor.memory", "4g")
```

在 PySpark SQL 中创建 SparkSession 对象的常用模板代码如下：

```
from pyspark.conf import SparkConf
from pyspark.sql import SparkSession

conf = SparkConf()

conf.set("spark.app.name", "demo")
conf.set("spark.master", "spark://xueai8:7077")
conf.set("spark.executor.cores", 2)
conf.set("spark.executor.memory", "512m")
conf.set("spark.executor.instances", 1)
conf.set("spark.locality.wait", "0")
conf.set("spark.serializer",
```

```
"org.apache.spark.serializer.KryoSerializer")

print(spark.conf.get("spark.executor.memory"))

spark = SparkSession.builder.config(conf=SparkConf()).getOrCreate()
```

可以使用 SparkSession 对象从各种源读取数据，例如 CSV、JSON、JDBC、Stream 等。此外，它还可以用来执行 SQL 语句、注册用户定义函数（UDF）和使用 DataFrame。

注意：Spark 2.0 中的 SparkSession 为 Hive 特性提供了内置支持，包括使用 HiveQL 编写查询、访问 Hive UDF 及从 Hive 表中读取数据的能力。要使用这些特性，不需要已有的 Hive 安装。

4.4 PySpark SQL 中的模式和对象

PySpark DataFrame 就像分布在内存中的表，具有指定的列和模式，其中每列都有特定的数据类型：整数、字符串、数组、map、real、日期、时间戳等。在我们看来，PySpark DataFrame 就像一张具有行和列的二维表。

4.4.1 模式

PySpark 中的模式（Schema）为一个 DataFrame 定义了列名和关联数据类型。最常见的情况是，当用户从外部数据源读取结构化数据时，需要用到模式。预先定义模式，而不是采用"读时模式"方法，有以下 3 个好处：

（1）减轻了 Spark 推断数据类型的责任。

（2）可以防止 Spark 仅为了读取文件的大部分内容以确定模式而创建单独的作业，这对于大型数据文件来讲是昂贵和耗时的。

（3）如果数据与模式不匹配，则可以尽早发现错误。

因此，当想从数据源读取大文件时，最好是预先定义模式。PySpark 支持以两种方式定义模式：

（1）第 1 种是通过编程方式定义 Schema。

（2）第 2 种是使用数据定义语言（Data Definition Language, DDL）字符串。

其中，第 2 种模式定义方式要简单得多，也更容易阅读。

要以编程方式为 DataFrame 定义一个模式，假设这个 DataFrame 有 3 个命名列，author、title 和 pages，可以使用 PySpark DataFrame API，代码如下：

```
from pyspark.sql.types import *

schema = StructType([
```

```
    StructField("author", StringType(), True),
    StructField("title", StringType(), False),
    StructField("pages", IntegerType(), True)
])
```

如果用 DDL 来定义相同的 Schema，则要简单得多。例如，用 DDL 定义同样的 Schema，代码如下：

```
schema = "author STRING, title STRING, pages INT"
```

4.4.2　列对象和行对象

在 PySpark SQL 中，列由 Column 类型表示。可以按名称列出所有列，并且可以使用关系表达式或计算表达式对列的值进行操作。还可以在列上使用逻辑或数学表达式代码如下：

```
from pyspark.sql.functions import *

blogsDF.select(col("Hits") * 2).show(2)
```

DataFrame 中的 Column 对象不能单独存在；每列都是记录（record）中一行的一部分，所有的行一起构成一个 DataFrame。

PySpark 中的行是一个通用的 Row 对象，包含一个或多个列。每列可以是相同的数据类型（例如，整数或字符串），也可以有不同的类型（整数、字符串、映射、数组等）。因为 Row 是 PySpark 中的一个对象，是字段的一个有序集合，因此可以在 PySpark 中实例化 Row，并通过从 0 开始的索引访问它的字段。在 PySpark 中使用 Row 对象的代码如下：

```
from pyspark.sql import Row

#创建一个 Row
blogRow = Row(6, "Reynold", "Xin", "https://tinyurl.6", 255568, "3/2/2015",
["twitter", "LinkedIn"])

#使用索引访问单个项
blogRow[1]            #Reynold
```

当需要快速交互和探索数据时，可以使用 Row 对象来创建 DataFrame，代码如下：

```
rows = [("Matei Zaharia", "CA"), ("Reynold Xin", "CA")]
authorsDF = spark.sparkContext.parallelize(rows).toDF(["Author", "State"])

authorsDF.show()
```

执行过程如图 4-5 所示。

但在实践中，用户通常希望从文件中读取 DataFrame。在大多数情况下，因为数据文件非常大，所以定义一个模式并使用它是创建 DataFrame 的一种更快、更有效的方法。

```
rows = [("Matei Zaharia", "CA"), ("Reynold Xin", "CA")]
authorsDF = spark.sparkContext.parallelize(rows).toDF(["Author", "State"])

authorsDF.show()
```

```
+-------------+-----+
|       Author|State|
+-------------+-----+
|Matei Zaharia|   CA|
|  Reynold Xin|   CA|
+-------------+-----+
```

图 4-5 使用 Row 对象来创建 DataFrame

4.5 简单构造 DataFrame

37min

作为 PySpark 2.0 及以后的应用程序统一入口，开发 PySpark 应用程序的第 1 步就是要创建一个 SparkSession 实例。为了访问 DataFrame API，需要 SparkSession 作为入口点，如图 4-6 所示。

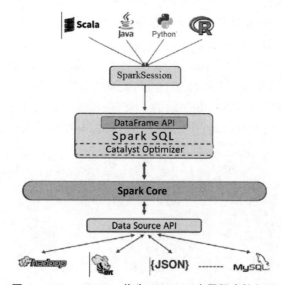

图 4-6 SparkSession 作为 PySpark 应用程序的入口

PySpark DataFrame 的创建类似于 RDD 的创建。在本节中，将演示如何从各种数据源创建 DataFrame。有多种方式可用来创建 DataFrame，包括以下几种：

(1) 简单创建单列和多列 DataFrame。
(2) 将已经存在的 RDD 转换为一个 DataFrame。
(3) 运行 SQL 查询返回一个 DataFrame。
(4) 将外部数据源的数据加载到一个 DataFrame。

4.5.1 简单创建单列和多列 DataFrame

SparkSession 有一个函数叫 range()，可以很容易地创建带有列名 id 和类型 LongType 的单列 DataFrame，代码如下：

```
df = spark.range(5)
df.printSchema()
df.show()
```

执行过程和输出结果如图 4-7 所示。

```
df = spark.range(5)
df.printSchema()
df.show()
```

```
root
 |-- id: long (nullable = false)

+---+
| id|
+---+
|  0|
|  1|
|  2|
|  3|
|  4|
+---+
```

图 4-7　创建单列的 DataFrame，带有默认列名 id

还可以在创建 DataFrame 时指定列名，代码如下：

```
val df1 = spark.range(5).toDF("num")
df1.printSchema
df1.show
```

执行过程和输出结果如图 4-8 所示。

另外，还可以指定范围的起始（含）和结束值（不含），代码如下：

```
val df2 = spark.range(5,10).toDF("num")
df2.show
```

执行过程和输出结果如图 4-9 所示。

另外，还可以指定步长，代码如下：

```
val df3 = spark.range(5,15,2).toDF("num")
df3.show
```

执行过程和输出结果如图 4-10 所示。

```
df1 = spark.range(5).toDF("num")

df1.printSchema()
df1.show()
```

```
root
 |-- num: long (nullable = false)

+---+
|num|
+---+
|  0|
|  1|
|  2|
|  3|
|  4|
+---+
```

图 4-8　创建单列的 DataFrame，指定列名

```
df2 = spark.range(5,10).toDF("num")
df2.show()
```

```
+---+
|num|
+---+
|  5|
|  6|
|  7|
|  8|
|  9|
+---+
```

图 4-9　创建单列的 DataFrame，指定范围的起始（含）和结束值（不含）

```
df3 = spark.range(5,15,2).toDF("num")
df3.show()
```

```
+---+
|num|
+---+
|  5|
|  7|
|  9|
| 11|
| 13|
+---+
```

图 4-10　创建单列的 DataFrame，指定步长

注意：toDF()方法采用的是元组列表，而不是标量元素。

通过将一个元组集合转换为一个 DataFrame，可以创建多列 DataFrame。这需要使用 SparkSession 对象的 toDF()方法。toDF()方法将列标签列表作为可选的参数，以指定转换后

的 DataFrame 的标题行，代码如下：

```
#一个list，元素为元组类型
movies = [
    ("马特•达蒙", "谍影重重:极限伯恩", 2007),
    ("马特•达蒙", "心灵捕手", 1997)
]

#将元组转换为DataFrame
moviesDF = spark.createDataFrame(movies, schema=['actor', 'title', 'year'])

#输出模式
moviesDF.printSchema()

#显示
moviesDF.show()
```

执行过程和输出结果如图 4-11 所示。

```
# 一个list,元素为元组类型
movies = [
    ("马特•达蒙", "谍影重重:极限伯恩", 2007),
    ("马特•达蒙", "心灵捕手", 1997)
]

# 将元组转为DataFrame
moviesDF = spark.createDataFrame(movies, schema=['actor', 'title', 'year'])

# 输出模式
moviesDF.printSchema()

# 显示
moviesDF.show()
```

```
root
 |-- actor: string (nullable = true)
 |-- title: string (nullable = true)
 |-- year: long (nullable = true)

+---------+----------------+----+
|    actor|           title|year|
+---------+----------------+----+
|马特•达蒙|谍影重重:极限伯恩|2007|
|马特•达蒙|        心灵捕手|1997|
+---------+----------------+----+
```

图 4-11　通过将一个元组集合转换为一个 DataFrame，可创建多列 DataFrame

通过元组来创建单列或多列 DataFrame，每个元组类似于一行。可以选择标题列；否则 PySpark 会创建一些模糊的名称，例如_1、_2。列的类型推断是隐式的。例如，与上面的代码类似，但是不指定标题列，执行过程和执行结果如图 4-12 所示。

```
# 一个list, 元素为元组类型
movies = [
    ("马特·达蒙", "谍影重重:极限伯恩", 2007),
    ("马特·达蒙", "心灵捕手", 1997)
]

# 将元组转为DataFrame
moviesDF = spark.createDataFrame(movies)

# 输出模式
moviesDF.printSchema()

# 显示
moviesDF.show()
```

```
root
 |-- _1: string (nullable = true)
 |-- _2: string (nullable = true)
 |-- _3: long (nullable = true)

+---------+---------------------+----+
|       _1|                   _2|  _3|
+---------+---------------------+----+
|马特·达蒙|谍影重重:极限伯恩    |2007|
|马特·达蒙|           心灵捕手  |1997|
+---------+---------------------+----+
```

图 4-12 创建多列 DataFrame，但不指定标题列

4.5.2 从 RDD 创建 DataFrame

在 PySpark 中有两种方法可以将 RDD 转换为 DataFrame：toDF()和 createDataFrame(rdd, schema)。

1. 方法一：使用 toDF()

通过显式调用 RDD 的 toDF()函数，将 RDD 转换到 DataFrame，并使用指定的列名，列的类型是从 RDD 中的数据推断出来的，代码如下：

```
#list
persons = [("张三",23),("李四",18),("王老五",35)]

#RDD
personRDD = spark.sparkContext.parallelize(persons)

#from RDD to DataFrame
personsDF = personRDD.toDF(["name", "age"])

#查看模式和数据
personsDF.printSchema()
personsDF.show()

print(personsDF.dtypes)              #注意其中的隐式类型推断
```

执行过程和结果如图 4-13 所示。

```
# list
persons = [("张三",23),("李四",18),("王老五",35)]

# RDD
personRDD = spark.sparkContext.parallelize(persons)

# from RDD to DataFrame
personsDF = personRDD.toDF(["name","age"])

# 查看模式和数据
personsDF.printSchema()
personsDF.show()

print(personsDF.dtypes)          # 注意其中的隐式类型推断
```

```
root
 |-- name: string (nullable = true)
 |-- age: long (nullable = true)

+------+---+
|  name|age|
+------+---+
|  张三| 23|
|  李四| 18|
|王老五| 35|
+------+---+

[('name', 'string'), ('age', 'bigint')]
```

图 4-13　显式调用 RDD 的 toDF()函数，将 RDD 转换到 DataFrame

在上面的代码中，创建了一个 RDD，它包含元组元素，然后调用它的 toDF()方法。需要注意，toDF()采用的是元组列表，而不是标量元素。每个元组类似于一行。可以指定列名（像在上面的代码中所做的那样）。如果不指定列名，则 PySpark 会自行创建一些模糊的名称，例如_1、_2，列的类型推断是隐式的。例如，将一个 RDD 转换为 DataFrame，通过 toDF()方法，但是不指定列名，代码如下：

```
#list
persons = [("张三",23),("李四",18),("王老五",35)]

#RDD
personRDD = spark.sparkContext.parallelize(persons)

#from RDD to DataFrame
personsDF = personRDD.toDF()

#查看模式和数据
personsDF.printSchema()
```

```
personsDF.show()

print(personsDF.dtypes)          #注意其中的隐式类型推断
```

执行过程和结果如图 4-14 所示。

```
# list
persons = [("张三",23),("李四",18),("王老五",35)]
# RDD
personRDD = spark.sparkContext.parallelize(persons)

# from RDD to DataFrame
personsDF = personRDD.toDF()

# 查看模式和数据
personsDF.printSchema()
personsDF.show()

print(personsDF.dtypes)  ———— # 注意其中的隐式类型推断
```

```
root
 |-- _1: string (nullable = true)
 |-- _2: long (nullable = true)

+-----+---+
|   _1| _2|
+-----+---+
| 张三| 23|
| 李四| 18|
|王老五| 35|
+-----+---+

[('_1', 'string'), ('_2', 'bigint')]
```

图 4-14 将 RDD 转换到 DataFrame 时不指定列名

也可以使用反射来推断包含特定对象类型的 RDD 的模式。PySpark SQL 可以将包含 Row 对象的 RDD 转换为 DataFrame，从而推断数据类型。Row 是通过将一组键-值对作为 kwargs 传递给 Row 类来构造的。这个列表的 key 用于定义表的列名，类型通过对整个数据集进行采样来推断，类似于对 JSON 文件执行的推断，代码如下：

```
from pyspark.sql.types import Row

#list
persons = [("张三",23),("李四",18),("王老五",35)]

#RDD[Row]
personRDD = spark.sparkContext \
    .parallelize(persons) \
    .map(lambda t: Row(stu_name=t[0],stu_age=t[1]))
```

```
#from RDD to DataFrame
personsDF = personRDD.toDF()

#查看模式和数据
personsDF.printSchema()
personsDF.show()

print(personsDF.dtypes)                    #注意其中的隐式类型推断
```

执行过程和结果如图 4-15 所示。

```
from pyspark.sql.types import Row

# list
persons = [("张三",23),("李四",18),("王老五",35)]

# RDD
personRDD = spark.sparkContext.parallelize(persons).map(lambda t: Row(stu_name=t[0],stu_age=t[1]))

# from RDD to DataFrame
personsDF = personRDD.toDF()

# 查看模式和数据
personsDF.printSchema()
personsDF.show()

print(personsDF.dtypes) ————# 注意其中的隐式类型推断

root
 |-- stu_name: string (nullable = true)
 |-- stu_age: long (nullable = true)

+--------+-------+
|stu_name|stu_age|
+--------+-------+
|    张三|     23|
|    李四|     18|
|  王老五|     35|
+--------+-------+

[('stu_name', 'string'), ('stu_age', 'bigint')]
```

图 4-15　使用反射来推断包含特定对象类型的 RDD 的模式

这种基于反射的方法使代码更简洁，如果在编写 PySpark 应用程序时已经了解了模式，则这种方法可以很好地工作。

2. 方法二：使用 createDataFrame(rdd, schema)

第 2 种方法是通过一个编程接口，先构造出一个模式（Schema），然后将其应用到现有的 RDD 中创建一个 DataFrame。这需要使用 SparkSession 的方法 createDataFrame 来创建。虽然这种方法比较冗长，但是在事先不知道列类型的情况下，可以通过这种方法自行构造 DataFrame。

可以通过以下 3 个步骤以编程的方式创建 DataFrame。

（1）从原始的 RDD 创建一个元组或列表的 RDD。

（2）创建由 StructType 表示的模式，该模式与在步骤 1 中创建的 RDD 中的元组或列表结构相匹配。

（3）通过 SparkSession 提供的 createDataFrame 方法将模式应用到 RDD。

例如，通过一个 RDD 应用指定的 Schema 模式，将其转换为一个 DataFrame，代码如下：

```python
from pyspark.sql.types import *

#指定一个 Schema(模式)
fields = [
    StructField("id", LongType(), True),
    StructField("name", StringType(), True),
    StructField("age", LongType(), True)
]
schema = StructType(fields)

#构造一个 RDD
peopleRDD = sc.parallelize([(1,"张三",30),(2, "李小四", 25),(3, "王老五", 35)])

#从给定的 RDD 应用给定的 Schema 创建一个 DataFrame
peopleDF = spark.createDataFrame(peopleRDD, schema)

#查看 DataFrame Schema
peopleDF.printSchema()

#输出
peopleDF.show()
```

执行过程和结果如图 4-16 所示。

4.5.3 读取外部数据源创建 DataFrame

PySpark 提供了一个接口 DataFrameReader，用来以各种格式（如 JSON、CSV、Parquet、Text、Avro、ORC 等）从众多的数据源将数据读到 DataFrame。同样，要将 DataFrame 以特定格式写回数据源，PySpark 使用 DataFrameWriter 接口。

要将存储系统中的文件加载到 DataFrame，可以通过 SparkSession 的 read 字段实现，它是 DataFrameReader 的一个实例。它有个 load()方法，可以直接从配置的数据源加载数据。另外它还有 5 个快捷方法：text、csv、json、orc 和 parquet，相当于先调用 format()方法再调用 load()方法。

为了演示如何将外部数据源加载到 DataFrame，本节使用 PySpark 自带的各种格式的

```python
from pyspark.sql.types import *

# 指定一个Schema(模式)
fields = [
    StructField("id", LongType(), True),
    StructField("name", StringType(), True),
    StructField("age", LongType(), True)
]
schema = StructType(fields)

# 构造一个RDD
peopleRDD = sc.parallelize([(1,"张三",30),(2,"李小四", 25),(3,"王老五",35)])

# 从给定的RDD应用给定的Schema创建一个DataFrame
peopleDF = spark.createDataFrame(peopleRDD, schema)

# 查看DataFrame Schema
peopleDF.printSchema()

# 输出
peopleDF.show()
```

```
root
 |-- id: long (nullable = true)
 |-- name: string (nullable = true)
 |-- age: long (nullable = true)

+---+------+---+
| id|  name|age|
+---+------+---+
|  1|  张三| 30|
|  2|李小四| 25|
|  3|王老五| 35|
+---+------+---+
```

图 4-16 使用反射来推断包含特定对象类型的 RDD 的模式

数据源文件。这些数据源文件位于 $SPARK_HOME/examples/src/main/resources/ 目录下。首先将它们上传到 HDFS 的 /data/spark/resources/ 目录下。在命令行终端执行 HDFS Shell 命令，命令如下：

```
$ hdfs dfs -put ~/bigdata/spark-3.1.2/examples/src/main/resources /data/spark/resources
```

1. 读取文本文件创建 DataFrame

文本文件是最常见的数据存储文件。PySpark DataFrame API 允许开发者将文本文件的内容转换为 DataFrame。

例如，将 PySpark 自带的文本文件 people.txt 读到一个 DataFrame 中，代码如下：

```
file = "/data/spark/resources/people.txt"
```

```
txtDF = spark.read.format("text").load(file)   #加载文本文件
#txtDF = spark.read.text(file)                 #等价于上一句，快捷方法

txtDF.printSchema()                            #打印 schema
txtDF.show()                                   #输出
```

执行过程和结果如图 4-17 所示。

```
file = "file:///home/hduser/data/spark/resources/people.txt"
txtDF = spark.read.format("text").load(file)   # 加载文本文件
# txtDF = spark.read.text(file)                # 等价于上一句，快捷方法

txtDF.printSchema()                            # 打印 schema
txtDF.show()                                   # 输出

root
 |-- value: string (nullable = true)

+-----------+
|      value|
+-----------+
|Michael, 29|
|   Andy, 30|
| Justin, 19|
+-----------+
```

图 4-17　读取文本文件创建 DataFrame

PySpark 会自动推断出模式，并相应地创建一个单列（列名为 value）的 DataFrame，因此，没有必要为文本数据定义模式。不过，当加载大数据文件时，定义一个 Schema 要比让 PySpark 进行推断效率更高。

2. 读取 CSV 文件创建 DataFrame

在 PySpark 3.x 中，加载 CSV 文件是非常简单的。例如，要将 Spark 自带的 people.csv 文件加载到一个 DataFrame 中，代码如下：

```
file = "/data/spark/resources/people.csv"
people_df = spark.read.load(file, format="csv", sep=";",
inferSchema = "true", samplingRatio=0.001, header="true")

#以下用法与上一种用法等价
#people_df = spark.read.format("csv").load(file, sep=";",
inferSchema = "true", samplingRatio=0.001, header="true")

people_df.printSchema()                        #打印 schema
people_df.show()                               #显示
```

在以上代码中，加载 CSV 文件时指定的 option 选项含义可参考表 4-3。

表 4-3 加载 CSV 文件时指定的 option 选项

Option	默认值	说明	引入版本
sep	,	设置单个字符作为每个字段和值的分隔符	v2.0.0
encoding	UTF-8	根据给定的编码类型解码 CSV 文件	v2.0.0
quote	"	设置用于转义引用值的单个字符，其中分隔符可以是值的一部分。如果要关闭引号，则需要设置的不是 null，而是一个空字符串	v2.0.0
escape	\	设置用于转义已引用值中的引号的单个字符	v2.0.0
comment	空字符串	设置用于跳过以该字符开头的行的单个字符。默认情况下，它是禁用的	v2.0.0
header	false	是否使用第 1 行作为列的名称。不支持两行标题	v2.0.0
inferSchema	false	告诉 Spark 是否尝试基于列值推断列类型（从数据自动推断输入模式）。它需要额外传递一次数据	v2.0.0
samplingRatio		根据抽样进行模式推断	
ignoreLeadingWhiteSpace	false	指示是否跳过正在读取的值中的前导空格	v2.0.0
ignoreTrailingWhiteSpace	false	指示是否跳过正在读取的值中的尾部空格	v2.0.0
nullValue	空字符串	设置 null 值的字符串表示形式。从 2.0.1 开始，这适用于所有受支持的类型，包括字符串类型	v2.0.0
nanValue	NaN	设置非数字"值"的字符串表示形式	v2.0.0
positiveInf	Inf	设置正无穷值的字符串表示形式	v2.0.0
negativeInf	-Inf	设置负无穷值的字符串表示形式	v2.0.0
dateFormat	yyyy-MM-dd	设置指示日期格式的字符串。自定义日期格式遵循 Java.text.SimpleDateFormat 的格式。这适用于日期（date）类型	v2.0.0
timestampFormat	yyyy-MM-dd'T'HH:mm:ss.SSSXXX	设置指示时间戳格式的字符串。自定义日期格式遵循 Java.text.SimpleDateFormat 的格式。这适用于时间戳（timestamp）类型	v2.1.0
maxColumns	20480	定义一个记录可以有多少列的硬限制	
maxCharsPerColumn	−1	定义允许读取任何给定值的最大字符数。默认值为 1 000 000	
mode	PERMISSIVE	允许在解析期间处理损坏记录的模式。它支持以下不区分大小写的模式。 - PERMISSIVE：当遇到损坏的记录时，将其他字段设置为 null，并将格式错误的字符串放入由 columnNameOfCorruptRecord 配置的字段中。要保存损坏的记录，用户可以在用户定义的模式中设置一个名为 columnNameOfCorruptRecord 的字符串类型字段。如果一个模式没有该字段，它将在解析期间删除损坏的记录。当解析后的 CSV tokens 长度小于模式的预期长度时，它会为额外字段设置 null。 - DROPMALFORMED：忽略整个损坏的记录。 - FAILFAST：当遇到损坏的记录时抛出异常	v2.0.0

续表

Option	默认值	说明	引入版本
columnNameOfCorruptRecord		允许重命名新字段，该字段具有由 PERMISSIVE 模式创建的格式错误字符串。这将覆盖 spark.sql.columnNameOfCorruptRecord。默认值为在 spark.sql.columnNameOfCorruptRecord 中指定的值	v2.2.0
multiLine		解析一条记录，它可能跨越多行	v2.2.0
wholeFile	false	解析一条记录，它可能跨越多行	

以上代码的执行过程和结果如图 4-18 所示。

```
file = "file:///home/hduser/data/spark/resources/people.csv"
people_df = spark.read.load(file,
                            format="csv",
                            sep=";",
                            inferSchema="true",
                            header="true")

people_df.printSchema()
people_df.show()
```

```
root
 |-- name: string (nullable = true)
 |-- age: integer (nullable = true)
 |-- job: string (nullable = true)

+-----+---+---------+
| name|age|      job|
+-----+---+---------+
|Jorge| 30|Developer|
|  Bob| 32|Developer|
+-----+---+---------+
```

图 4-18 读取 CSV 文件创建 DataFrame

或者，也可以使用快捷方法 csv()，代码如下：

```
file = "/data/spark/resources/people.csv"

people_df = spark.read.options(sep=";", inferSchema="true",
samplingRatio=0.001, header="true").csv(file)

#以下用法与上一种用法等价
#people_df = spark.read.csv(file, sep=";", inferSchema="true",
samplingRatio=0.001, header="true")

people_df.printSchema()
people_df.show()
```

执行过程和结果如图 4-19 所示。

```
file = "file:///home/hduser/data/spark/resources/people.csv"
people_df = spark.read.options(sep=";",inferSchema="true",header="true").csv(file)

people_df.printSchema()
people_df.show()
```

```
root
 |-- name: string (nullable = true)
 |-- age: integer (nullable = true)
 |-- job: string (nullable = true)

+-----+---+---------+
| name|age|      job|
+-----+---+---------+
|Jorge| 30|Developer|
|  Bob| 32|Developer|
+-----+---+---------+
```

图 4-19　读取 CSV 文件创建 DataFrame 的简捷方式

在上面的代码中，使用了模式自动推断。对于大型的数据源，指定一个 Schema 要比让 PySpark 进行模式推断效率更高。修改上面的代码，明确提供一个 Schema，代码如下：

```
from pyspark.sql.types import *

file = "/data/spark/resources/people.csv"

#指定一个Schema(模式)
fields = [
    StructField("p_name", StringType(), True),
    StructField("p_age", IntegerType(), True),
    StructField("p_job", StringType(), True)
]
schema = StructType(fields)

#将CSV文件加载到DataFrame，并指定schema
people_df2 = spark.read \
    .options(sep=";", header="true") \
    .schema(schema) \
    .csv(file)

people_df2.printSchema()
people_df2.show()
```

执行过程和结果如图 4-20 所示。

```
from pyspark.sql.types import *

file = "file:///home/hduser/data/spark/resources/people.csv"

# 指定一个Schema(模式)
fields = [
    StructField("p_name", StringType(), True),
    StructField("p_age", IntegerType(), True),
    StructField("p_job", StringType(), True)
]
schema = StructType(fields)

# 将CSV文件加载到DataFrame，并指定Schema
people_df2 = spark.read.options(sep=",",header="true").schema(schema).csv(file)

people_df2.printSchema()
people_df2.show()
```

```
root
 |-- p_name: string (nullable = true)
 |-- p_age: integer (nullable = true)
 |-- p_job: string (nullable = true)

+------+-----+---------+
|p_name|p_age|    p_job|
+------+-----+---------+
| Jorge|   30|Developer|
|   Bob|   32|Developer|
+------+-----+---------+
```

图 4-20　读取 CSV 文件创建 DataFrame，指定 schema

可以看出，它返回了由行和命名列组成的 DataFrame，该 DataFrame 具有模式中指定的类型。

注意：如果一个文件是由 DataFrame 以 Parquet 格式写出存储的，则模式被保留为 Parquet 元数据的一部分。在这种情况下，随后读入 DataFrame 不需要手动提供模式。Parquet 是一种流行的柱状格式，是 Spark/PySpark 的默认格式；它使用 snappy 来压缩数据。

也可以使用 CSV 格式读取 TSV 文件。所谓 TSV 文件，指的是以制表符（Tab）作为字段分隔符的文件。例如，将 Spark 自带的 people.tsv 文件加载到一个 DataFrame 中，代码如下：

```
file = "/data/spark/resources/people.tsv"

people_df3 = spark.read \
    .options(sep="\t",inferSchema="true",header="true") \
    .csv(file)

people_df3.printSchema()
people_df3.show()
```

执行过程和结果如图 4-21 所示。

```
file = "file:///home/hduser/data/spark/resources/people.tsv"
people_df3 = spark.read.options(sep="\t",inferSchema="true",header="true").csv(file)
people_df3.printSchema()
people_df3.show()
```

```
root
 |-- name: string (nullable = true)
 |-- age: integer (nullable = true)
 |-- job: string (nullable = true)

+-----+---+---------+
| name|age|      job|
+-----+---+---------+
|Jorge| 30|Developer|
|  Bob| 32|Developer|
+-----+---+---------+
```

图 4-21　读取 TSV 文件创建 DataFrame

在使用 SparkSession 对象的 read 字段读取 CSV 文件时，可指定的 option 选项参数和值的列表很长，见表 4-3。

3. 读取 JSON 文件创建 DataFrame

PySpark SQL 可以自动推断 JSON DataFrame 的模式，并将其加载为 DataFrame。这种转换可以在 JSON 文件上使用 SparkSession.read.json()完成。

注意：作为 JSON 文件提供的文件实际上并不是标准意义上的 JSON 文件，而是要求每行必须包含一个单独的、自包含的有效 JSON 对象。对于常规的多行 JSON 文件，将 multiLine 选项设置为 true。

读取 JSON 数据源文件时，PySpark 会从 key 中自动推断模式，并相应地创建一个 DataFrame，因此，没有必要为 JSON 数据定义模式。此外，PySpark 极大地简化了访问复杂 JSON 数据结构中的字段所需的查询语法。例如，将 Spark 自带的 people.json 文件内容读取到一个 DataFrame 中，代码如下：

```
file = "/data/spark/resources/people.json"

#JSON 解析；列名和数据类型隐式地推断
people_df4 = spark.read.load(file, format="json")

#people_df4 = spark.read.json(file)     #简捷方法

people_df4.printSchema()
people_df4.show()
```

执行过程和结果如图4-22所示。

```
file = "file:///home/hduser/data/spark/resources/people.json"

# JSON解析：列名和数据类型隐式地推断
people_df4 = spark.read.load(file, format="json")

# people_df4 = spark.read.json(file)    # 简捷方法

people_df4.printSchema()
people_df4.show()
```

```
root
 |-- age: long (nullable = true)
 |-- name: string (nullable = true)

+----+-------+
| age|   name|
+----+-------+
|null|Michael|
|  30|   Andy|
|  19| Justin|
+----+-------+
```

图 4-22　读取 JSON 文件创建 DataFrame

当然，也可以明确指定一个 schema，覆盖 PySpark 的推断 Schema，代码如下：

```
from pyspark.sql.types import *

file = "/data/spark/resources/people.json"

#指定一个 Schema(模式)。字段名称要与 JSON 对象的 key 名称保持一致
fields = [
    StructField("name", StringType(), True),
    StructField("age", IntegerType(), True)
]
schema = StructType(fields)

#将数据读到 DataFrame，并指定 schema
people_df5 = spark.read.schema(schema).json(file)

people_df5.printSchema()
people_df5.show()
```

执行过程和结果如图4-23所示。

如果在加载 JSON 文件时 JSON 格式解析错误，则默认创建的 DataFrame 的相应各行各列值都设为 null。例如，在 Schema 中将姓名这一列（name）对应的数据类型错误地设

```
from pyspark.sql.types import *

file = "file:///home/hduser/data/spark/resources/people.json"

# 指定一个Schema(模式)
fields = [
    StructField("name", StringType(), True),
    StructField("age", IntegerType(), True)
]
schema = StructType(fields)

# 将数据读到DataFrame，并指定Schema
people_df5 = spark.read.schema(schema).json(file)

people_df5.printSchema()
people_df5.show()
```

```
root
 |-- name: string (nullable = true)
 |-- age: integer (nullable = true)

+-------+----+
|   name| age|
+-------+----+
|Michael|null|
|   Andy|  30|
| Justin|  19|
+-------+----+
```

图 4-23　读取 JSON 文件创建 DataFrame，并指定一个 schema

为 boolean 类型，因此 SparkSession 在解析时出现解析错误，因而整个 DataFrame 中的各行各列值均为 null，代码如下：

```
from pyspark.sql.types import *

file = "/data/spark/resources/people.json"

#指定一个Schema(模式)。字段名称要与JSON对象的key名称保持一致
fields = [
    StructField("name", BooleanType(), True),
    StructField("age", IntegerType(), True)
]
schema = StructType(fields)

#将数据读到DataFrame，并指定schema
people_df6 = spark.read.schema(schema).json(file)

people_df6.printSchema()
people_df6.show()
```

执行以上代码,并不会出现任何异常。执行过程和结果如图 4-24 所示。

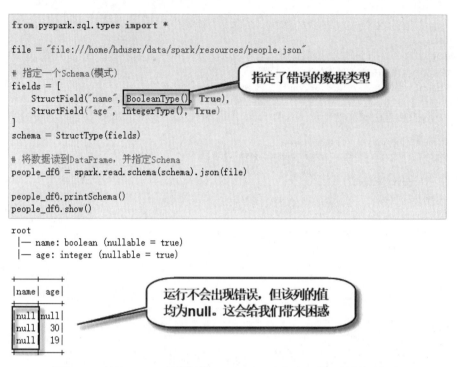

图 4-24　读取 JSON 文件创建 DataFrame,指定了错误的解析类型

可以看到,name 列的值全部被解析为 null,但是这样处理 JSON 格式解析错误并不是一种好的方式,因为它经常会掩盖错误事实,让用户迷惑,所以最好的方式是,如果 JSON 格式解析错误,就直接抛出异常(快速失败),而不是全都设为 null 值。例如,在将 Spark 自带的 people.json 文件读到一个 DataFrame 时,指定快速失败的处理方式,代码如下:

```
from pyspark.sql.types import *

file = "/data/spark/resources/people.json"

#指定一个 Schema(模式)。字段名称要与 JSON 对象的 key 名称保持一致
fields = [
   StructField("name", BooleanType(), True),
   StructField("age", IntegerType(), True)
]
schema = StructType(fields)

#将数据读到 DataFrame,并指定 schema
people_df6 = spark.read \
   .option("mode","failFast") \
```

```
    .schema(schema) \
    .json(file)

#输出模式和显示内容
people_df6.printSchema()
people_df6.show()
```

在上面的代码中,将 mode 指定为 failFast 模式,这是用来告诉 PySpark,当遇到解析错误时,快速失败,而不是生成 null 值。执行以上代码,会抛出异常信息,执行过程和异常信息如图 4-25 所示。

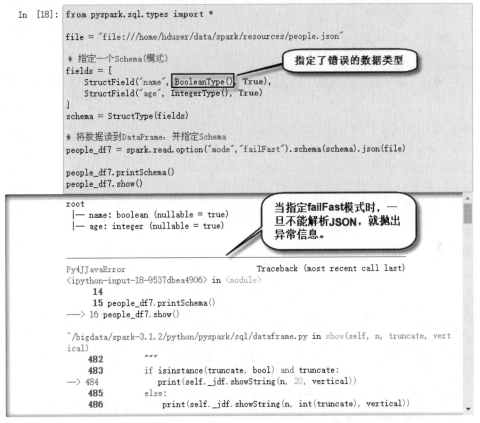

图 4-25 读取 JSON 文件创建 DataFrame,指定 failFast 模式

可以看到,在 JSON 格式解析错误时会立即抛出异常信息,从而不会误导用户。

4. 读取 Parquet 文件创建 DataFrame

Apache Parquet 是一种高效的、压缩的、面向列的开源数据存储格式。它提供了多种存储优化,允许读取单独的列而非整个文件,这不仅节省了存储空间而且提升了读取效率。

它是Spark/PySpark 默认的文件格式，支持非常有效的压缩和编码方案，也可用于Hadoop生态系统中的任何项目，可以大大提高这类应用程序的性能。

Apache Spark 提供了对读取和写入 Parquet 文件的支持，这些文件拥有自动保存原始数据的模式。Parquet 是一种非常流行的格式，在 PySpark 中有一些额外的选项可以用于读写 Parquet 文件。在写入 Parquet 文件时，出于兼容性考虑，所有列都会被自动转换为 nullable。

例如，将 Spark 自带的 Parquet 文件 users.parquet 的内容读到一个 DataFrame 中，然后打印其 Schema 并输出数据，代码如下：

```
file = "/data/spark/resources/users.parquet"

#读取 Parquet 文件
#Parquet 文件是自描述的，因此模式得以保留
#加载 Parquet 文件的结果也是一个 DataFrame
parquet_df = spark.read.load(file,format="parquet")
#parquet_df = spark.read.parquet(file)      #简捷写法

#输出模式和内容
parquet_df.printSchema()
parquet_df.show()
```

执行过程和结果如图 4-26 所示。

```
file = "file:///home/hduser/data/spark/resources/users.parquet"

# 读取Parquet文件
# Parquet文件是自描述的，因此模式得以保留
# 加载Parquet文件的结果也是一个DataFrame
parquet_df = spark.read.load(file,format="parquet")
# parquet_df = spark.read.parquet(file)      # 简捷写法

# 输出模式和内容
parquet_df.printSchema()
parquet_df.show()
```

```
root
 |-- name: string (nullable = true)
 |-- favorite_color: string (nullable = true)
 |-- favorite_numbers: array (nullable = true)
 |    |-- element: integer (containsNull = true)

+------+--------------+----------------+
|  name|favorite_color|favorite_numbers|
+------+--------------+----------------+
|Alyssa|          null|  [3, 9, 15, 20]|
|   Ben|           red|              []|
+------+--------------+----------------+
```

图 4-26　读取 Parquet 文件创建 DataFrame

5. 读取 ORC 文件创建 DataFrame

ORC (Optimized Row Columnar)是一种流行的大数据文件存储格式。ORC 文件是一种自描述的、类型感知的列存储格式数据文件，它针对大型数据的读写进行了优化，具有高性能、可压缩性强等特点，被许多顶级 Apache 产品支持，例如 Hive、Crunch、Cascading、Spark 等，是大数据常用的文件格式。Apache Spark 提供了对读取和写入 ORC 文件的支持。

例如，将 Spark 自带的 ORC 文件 users.orc 读到一个 DataFrame 中，代码如下：

```
#数据源文件
orcFile = "/data/spark/resources/users.orc"

#读取 ORC 文件，构造 DataFrame
orc_df = spark.read.format("orc").load(orcFile)
#orc_df = spark.read.load(orcFile,format="orc")
#orc_df = spark.read.orc(orcFile)    #简捷写法

orc_df.printSchema()
orc_df.show()
```

执行过程和结果如图 4-27 所示。

```
# 数据源文件
orcFile = "file:///home/hduser/data/spark/resources/users.orc"

# 读取ORC文件，构造DataFrame
orc_df = spark.read.format("orc").load(orcFile)
# orc_df = spark.read.load(orcFile,format = "orc")
# orc_df = spark.read.orc(orcFile)   # 简捷写法

orc_df.printSchema()
orc_df.show()
```

```
root
 |-- name: string (nullable = true)
 |-- favorite_color: string (nullable = true)
 |-- favorite_numbers: array (nullable = true)
 |    |-- element: integer (containsNull = true)

+------+--------------+----------------+
|  name|favorite_color|favorite_numbers|
+------+--------------+----------------+
|Alyssa|          null| [3, 9, 15, 20]|
|   Ben|           red|             []|
+------+--------------+----------------+
```

图 4-27 读取 ORC 文件创建 DataFrame

从 Spark 2.3 开始，PySpark 支持一个带有新的 ORC 文件格式的向量化 ORC reader，配置见表 4-4。

表 4-4 读取 ORC 文件时的配置

属 性 名	默认值	含 义
spark.sql.orc.impl	native	The name of ORC 实现的名称。可以是 native 或 hive。native 意味着本地 ORC 支持，它构建在 Apache ORC 1.4 之上。hive 意味着 Hive 1.2.1 中的 ORC 库
spark.sql.orc.enableVectorizedReader	true	在 native 实现中启用向量化 orc 解码。如果值为 false，则在 native 实现中使用新的非向量化 ORC reader

当 spark.sql.orc.impl 设置为 native 及 spark.sql.orc.enableVectorizedReader 设置为 true 时，会将向量化的 reader 用于本地 ORC 表（例如，使用 USING ORC 子句创建的表）。对于 Hive ORC serde 表，当 spark.sql.hive.convertMetastoreOrc 也被设置为 true 时，也会使用向量化的 reader。

6. 使用 JDBC 从数据库创建 DataFrame

PySpark SQL 还包括一个可以使用 JDBC 从其他关系型数据库读取数据的数据源。开发人员可以使用 JDBC 创建来自其他数据库的 DataFrame，只要确保预定数据库的 JDBC 驱动程序是可访问的（需要在 Spark 类路径中包含特定数据库的 JDBC 驱动程序）。

注意：Spark 安装程序默认为没有提供数据库驱动，所以在使用前需要将相应的数据库驱动上传到$SPARK_HOME 目录下的 jars 目录中。本书示例使用的是 MySQL 数据库，所以使用前需要将相应的 mysql-connector-Java-x.x.x.jar 包上传到$SPARK_HOME/jars 目录下。

在下面的示例中，通过 JDBC 读取 MySQL 数据库中的一个 people 数据表，并创建 DataFrame。为此，先在 MySQL 中执行如下脚本，创建一个名为 xueai8 的数据和一个名为 people 的数据表，并向表中插入一些样本记录，命令如下：

```
mysql> create database xueai8;
mysql> use xueai8;
mysql> create table people(id int not null primary key, name varchar(20), age int);
mysql> insert into people values(1,"张三",23),(2,"李四",18),(3,"王老五",35);
mysql> select * from people;
```

然后编写代码，从 MySQL 中将 people 表中的数据读到 DataFrame 中，代码如下：

```
jdbc_df1 = spark.read \
    .format('jdbc') \
    .options(url = 'jdbc:mysql://localhost:3306/xueai8',
             dbtable = 'people',
             user = 'root',
             password = '123456') \
    .load()

#获取表数据作为一个 DataFrame
jdbc_df1.show()
```

注意：数据库连接账号和密码应修改为自己本机数据的账号和密码。

执行过程和结果如图 4-28 所示。

```
jdbc_df1 = spark.read \
    .format('jdbc') \
    .options(url = 'jdbc:mysql://localhost:3306/xueai8',
            dbtable = 'people',
            user = 'root',
            password = 'admin') \
    .load()
# 获取表数据作为一个DataFrame
jdbc_df1.show()
```

```
| id| name|age|
|  1|  张三| 23|
|  2|  李四| 18|
|  3|王老五| 35|
```

图 4-28　读取 JDBC 数据库创建 DataFrame

也可以使用快捷方法 jdbc(url, table, properties={"user": "username", "password": "password"})从关系型数据库中加载数据。例如，改写上面的示例，代码如下：

```
jdbc_df3 = spark.read \
    .jdbc("jdbc:mysql://localhost:3306/xueai8",
        "people",
        properties={"user": "root", "password": "admin"})

#获取表数据作为一个 DataFrame
jdbc_df3.show()
```

执行过程和结果如图 4-29 所示。

```
jdbc_df3 = spark.read \
    .jdbc("jdbc:mysql://localhost:3306/xueai8",
        "people",
        properties={"user": "root", "password": "admin"})
# 获取表数据作为一个DataFrame
jdbc_df3.show()
```

```
| id| name|age|
|  1|  张三| 23|
|  2|  李四| 18|
|  3|王老五| 35|
```

图 4-29　使用快捷方式读取 JDBC 数据库创建 DataFrame

在读取 JDBC 关系型数据库中的表数据时，也可以指定相应的 DataFrame 的列数据类型，代码如下：

```
#指定读取时的DataFrame列数据类型
jdbc_df2 = spark.read \
    .format("jdbc") \
    .option("url", "jdbc:mysql://localhost:3306/xueai8") \
    .option("dbtable", "people") \
    .option("user", "root") \
    .option("password", "admin") \
    .option("customSchema", "id DECIMAL(38, 0), name STRING, age LONG") \
    .load()

jdbc_df2.printSchema()
jdbc_df2.show()
```

执行过程和结果如图 4-30 所示。

```
# 指定读取时的DataFrame列数据类型
jdbc_df2 = spark.read \
    .format("jdbc") \
    .option("url", "jdbc:mysql://localhost:3306/xueai8") \
    .option("dbtable", "people") \
    .option("user", "root") \
    .option("password", "admin") \
    .option("customSchema", "id DECIMAL(38, 0), name STRING, age LONG") \
    .load()

jdbc_df2.printSchema()
jdbc_df2.show()
```

```
root
 |-- id: decimal(38,0) (nullable = true)
 |-- name: string (nullable = true)
 |-- age: long (nullable = true)

+---+------+---+
| id|  name|age|
+---+------+---+
|  1|  张三| 23|
|  2|  李四| 18|
|  3|王老五| 35|
+---+------+---+
```

图 4-30　读取 JDBC 数据库创建 DataFrame，并指定相应的列数据类型

还可以使用 query 选项指定用于将数据读入 PySpark 的查询语句。指定的查询将被圆括号括起来，并在 FROM 子句中用作子查询。PySpark 还将为子查询子句分配一个别名。例如，PySpark 将向 JDBC 源发出查询，模板代码如下：

```
SELECT <columns> FROM (<user_specified_query>) spark_gen_alias
```

使用此选项时有一些限制,包括以下限制:

(1)不允许同时指定 dbtable 和 query 选项。

(2)不允许同时指定 query 和 partitionColumn 选项。当需要指定 partitionColumn 选项时,可以使用 dbtable 选项指定子查询,分区列可以使用作为 dbtable 的一部分提供的子查询别名进行限定。

例如,通过 PySpark SQL 读取 MySQL 中的数据,使用 query 选项,代码如下:

```
jdbc_df4 = spark.read \
    .format("jdbc") \
    .option("url", "jdbc:mysql://localhost:3306/xueai8") \
    .option("query", "select name,age from peoples") \
    .option("user", "root") \
    .option("password", "admin") \
    .load()

jdbc_df4.printSchema()
jdbc_df4.show()
```

执行过程和结果如图 4-31 所示。

```
jdbc_df4 = spark.read \
    .format("jdbc") \
    .option("url", "jdbc:mysql://localhost:3306/xueai8") \
    .option("query", "select name,age from peoples") \
    .option("user", "root") \
    .option("password", "admin") \
    .load()

jdbc_df4.printSchema()
jdbc_df4.show()

root
 |-- name: string (nullable = true)
 |-- age: integer (nullable = true)

+------+---+
|  name|age|
+------+---+
|  张三| 23|
|  李四| 18|
|王老五| 35|
+------+---+
```

图 4-31 读取 JDBC 数据库创建 DataFrame,使用 query 选项

7. 读取图像文件创建 DataFrame

随着用于图像分类和对象检测的深度学习框架的进展,Apache Spark 中对标准图像处理的需求也越来越大。图像处理和预处理有其特定的挑战,例如,图像有不同的格式(如 jpeg、png 等)、大小和颜色方案,而且没有简单的方法来测试正确性(静默失败)。图像数据源通

过提供标准表示来解决这些问题,用户可以针对特定图像表示的细节进行编码和抽象。

Apache Spark 2.3 提供了 ImageSchema.readImages API,它最初是在 MMLSpark 库中开发的,在 Apache Spark 2.4 中成为一个内置的数据源,因此更容易使用。ImageDataSource 实现了一个 PySpark SQL 数据源 API,用于将图像数据作为 DataFrame 加载。

例如,将 images 目录下的所有图像读到 DataFrame 中,代码如下:

```
#加载图像数据源
image_file = "/data/spark/resources/images/*"
image_df = spark.read \
    .format("image") \
    .option("dropInvalid", "true") \
    .load(image_file)

image_df.printSchema()
image_df \
    .select("image.origin","image.width","image.height","image.nChannels","image.mode") \
    .show(truncate=False)
```

执行过程和结果如图 4-32 所示。

```
image_file = "file:///home/hduser/data/spark/resources/images/*"
image_df = spark.read.format("image").option("dropInvalid", "true").load(image_file)

image_df.printSchema()
image_df \
    .select("image.origin","image.width","image.height","image.nChannels","image.mode") \
    .show(truncate=False)
```

```
root
 |-- image: struct (nullable = true)
 |    |-- origin: string (nullable = true)
 |    |-- height: integer (nullable = true)
 |    |-- width: integer (nullable = true)
 |    |-- nChannels: integer (nullable = true)
 |    |-- mode: integer (nullable = true)
 |    |-- data: binary (nullable = true)

+--------------------------------------------------------------------+-----+------+---------+----+
|origin                                                              |width|height|nChannels|mode|
+--------------------------------------------------------------------+-----+------+---------+----+
|file:///home/hduser/data/spark/resources/images/white_lines.jpg     |2436 |1366  |3        |16  |
|file:///home/hduser/data/spark/resources/images/bridge_trees_example.jpg|1280 |720   |3        |16  |
|file:///home/hduser/data/spark/resources/images/udacity_sdc.png     |320  |213   |4        |24  |
|file:///home/hduser/data/spark/resources/images/curved_lane.jpg     |1280 |720   |3        |16  |
+--------------------------------------------------------------------+-----+------+---------+----+
```

图 4-32　读取 TSV 文件创建 DataFrame

从上面的示例可以看出,使用图像数据源可以从目录加载图像文件,加载的 DataFrame 有一个 StructType 列 image,其中包含作为 image 的 Schema 存储的图像数据。该 image 列的 Schema 构成如下:

(1) origin：StringType（代表该头像的文件路径）。
(2) height：IntegerType（图像的高度）。
(3) width：IntegerType（图像的宽度）。
(4) nChannels：IntegerType（图像通道的数量）。
(5) mode：IntegerType（OpenCV 兼容的类型）。
(6) data：BinaryType（以 OpenCV 兼容顺序的图像字节：在大多数情况下按行排列的 BGR）。

关于 nChannels，指的是颜色通道的数量。典型值：1 为灰度图像，3 为彩色图像（例如，RGB），4 为带有 alpha 通道的彩色图像。

关于 mode，这是一个整数标志，提供如何解释数据字段的信息。它用于指定数据存储的数据类型和通道顺序。该字段的值被期望（但不是强制的）映射到 OpenCV 类型之一，OpenCV 类型定义为 1、2、3 或 4 个通道和像素值的一些数据类型，见表 4-5。

表 4-5 OpenCV 中类型到数字的映射(数据类型和通道数量)

类型	C1	C2	C3	C4
CV_8U	0	8	16	24
CV_8S	1	9	17	25
CV_16U	2	10	18	26
CV_16S	3	11	19	27
CV_32U	4	12	20	28
CV_32S	5	13	21	29
CV_64F	6	14	22	30

关于 data，这是以二进制格式存储的图像数据。图像数据表示为一个三维数组，其维度形状（高度、宽度、nChannels）和类型为 t 的数组值由 mode 字段指定。数组按行主顺序存储。

对于可加载的图像，PySpark 规范明确针对当前行业中最常见的用例：二进制、int32、int64、浮点或双数据的多通道矩阵，可以轻松地放入 JVM 堆中。以下是一些限制：
(1) 图像的总大小应该限制在 2GB 以下（大致）。
(2) 颜色通道的含义是特定于应用程序的，不受标准的约束（与 OpenCV 标准一致）。
(3) 不支持气象学、医学等领域使用的专门格式。
(4) 这种格式专门用于图像，并不试图解决 Spark 中表示 n 维张量的更一般的问题。

8. 读取 Avro 创建 DataFrame

从 Apache Spark 2.4 开始，Spark SQL 为读取和写入 Apache Avro 数据提供了内置支持，新增一个基于 Databricks 的 spark-avro 模块的原生 Avro 数据源。

因为 spark-avro 模块是外部的,所以 DataFrameReader 或 DataFrameWriter 中没有.avro API。要想使用 Avro 数据源,需要添加相关的依赖。

如果使用 spark-submit 命令启动应用程序,则可使用--packages 选项添加 spark-avro_2.12 及其依赖,代码如下:

```
$ ./bin/spark-submit --packages org.apache.spark:spark-avro_2.12:3.1.2 ...
```

同样,如果使用 PySpark Shell 命令交互式执行程序代码,也可以使用--packages 选项添加 spark-avro_2.12 及其依赖,代码如下:

```
$ ./bin/pyspark --packages org.apache.spark:spark-avro_2.12:3.1.2 ...
```

在下面的示例中,将 PySpark 自带的 user.avro 数据源文件读到 DataFrame 中。要以 Avro 格式加载/保存数据,需要将数据源的 format 选项指定为 avro,代码如下:

```
#数据源文件
avro_file = "/data/spark/resources/users.avro"

#读取 Avro 文件,创建 DataFrame
users_df = spark.read.format("avro").load(avro_file)

users_df.printSchema()
users_df.show()
```

执行过程和结果如图 4-33 所示。

```
# 数据源文件
avro_file = "file:///home/hduser/data/spark/resources/users.avro"

# 读取Avro文件,创建DataFrame
users_df = spark.read.format("avro").load(avro_file)

users_df.printSchema()
users_df.show()
```

```
root
 |-- name: string (nullable = true)
 |-- favorite_color: string (nullable = true)
 |-- favorite_numbers: array (nullable = true)
 |    |-- element: integer (containsNull = true)

+------+--------------+----------------+
|  name|favorite_color|favorite_numbers|
+------+--------------+----------------+
|Alyssa|          null| [3, 9, 15, 20]|
|   Ben|           red|             []|
+------+--------------+----------------+
```

图 4-33 读取 Avro 文件创建 DataFrame

Avro 的数据源选项可以使用 DataFrameReader 或 DataFrameWriter 上的.option 方法进行设置，可使用的配置项见表 4-6。

表 4-6　Avro 数据源配置选项

属　性　名	默 认 值	含　义	应用范围
avroSchema	None	用户以 JSON 格式提供的可选 Avro 模式。记录字段的数据类型和命名在从 Avro 读取时应与 Avro 数据类型匹配，在写入 Avro 文件时应与 Spark 的内部数据类型（例如 StringType、IntegerType）匹配；否则读/写操作将失败	read write
recordName	topLevelRecord	写入结果中的顶级记录名，这在 Avro 规范中是必需的	write
recordNamespace	""	在写入结果中的记录命名空间	write
ignoreExtension	true	该选项控制在 read 中，忽略没有.avro 扩展名的文件。如果启用了该选项，则加载所有文件(有和没有.avro 扩展名)	read
compression	snappy	compression 选项允许指定写入中使用的压缩编解码器。目前支持的编解码器有 uncompressed、snappy、deflate、bzip2 和 xz。 如果未设置此选项，则会考虑配置 spark.sql.avro.compression.codec 配置项	write

Avro 的配置可以在 SparkSession 上使用 setConf()方法进行设置，或者使用 SQL 运行 SET key=value 命令进行设置，可设置的配置见表 4-7。

表 4-7　Avro 配置

属　性　名	默认值	含　义
spark.sql.legacy.replaceDatabricksSparkAvro.enabled	true	如果将其设置为 true，则数据源提供程序 com.databricks.spark.avro 被映射到内置但属于外部的 Avro 数据源模块，以便向后兼容
spark.sql.avro.compression.codec	snappy	用于写 Avro 文件的压缩编解码器。支持的编解码器：uncompressed、deflate、snappy、bzip2 和 xz。默认的编解码器是 snappy
spark.sql.avro.deflate.level	–1	写 Avro 文件中使用的 deflate codec 的压缩水平（级别）。有效值必须在 1 到 9（含 9）或 –1。默认值为 –1，对应于当前实现中的 6 级

9. 读取二进制文件创建 DataFrame

从 Spark 3.0 开始，PySpark 支持二进制文件数据源，它读取二进制文件，并将每个文件转换成一条记录，其中包含文件的原始内容和元数据。它生成一个 DataFrame，包含以下列和可能的分区列。

(1) path：StringType。
(2) modificationTime：TimestampType。
(3) length：LongType。
(4) content：BinaryType。

要读取整个二进制文件,需要将数据源 format 指定为 binaryFile。要加载具有匹配给定全局模式的路径的文件,同时保持分区发现的行为,可以使用通用数据源选项 pathGlobFilter。例如,从输入目录中读取所有 JPG 文件,代码如下：

```
#二进制文件路径
filePath = "/data/spark/resources/images/*"

#读取 JPG 文件
binDf = spark \
    .read \
    .format("binaryFile") \
    .option("pathGlobFilter", "*.jpg") \
    .load(filePath)

#输出
binDf.printSchema()
binDf.show()
```

执行以上代码,输出结果如下：

```
root
 |-- path: string (nullable = true)
 |-- modificationTime: timestamp (nullable = true)
 |-- length: long (nullable = true)
 |-- content: binary (nullable = true)

+--------------------+--------------------+------+--------------------+
|                path|    modificationTime|length|             content|
+--------------------+--------------------+------+--------------------+
|hdfs://xueai8:802...|2022-02-11 10:14:...|358076|[FF D8 FF E0 00 1...|
|hdfs://xueai8:802...|2022-02-11 10:14:...|232093|[FF D8 FF E0 00 1...|
|hdfs://xueai8:802...|2022-02-11 10:14:...|128326|[FF D8 FF E0 00 1...|
+--------------------+--------------------+------+--------------------+
```

4.6 操作 DataFrame

在 PySpark 2.0 中,DataFrame 为结构化数据操作提供了一种特定于领域的语言(Domain-Specific Language,DSL)。这些操作被分为两类,即 Transformation 和 Action。开发人员

连接多个操作来选择、过滤、转换、聚合和排序在 DataFrame 中的数据。底层的 Catalyst 优化器确保了这些操作的高效执行。

4.6.1 列的多种引用方式

在学习操作 DataFrame 之前，需要掌握 PySpark 所提供的引用 DataFrame 列的多种方式。首先创建一个 DataFrame，代码如下：

```
#将元组转换为DataFrame
kvDF = spark.createDataFrame([(1,2),(3,4)],["key","value"])

kvDF.printSchema()
kvDF.show()
```

执行以上代码，输出内容如下：

```
root
 |-- key: integer (nullable = false)
 |-- value: integer (nullable = false)

+---+-----+
|key|value|
+---+-----+
|  1|    2|
|  3|    4|
+---+-----+
```

要显示一个 DataFrame 的所有列名，可以调用其 columns 属性，代码如下：

```
kvDF.columns              #输出：['key', 'value']
```

如果仅仅是为了获取列值，而不对列值做任何计算和比较，则可直接引用列名字符串，代码如下：

```
kvDF.select("key").show()            #列为字符串类型
```

执行以上代码，输出内容如下：

```
+---+
|key|
+---+
|  1|
|  3|
+---+
```

如果要对列做任何计算和比较等操作，则需要获得列的对象类型，这需要使用内置的 col 函数来选择列（这时获得的是一个 pyspark.sql.Column 对象），代码如下：

```
from pyspark.sql.functions import col
```

```
kvDF.select(col("key")).show()            #列为Column 类型
kvDF.where(col("key") > 1).show()         #列为Column 类型
```

执行以上代码,输出内容如下:

```
+---+
|key|
+---+
|  1|
|  3|
+---+


+---+-----+
|key|value|
+---+-----+
|  3|    4|
+---+-----+
```

也可以使用 expr()函数,它与 col()函数相同,返回一个 pyspark.sql.Column 对象,不过它允许使用表达式作为参数,代码如下:

```
from pyspark.sql.functions import expr

kvDF.select(expr("key")).show()
kvDF.select(expr("key > 1")).show()
kvDF.select(expr("key") > 1).show()
```

执行以上代码,输出内容如下:

```
+---+
|key|
+---+
|  1|
|  3|
+---+


+---------+
|(key > 1)|
+---------+
|    false|
|     true|
+---------+


+---------+
|(key > 1)|
+---------+
|    false|
|     true|
+---------+
```

也可以直接通过 DataFrame 引用，代码如下：

```
kvDF.select(kvDF.key).show()
kvDF.select(kvDF["key"]).show()
```

执行以上代码，输出内容如下：

```
+---+
|key|
+---+
|  1|
|  3|
+---+

+---+
|key|
+---+
|  1|
|  3|
+---+
```

下面的代码选择 key 列，并增加一个新的列，新列的值由 key 列计算得来，为 boolean 值，表示 key 列的值是否大于 1，代码如下：

```
from pyspark.sql.functions import col

kvDF.select(col("key"), col("key") > 1).show()
```

执行以上代码，输出内容如下：

```
+---+---------+
|key|(key > 1)|
+---+---------+
|  1|    false|
|  3|     true|
+---+---------+
```

也可以给列取一个别名，代码如下：

```
kvDF.select(col("key"), (col("key") > 1).alias("a")).show()
#或
kvDF.select(kvDF.key, (kvDF.key > 1).alias("a")).show()
```

执行以上代码，输出内容如下：

```
+---+------+
|key|     a|
+---+------+
|  1| false|
```

```
|  3|  true|
+---+------+
```

也可以使用正则表达式来选择列,代码如下:

```
kvDF.select(kvDF.colRegex("`^k.*`")).show()
```

执行以上代码,输出内容如下:

```
+---+
|key|
+---+
|  1|
|  3|
+---+
```

对于列的引用方式,主要区别如下:
(1) 如果只是为了获取特定列的值,则直接以字符串类型引用列名即可。
(2) 如果引用列是为了任何形式的计算,包括排序、类型转换、别名、比较、计算列等,则需要应用上述任一函数将列转换为 Column 对象。

4.6.2 对 DataFrame 执行 Transformation 转换操作

DataFrame API 提供了许多函数用来执行关系运算,这些函数模拟了 SQL 关系操作:
(1) 选择数据:select()。
(2) 删除某列:drop()。
(3) 过滤数据:where()和 filter()(两者等价)。
(4) 限制返回的数量:limit()。
(5) 重命名列:withColumnRenamed()。
(6) 增加一个新的列:withColumn()。
(7) 数据分组:groupBy()。
(8) 数据排序:orderBy()和 sort()(两者等价)。

在进一步演示 DataFrame 的各种操作方法之前,先准备好要用到的数据。这里将使用一部电影数据集 movies.csv,需要将它上传到 HDFS 的/data/spark/目录下,其中部分数据格式如下:

```
actor,title,year
"McClure, Marc (I)",Freaky Friday,2003
"McClure, Marc (I)",Coach Carter,2005
"McClure, Marc (I)",Superman II,1980
"McClure, Marc (I)",Apollo 13,1995
"McClure, Marc (I)",Superman,1978
"McClure, Marc (I)",Back to the Future,1985
"McClure, Marc (I)",Back to the Future Part III,1990
```

```
"Cooper, Chris (I)","Me, Myself & Irene",2000
"Cooper, Chris (I)",October Sky,1999
"Cooper, Chris (I)",Capote,2005
"Cooper, Chris (I)",The Bourne Supremacy,2004
```

可以看到，数据集有 3 个字段：actor、title 和 year。数据集的第 1 行是标题行。

首先，将数据集加载到 DataFrame 中，代码如下：

```python
from pyspark.sql import SparkSession

#构建 SparkSession 实例
spark = SparkSession.builder \
    .master("spark://localhost:7077") \
    .appName("pyspark sql demo") \
    .getOrCreate()

#将电影数据集文件加载到 DataFrame 中
file = "/data/spark/movies2/movies.csv"
movies = spark.read \
      .option("header","true") \
      .option("inferSchema","true") \
      .csv(file)

movies.printSchema()
movies.show(5)
```

执行以上代码，输出内容如下：

```
root
 |-- actor: string (nullable = true)
 |-- title: string (nullable = true)
 |-- year: integer (nullable = true)

+-----------------+-------------+----+
|            actor|        title|year|
+-----------------+-------------+----+
|McClure, Marc (I)| Freaky Friday|2003|
|McClure, Marc (I)|  Coach Carter|2005|
|McClure, Marc (I)|   Superman II|1980|
|McClure, Marc (I)|     Apollo 13|1995|
|McClure, Marc (I)|      Superman|1978|
+-----------------+-------------+----+
only showing top 5 rows
```

接下来，学习 DataFrame 的各种转换操作函数。

1) select()函数

类似于 SQL 中的 select 子句，用来选择指定的列。例如，选取 movies 中的 title 和 year

两列,并返回一个新的 DataFrame,代码如下:

```
from pyspark.sql.functions import *

movies.select("title","year").show(5)
```

执行以上代码,输出内容如下:

```
+-------------+----+
|        title|year|
+-------------+----+
|Freaky Friday|2003|
| Coach Carter|2005|
|  Superman II|1980|
|    Apollo 13|1995|
|     Superman|1978|
+-------------+----+
only showing top 5 rows
```

将电影上演的年份转换到年代,使用 col()函数,并赋予一个别名,代码如下:

```
from pyspark.sql.functions import *

#将年份列转换到年代列
movies2 = movies.select(col('title') ,(col('year') - col('year') % 10).alias("decade"))
movies2.show(5)
```

执行以上代码,输出内容如下:

```
+-------------+------+
|        title|decade|
+-------------+------+
|Freaky Friday|  2000|
| Coach Carter|  2000|
|  Superman II|  1980|
|    Apollo 13|  1990|
|     Superman|  1970|
+-------------+------+
only showing top 5 rows
```

2) selectExp()函数

用来选择一组 SQL 表达式,即使用 SQL 表达式来选择列。例如,在下面的代码中,用通配符星号(*)来表示选择所有的列,并增加一个新的列 decade,新列的值是通过对 year 列值计算得到的电影上映的年代,代码如下:

```
from pyspark.sql.functions import *

movies.selectExpr("*","(year - year % 10) as decade").show(5)
```

执行以上代码，输出内容如下：

```
+----------------+-------------+----+------+
|           actor|        title|year|decade|
+----------------+-------------+----+------+
|McClure, Marc (I)|Freaky Friday|2003|  2000|
|McClure, Marc (I)| Coach Carter|2005|  2000|
|McClure, Marc (I)|  Superman II|1980|  1980|
|McClure, Marc (I)|    Apollo 13|1995|  1990|
|McClure, Marc (I)|     Superman|1978|  1970|
+----------------+-------------+----+------+
only showing top 5 rows
```

在 selectExpr()方法中，不但支持使用 SQL 表达式，还支持直接使用 SQL 内置函数。例如，使用 SQL 表达式和内置函数，来查询电影数量和演员数量这两个值，代码如下：

```
from pyspark.sql.functions import *

movies.selectExpr("count(distinct(title)) as movies",
                  "count(distinct(actor)) as actors").show()
```

执行以上代码，输出内容如下：

```
+------+------+
|movies|actors|
+------+------+
|  1409|  6527|
+------+------+
```

可以看出，数据集中的电影共有 1409 部，演员共有 6527 名。

3）filter()和 where()函数

使用给定的条件过滤 DataFrame 中的行。这两个函数是等价的，相当于 SQL 中的 where 子句。例如，要找出 2000 年以前上映的电影，需要在 filter()函数或 where()函数中指定过滤条件，代码如下：

```
from pyspark.sql.functions import *

movies.filter('year < 2000').show(5)
#movies.where('year < 2000').show(5)    #等价
```

执行以上代码，输出内容如下：

```
+----------------+-------------------+----+
|           actor|              title|year|
+----------------+-------------------+----+
|McClure, Marc (I)|        Superman II|1980|
|McClure, Marc (I)|          Apollo 13|1995|
```

```
|McClure, Marc (I)|            Superman|1978|
|McClure, Marc (I)| Back to the Future|1985|
|McClure, Marc (I)|Back to the Futur...|1990|
+-----------------+--------------------+----+
only showing top 5 rows
```

找出2000年及以后上映的电影,代码如下:

```
from pyspark.sql.functions import *

movies.filter('year >= 2000').show(5)
#movies.where('year >= 2000').show(5)    #等价
```

执行以上代码,输出内容如下:

```
+-----------------+--------------------+----+
|            actor|               title|year|
+-----------------+--------------------+----+
|McClure, Marc (I)|        Freaky Friday|2003|
|McClure, Marc (I)|        Coach Carter|2005|
|Cooper, Chris (I)| Me, Myself & Irene|2000|
|Cooper, Chris (I)|              Capote|2005|
|Cooper, Chris (I)|The Bourne Supremacy|2004|
+-----------------+--------------------+----+
only showing top 5 rows
```

找出2000年上映的电影,代码如下:

```
from pyspark.sql.functions import *

#相等比较
movies.filter('year = 2000').show(5)
#movies.where('year = 2000').show(5)    #等价
```

执行以上代码,输出内容如下:

```
+-----------------+--------------------+----+
|            actor|               title|year|
+-----------------+--------------------+----+
|Cooper, Chris (I)| Me, Myself & Irene|2000|
|Cooper, Chris (I)|         The Patriot|2000|
|   Jolie, Angelina|Gone in Sixty Sec...|2000|
|   Yip, Françoise|    Romeo Must Die|2000|
|   Danner, Blythe|   Meet the Parents|2000|
+-----------------+--------------------+----+
only showing top 5 rows
```

不等比较使用的操作符是!=。找出非2000年上映的电影,代码如下:

```
from pyspark.sql.functions import *
```

```
#不等比较使用的操作符是 !=
movies.select("title","year").filter('year != 2000').show(5)
#movies.select("title","year").where('year != 2000').show(5)     #等价
#movies.select("title","year").filter(col('year') != 2000).show(5)     #等价
```

执行以上代码,输出内容如下:

```
+--------------+----+
|         title|year|
+--------------+----+
| Freaky Friday|2003|
|  Coach Carter|2005|
|   Superman II|1980|
|     Apollo 13|1995|
|      Superman|1978|
+--------------+----+
only showing top 5 rows
```

支持离散值匹配。例如,找出2001—2002年间上映的电影,代码如下:

```
from pyspark.sql.functions import *

movies.filter(col("year").isin([2001,2002])).show(5, truncate=False)

#等价
#movies.where(col("year").isin([2001,2002])).show(5, truncate=False)
```

执行以上代码,输出内容如下:

```
+-------------------+----------------------------------------+----+
|actor              |title                                   |year|
+-------------------+----------------------------------------+----+
|Cooper, Chris (I)  |The Bourne Identity                     |2002|
|Cassavetes, Frank  |John Q                                  |2002|
|Knight, Shirley (I)|Divine Secrets of the Ya-Ya Sisterhood  |2002|
|Jolie, Angelina    |Lara Croft: Tomb Raider                 |2001|
|Cueto, Esteban     |Collateral Damage                       |2002|
+-------------------+----------------------------------------+----+
only showing top 5 rows
```

可使用OR和AND运算符组合一个或多个比较表达式。例如,要找出2000年及以后上映并且名称长度少于5个字符的电影,代码如下:

```
from pyspark.sql.functions import *

movies.filter('year >= 2000' and length('title') < 5).show(5)
#movies.where('year >= 2000' and length('title') < 5).show(5)     #等价
```

执行以上代码,输出内容如下:

```
+---------------+-----+----+
|          actor|title|year|
+---------------+-----+----+
|Jolie, Angelina| Salt|2010|
| Cueto, Esteban|  xXx|2002|
| Butters, Mike |  Saw|2004|
| Franko, Victor|   21|2008|
| Ogbonna, Chuk | Salt|2010|
+---------------+-----+----+
only showing top 5 rows
```

另一种实现相同结果的方法是调用 filter 函数两次,代码如下:

```
from pyspark.sql.functions import *

movies.filter('year >= 2000').filter(length('title') < 5).show(5)
#movies.where('year >= 2000').where(length('title') < 5).show(5)    #等价
```

4) distinct()和 dropDuplicates()函数

返回一个新数据集,其中仅包含此数据集中的唯一行。这是 dropDuplicates()的别名。例如,想知道数据集中共有多少条唯一行(去重),代码如下:

```
from pyspark.sql.functions import *

movies.distinct().count()
movies.dropDuplicates().count()     #与上句等价
```

执行以上代码,输出内容如下:

```
31394
```

但是,如果想要知道数据集中共包含多少部电影,则需要基于 title 字段进行唯一值统计,代码如下:

```
from pyspark.sql.functions import *

movies.select("title").distinct().count()

#其实也可以使用 SQL 的 distinct 函数
#movies.selectExpr("count(distinct(title)) as movies").show()
```

执行以上代码,输出内容如下:

```
1409
```

5) dropDuplicates()函数

仅考虑列的子集,返回删除(按列的子集)重复行的新数据集。例如,同样需要统计

数据集中共包含多少部电影,代码如下:

```
movies.dropDuplicates(["title"]).count()
```

执行以上代码,输出内容如下:

```
1409
```

6) sort()和 orderBy()函数

相当于 SQL 中的 order by 子句,它返回按指定的列排序后的新数据集。例如,要按电影名称长度顺序及上映年份倒序显示,代码如下:

```
from pyspark.sql.functions import *

movies.dropDuplicates(["title", "year"]) \
    .selectExpr("title", "length(title) as title_length", "year") \
    .orderBy(col("title_length").asc(), col("year").desc()) \
    .show(10,False)
```

orderBy()函数与 sort()函数是等价的,所以在上面代码中的 orderBy()也可以替换为 sort(),结果是相同的。执行以上代码,输出内容如下:

```
+-----+------------+----+
|title|title_length|year|
+-----+------------+----+
|Up   |2           |2009|
|21   |2           |2008|
|12   |2           |2007|
|RV   |2           |2006|
|X2   |2           |2003|
|Rio  |3           |2011|
|Hop  |3           |2011|
|300  |3           |2006|
|Saw  |3           |2004|
|Elf  |3           |2003|
+-----+------------+----+
only showing Top 10 rows
```

7) groupBy()

相当于 SQL 中的 group by 子句,按指定的列对数据进行分组,以便执行聚合统计操作。例如,统计每年上映的电影数量并按数量倒序显示,代码如下:

```
from pyspark.sql.functions import *

movies.groupBy("year").count().orderBy(col("count").desc()).show(10)
```

执行以上代码,输出内容如下:

```
+----+-----+
|year|count|
+----+-----+
|2006| 2078|
|2004| 2005|
|2007| 1986|
|2005| 1960|
|2011| 1926|
|2008| 1892|
|2009| 1890|
|2010| 1843|
|2002| 1834|
|2001| 1687|
+----+-----+
only showing Top 10 rows
```

统计上映电影数量超过 2000 部的年份,代码如下:

```
from pyspark.sql.functions import *
movies.groupBy("year").count().where(col("count") > 2000).show()
```

执行以上代码,输出内容如下:

```
+----+-----+
|year|count|
+----+-----+
|2006| 2078|
|2004| 2005|
```

8) limit()

通过获取前 n 行数据返回一个新的 DataFrame,相当于 SQL 中的 limit 子句,限制返回的行数。通常与 orderBy 配合实现 Top N 算法。例如,统计上映电影数量最多的 5 个年份,代码如下:

```
from pyspark.sql.functions import *
movies.groupBy("year") \
    .count() \
    .orderBy(col("count").desc()) \
    .limit(5) \
    .show()
```

执行以上代码,输出内容如下:

```
+----+-----+
|year|count|
+----+-----+
```

```
|2006| 2078|
|2004| 2005|
|2007| 1986|
|2005| 1960|
|2011| 1926|
+----+-----+
```

查询电影名称最长的 5 部电影，代码如下：

```
from pyspark.sql.functions import *

movies.dropDuplicates(["title"]) \
      .selectExpr("title", "length(title) as title_length") \
      .orderBy(col("title_length").desc()) \
      .limit(5) \
      .show(truncate=False)
```

执行以上代码，输出内容如下：

```
+---------------------------------------------------------------------------------
--------------+------------+
|title                                                                           
              |title_length|
+---------------------------------------------------------------------------------
--------------+------------+
|Borat: Cultural Learnings of America for Make Benefit Glorious Nation of 
Kazakhstan|83          |
|The Chronicles of Narnia: The Lion, the Witch and the Wardrobe                  
              |62          |
|Hannah Montana & Miley Cyrus: Best of Both Worlds Concert                       
              |57          |
|The Chronicles of Narnia: The Voyage of the Dawn Treader                        
              |56          |
|Istoriya pro Richarda, milorda i prekrasnuyu Zhar-ptitsu                        
              |56          |
+---------------------------------------------------------------------------------
--------------+------------+
```

9）union()和 unionAll()

按列位置合并两个数据集的行，相当于 SQL 中的 union all。假设现在想在名称为"12"的电影中添加一个缺失的演员，可以采用如下的方法：新创建一个 DataFrame，其中包含所缺失的演员信息，然后将原数据集与这个含有缺失演员信息的新数据集执行一个 union() 操作，将两个数据集合并在一起。这样就将缺失的演员添加到了原数据集中。

首先获得电影"12"的数据集，代码如下：

```
shortNameMovieDF = movies.where('title == "12"')
shortNameMovieDF.show()
```

执行以上代码,输出内容如下:

```
+--------------------+-----+----+
|               actor|title|year|
+--------------------+-----+----+
|    Efremov, Mikhail|   12|2007|
|     Stoyanov, Yuriy|   12|2007|
|     Gazarov, Sergey|   12|2007|
|Verzhbitskiy, Viktor|   12|2007|
+--------------------+-----+----+
```

可以看到,该部电影有 4 名演员,但是缺少了演员 Edita Brychta 的信息。接下来,新创建一个 DataFrame,包含所缺失演员 Edita Brychta 的信息,然后将这个 DataFrame 和上面的 DataFrame 进行合并,这样就将演员 Edita Brychta 的信息合并到上面的数据集中了,代码如下:

```
from pyspark.sql.functions import *

#将缺失演员信息构造到一个DataFrame
forgottenActor = [("Brychta, Edita", "12", 2007)]
forgottenActorDF = spark \
    .createDataFrame(forgottenActor,shortNameMovieDF.schema)

#现在添加缺失的演员姓名
completeShortNameMovieDF = shortNameMovieDF.union(forgottenActorDF)
completeShortNameMovieDF.show()
```

执行以上代码,输出内容如下:

```
+--------------------+-----+----+
|               actor|title|year|
+--------------------+-----+----+
|    Efremov, Mikhail|   12|2007|
|     Stoyanov, Yuriy|   12|2007|
|     Gazarov, Sergey|   12|2007|
|Verzhbitskiy, Viktor|   12|2007|
|      Brychta, Edita|   12|2007|
+--------------------+-----+----+
```

如果要实现类 SQL 中 UNION 的功能,则需要在 union()操作之后使用 distinct()操作,这样就可以去掉重复的数据了。

10) withColumn()

通过添加列或替换具有相同名称的现有列返回一个新的数据集。例如,向 movies 数据集增加一个新列 decade,该列的值是基于 year 这个列的表达式,计算的结果是该电影上映的年代,代码如下:

```
from pyspark.sql.functions import *
movies.withColumn("decade", (col('year') - col('year') % 10)).show(5)
```

执行以上代码，输出内容如下：

```
+----------------+-------------+----+------+
|           actor|        title|year|decade|
+----------------+-------------+----+------+
|McClure, Marc (I)| Freaky Friday|2003|  2000|
|McClure, Marc (I)| Coach Carter|2005|  2000|
|McClure, Marc (I)|  Superman II|1980|  1980|
|McClure, Marc (I)|    Apollo 13|1995|  1990|
|McClure, Marc (I)|     Superman|1978|  1970|
+----------------+-------------+----+------+
only showing top 5 rows
```

如果传给 withColumn() 函数的列名与现有列名相同，则意味着用新值替换旧值。例如，将 year 列的值替换为年代（原值为年份），代码如下：

```
from pyspark.sql.functions import *
movies.withColumn("decade", (col('year') - col('year') % 10)).show(5)
```

执行以上代码，输出内容如下：

```
+----------------+-------------+----+
|           actor|        title|year|
+----------------+-------------+----+
|McClure, Marc (I)| Freaky Friday|2000|
|McClure, Marc (I)| Coach Carter|2000|
|McClure, Marc (I)|  Superman II|1980|
|McClure, Marc (I)|    Apollo 13|1990|
|McClure, Marc (I)|     Superman|1970|
+----------------+-------------+----+
only showing top 5 rows
```

11）withColumnRenamed()

返回一个重命名列的新数据集。如果模式不包含指定的列名，则不进行任何操作。例如，将 movies 数据集中的列名改为新的名称，代码如下：

```
from pyspark.sql.functions import *
movies.withColumnRenamed("actor", "actor_name") \
      .withColumnRenamed("title", "movie_title") \
      .withColumnRenamed("year", "produced_year") \
      .show(5)
```

执行以上代码，输出内容如下：

```
+----------------+-------------+-------------+
|      actor_name|  movie_title|produced_year|
+----------------+-------------+-------------+
|McClure, Marc (I)| Freaky Friday|         2003|
|McClure, Marc (I)| Coach Carter|         2005|
|McClure, Marc (I)|   Superman II|         1980|
|McClure, Marc (I)|     Apollo 13|         1995|
|McClure, Marc (I)|      Superman|         1978|
+----------------+-------------+-------------+
only showing top 5 rows
```

12) drop()

返回一个删除了指定列的新 DataFrame。如果要被删除的列不存在，则不进行任何操作。例如，删除 movies 数据集中指定的列，代码如下：

```
from pyspark.sql.functions import *

movies.drop("actor", "me").printSchema()
movies.drop("actor", "me").show(5)
```

执行以上代码，输出内容如下：

```
root
 |-- title: string (nullable = true)
 |-- year: integer (nullable = true)

+-------------+----+
|        title|year|
+-------------+----+
|Freaky Friday|2003|
| Coach Carter|2005|
|  Superman II|1980|
|    Apollo 13|1995|
|     Superman|1978|
+-------------+----+
only showing top 5 rows
```

从输出结果可以看出，actor 这一列被删除了，而 me 这一列在原 DataFrame 中并不存在，所以删除不存在的列没有任何影响。

13) sample()

数据抽样。通过随机种子对部分行进行抽样（不进行替换），返回一个新的 DataFrame。抽样比例由 fraction 参数决定，它指定要生成的行的比例，范围为[0.0,1.0]。这种方法有多个不同的重载版本。

（1）sample(fraction: Double, seed: Long)：指定种子。

（2）sample(withReplacement: Boolean, fraction: Double)：指定是否替换，使用随机种子。

（3）sample(withReplacement: Boolean, fraction: Double, seed: Long)：同上，但指定种子。

例如，指定无放回抽样和抽样比例，代码如下：

```
from pyspark.sql.functions import *
movies.sample(False, 0.0003).show(truncate=False)
```

执行以上代码，输出内容如下（随机的）：

```
+--------------------+-----------------------------------------+----+
|actor               |title                                    |year|
+--------------------+-----------------------------------------+----+
|Baskin, Elya        |Austin Powers: International Man of Mystery|1997|
|Rapace, Noomi       |Prometheus                               |2012|
|Hatcher, Teri       |Coraline                                 |2009|
|Damon, Matt         |The Bourne Supremacy                     |2004|
|Novak, B.J.         |Knocked Up                               |2007|
|Masamune, Tohoru    |Jumper                                   |2008|
|Cruise, Tom         |Collateral                               |2004|
|Harvey, Steve (I)   |Racing Stripes                           |2005|
|Giuliani, Rudolph W.|Anger Management                         |2003|
|Sloan, Amy          |The Aviator                              |2004|
|Krumholtz, David (I)|The Mexican                              |2001|
+--------------------+-----------------------------------------+----+
```

执行有放回抽样并指定比例因子、种子的抽样，代码如下：

```
from pyspark.sql.functions import *
movies.sample(True, 0.0003, 123456).show()
```

执行以上代码，输出内容如下（保持不变的）：

```
+-------------------+--------------------+----+
|              actor|               title|year|
+-------------------+--------------------+----+
|       Piddock, Jim|        The Prestige|2006|
|       Reed, Tanoai|  The Stepford Wives|2004|
|       Moyo, Masasa|     Angels & Demons|2009|
|   Zemeckis, Leslie|             Beowulf|2007|
|      Huston, Danny|X-Men Origins: Wo...|2009|
|      Pompeo, Ellen|          Old School|2003|
|   Utt, Kenneth (I)|        Philadelphia|1993|
|   Cannon, Kevin (I)|            Cop Out|2010|
+-------------------+--------------------+----+
```

14）randomSplit()

使用提供的权重随机分割数据集。如果分割的权重和不等于 1，则将被标准化。该操作返回一个切分后的 DataSet 集合数组。例如，将 movies 分割为 3 个数据集，所占数据比例分别为 0.6、0.3 和 0.1，代码如下：

```python
from pyspark.sql.functions import *

#权重需要是一个数组
smallerMovieDFs = movies.randomSplit([0.6, 0.3, 0.1])

#检查各部分计数之和是否等于1
print("总数据量: ", movies.count())

#分割后的第1个数据集的数量
print("第1个子数据集的数量: ", smallerMovieDFs[0].count())

#3个子数据集之和
total = smallerMovieDFs[0].count() + smallerMovieDFs[1].count() + smallerMovieDFs[2].count()
print("3个子数据集之和: ", total)
```

执行以上代码，输出内容如下：

```
总数据量：31394
第1个子数据集的数量：18906
3个子数据集之和：31394
```

4.6.3 对 DataFrame 执行 Action 操作

与 RDD 类似，DataFrame 上的 Transformation 转换操作都是延迟执行的。只有当在 DataFrame 上执行 Action 操作时，才会触发真正的计算。这些 Action 操作相对比较简单，关于这些 Action 操作方法的使用，代码如下：

```python
from pyspark.sql.functions import *

#查看前5条数据。第2个参数指定当列内容较长时，是否截断显示，False 为不截断
movies.show(5,truncate=False)

#返回数据集中的数量
movies.count

#返回数据集中的第1条数据
movies.first()

#等价于 first 方法
```

```
movies.head()

#以 Array 形式返回数据集中的前 3 条数据
movies.head(3)

#以 Array 形式返回数据集中的前 3 条数据
movies.take(3)

#返回一个包含数据集中所有行的数组
movies.collect()

#以列表形式返回数据集的列名
movies.columns            #['actor', 'title', 'year']
```

4.6.4 对 DataFrame 执行描述性统计操作

PySpark 还为 DataFrame 提供了一个 describe 函数，用来计算数字列和字符串列的基本统计信息，包括 count、mean、stddev、min 和 max。如果没有给定列，则此函数将计算所有数值列或字符串列的统计信息。该函数返回的也是一个 DataFrame，其方法签名如下：

```
DataFrame.describe(*cols: Union[str, List[str]]) →
pyspark.sql.dataframe.DataFrame
```

这种方法是一个 Action 操作，经常用于对数据集进行探索性数据分析，代码如下：

```
descDF = movies.describe()

descDF.printSchema()
descDF.show()
```

执行以上代码，输出内容如下：

```
root
 |-- summary: string (nullable = true)
 |-- actor: string (nullable = true)
 |-- title: string (nullable = true)
 |-- year: string (nullable = true)

+-------+------------------+--------------------+------------------+
|summary|             actor|               title|              year|
+-------+------------------+--------------------+------------------+
|  count|             31394|               31394|             31393|
|   mean|              null|   312.61538461538464| 2002.7964514382188|
| stddev|              null|    485.7043414390151|  6.377135379933117|
|    min|   Aaron, Caroline|'Crocodile' Dunde...|              1961|
|    max|  von Sydow, Max (I)|                 xXx|              2012|
+-------+------------------+--------------------+------------------+
```

计算 year 字段的基本统计信息，代码如下：

```
descDF = movies.describe("year")

descDF.printSchema()
descDF.show()
```

执行以上代码，输出内容如下：

```
root
 |-- summary: string (nullable = true)
 |-- year: string (nullable = true)

+-------+------------------+
|summary|              year|
+-------+------------------+
|  count|             31393|
|   mean| 2002.7964514382188|
| stddev| 6.377135379933117|
|    min|              1961|
|    max|              2012|
+-------+------------------+
```

计算 year 和 actor 这两列的基本统计信息，代码如下：

```
descDF = movies.describe("year","actor")

descDF.printSchema()
descDF.show()
```

执行以上代码，输出内容如下：

```
root
 |-- summary: string (nullable = true)
 |-- year: string (nullable = true)
 |-- actor: string (nullable = true)

+-------+------------------+------------------+
|summary|              year|             actor|
+-------+------------------+------------------+
|  count|             31393|             31394|
|   mean| 2002.7964514382188|              null|
| stddev| 6.377135379933117|              null|
|    min|              1961|    Aaron, Caroline|
|    max|              2012| von Sydow, Max (I)|
+-------+------------------+------------------+
```

注意：这个函数用于进行探索性数据分析，它不能保证结果数据集模式的向后兼容性。如果希望以编程方式计算汇总统计信息，则应使用 agg() 函数。

PySpark 还提供了一个与 describe()函数类似的 summary()函数，用于提供数据集的摘要信息。如果没有给出统计信息，则这个函数将计算 count、mean、stddev、min、近似四分位数（25%、50%和75%的百分位数）和 max。例如，对 movies 调用 summary()函数，代码如下：

```
summaryDF = movies.summary()

summaryDF.printSchema()
summaryDF.show()
```

执行以上代码，输出内容如下：

```
root
 |-- summary: string (nullable = true)
 |-- actor: string (nullable = true)
 |-- title: string (nullable = true)
 |-- year: string (nullable = true)

+-------+------------------+------------------+------------------+
|summary|             actor|             title|              year|
+-------+------------------+------------------+------------------+
|  count|             31394|             31394|             31393|
|   mean|              null| 312.61538461538464|2002.7964514382188|
| stddev|              null| 485.7043414390151| 6.377135379933117|
|    min|   Aaron, Caroline|'Crocodile' Dunde...|              1961|
|    25%|              null|              21.0|              1999|
|    50%|              null|              21.0|              2004|
|    75%|              null|             300.0|              2008|
|    max| von Sydow, Max (I)|               xXx|              2012|
+-------+------------------+------------------+------------------+
```

也可以指定想要的统计信息，代码如下：

```
summaryDF = movies.summary("count", "min", "25%", "75%", "max")

summaryDF.printSchema()
summaryDF.show()
```

执行以上代码，输出内容如下：

```
root
 |-- summary: string (nullable = true)
 |-- actor: string (nullable = true)
 |-- title: string (nullable = true)
 |-- year: string (nullable = true)
```

```
+-------+--------------------+--------------------+-----+
|summary|               actor|               title| year|
+-------+--------------------+--------------------+-----+
|  count|               31394|               31394|31393|
|    min|     Aaron, Caroline|'Crocodile' Dunde...| 1961|
|    25%|                null|                21.0| 1999|
|    75%|                null|               300.0| 2008|
|    max|    von Sydow, Max (I)|                 xXx| 2012|
+-------+--------------------+--------------------+-----+
```

要对指定的列做一个摘要，首先选择这些列，然后执行 summary() 方法，代码如下：

```
summaryDF = movies.select("year").summary()

summaryDF.printSchema()
summaryDF.show()
```

执行以上代码，输出内容如下：

```
root
 |-- summary: string (nullable = true)
 |-- year: string (nullable = true)

+-------+------------------+
|summary|              year|
+-------+------------------+
|  count|             31393|
|   mean| 2002.7964514382188|
| stddev| 6.377135379933117|
|    min|              1961|
|    25%|              1999|
|    50%|              2004|
|    75%|              2008|
|    max|              2012|
+-------+------------------+
```

4.6.5 提取 DataFrame Row 中特定字段

有时，用户需要从 DataFrame 中获取特定行的特定字段的值，这时可以通过索引的方式实现。

在下面的示例中，要求加载 PySpark 自带的数据文件 people.json，并输出每个字段的值。首先将文件 people.json 读到一个 DataFrame 中，代码如下：

```
#加载数据源，构造 DataFrame
input = "file://home/hduser/data/spark/resources/people.json"
df = spark.read.json(input)
```

```
df.printSchema()
df.show()
```

执行以上代码,输出内容如下:

```
root
 |-- age: long (nullable = true)
 |-- name: string (nullable = true)

+----+-------+
| age|   name|
+----+-------+
|null|Michael|
|  30|   Andy|
|  19| Justin|
+----+-------+
```

DataFrame 返回的每行是一个 Row 对象,可以按字段的位置取每个字段的值,代码如下:

```
#取每个字段的值
for row in df.collect():
    print(row[0],"\t",row[1])
```

执行以上代码,输出内容如下:

```
Michael  0
Andy     30
Justin   19
```

4.6.6 操作 DataFrame 示例

下面通过一个示例来进一步理解和掌握 DataFrame 的 Transformation 和 Action 操作。

【例 4-2】假设现在给出一个员工信息名单 employees.csv,包含员工薪资等信息,如下所示。

```
张三,paramedic i/c,fire,f,salary,,91080.00,
李四,lieutenant,fire,f,salary,,114846.00,
王老五,sergeant,police,f,salary,,104628.00,
赵六,police officer,police,f,salary,,96060.00,
钱七,clerk iii,police,f,salary,,53076.00,
周扒皮,firefighter,fire,f,salary,,87006.00,
吴用,law clerk,law,f,hourly,35,,14.51
```

其中,各个字段的含义依次为姓名,职业,部门,全职或兼职,固定薪资或计时薪资,工作时长,年薪,每小时工资。

现在要求找出其中收入最高的前 3 名员工(Top N 问题)。

需要注意数据集中，"吴用"没有年薪，是计时薪资（hourly，时薪），所以有最后一个"每小时工资"的字段，而其他人都是固定薪资（salary），所以有"年薪"字段，但是没有"每小时工资"字段。

首先，使用 HDFS Shell 命令将该数据文件 employees.csv 上传到 HDFS 上，命令如下：

```
$ hdfs dfs -put employees.csv /data/spark/
```

然后，编写 PySpark 程序，代码如下：

```
#第4章/dataframe_demo02.ipynb

from pyspark.sql import SparkSession
from pyspark.sql.functions import col

#构建SparkSession实例
spark = SparkSession.builder \
   .master("spark://localhost:7077") \
   .appName("pyspark demo") \
   .getOrCreate()

#定义文件路径
file = "/data/spark/employees.csv"

#指定列名
columns = ["uname", "designation", "department", "jobtype", "NA", "NA2", "salary", "NA3"]

#加载数据文件，并构造DataFrame
df = spark.read \
     .option("inferSchema", "true") \
     .option("header","true") \
     .csv(file) \
     .toDF(*columns)

#按salary列降序排列，并显示
df.orderBy(col("salary").desc()).limit(3).show()
```

执行以上代码，输出内容如下：

```
+------+--------------+----------+-------+------+------+--------+----+
| uname|   designation|department|jobtype|    NA|   NA2|  salary| NA3|
+------+--------------+----------+-------+------+------+--------+----+
|  李四|    lieutenant|      fire|      f|salary|  null|114846.0|null|
|王老五|      sergeant|    police|      f|salary|  null|104628.0|null|
|  赵六|police officer|    police|      f|salary|  null| 96060.0|null|
+------+--------------+----------+-------+------+------+--------+----+
```

4.7 存储 DataFrame

有时,需要将 DataFrame 中的数据写到外部存储系统中,例如,在一个典型的 ETL 数据处理作业中,处理结果通常需要被写到别的存储系统中,如本地文件系统、HDFS、Hive 或 Amazon S3。下面学习如何存储这些 DataFrame。

4.7.1 写出 DataFrame

在 PySpark SQL 中,pyspark.sql.DataFrameWriter 类负责将 DataFrame 中的数据写入外部存储系统。在 DataFrame 中有一个变量 write,它实际上就是 DataFrameWriter 类的一个实例。与 DataFrameWriter 交互的模式与 DataFrameReader 的交互模式有点类似。

与 DataFrameWriter 交互的常见模式,代码如下:

```
movies.write  \              #DataFrameWriter 实例对象
  .format(...) \             #指定存储格式
  .mode(...) \               #指定写出模式:append 或 overwrite
  .option(...) \             #指定选项
  .partitionBy(...) \        #指定分区列
  .bucketBy(...) \           #指定分桶
  .sortBy(...) \             #排序
  .save(path)                #保存到指定路径(文件夹)
```

与 DataFrameReader 相似,可以使用 json、orc 和 parquet 文件存储格式,默认格式是 parquet。需要注意的是,save()函数的输入参数是目录名,而不是文件名,它将数据直接保存到文件系统,如 HDFS、Amazon S3 或者一个本地路径 URL。

这些方法大多有相应的快捷方式,如 df.write.csv()、df.write.json()、df.write.orc()、df.write.parquet()、df.write.jdbc()等。这些方法相当于先调用 format()方法,再调用 save()方法。

在下面这个示例中,先读取 PySpark 自带的数据文件 people.json,然后进行简单计算,并把结果 DataFrame 保存到 CSV 格式的存储文件中,最后将这个结果存储文件加载到 RDD 中,代码如下:

```
#第4章/dataframe_demo04.ipynb

from pyspark.sql import SparkSession
from pyspark.sql.functions import col

#构建 SparkSession 实例
spark = SparkSession.builder \
  .master("spark://xueai8:7077") \
  .appName("pyspark sql demo") \
```

```python
        .getOrCreate()

#读取 JSON 文件源,创建 DataFrame
file = "/data/spark/resources/people.json"
df = spark.read.json(file)

df.printSchema
df.show()

#保存目录位置
output = "/data/spark/people-csv-output"
#找出 age 不是 null 的信息,保存到 CSV 文件中
df.where(col("age").isNotNull()).write.format("csv").save(output)
#df.where(col("age").isNotNull()).write.csv(output)   #等价上一句

#将保存的 CSV 数据再次加载到 RDD 中
textFile = spark.sparkContext.textFile(output)
for row in textFile.collect():
    print(row)
```

执行以上代码,输出内容如下:

```
30,Andy
19,Justin
```

打开浏览器,访问 HDFS 的 Web UI(http://localhost:9870),然后浏览 HDFS 文件系统中的结果文件存储目录(在上面的代码中指定的/data/spark/people-csv-output),可以看到存储内容如图 4-34 所示。

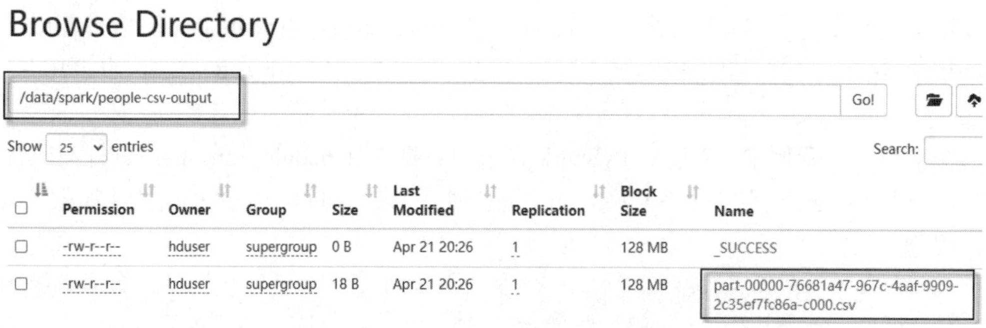

图 4-34 DataFrame 存储的结果文件

注意:write.format()支持输出 json、parquet、jdbc、orc、libsvm、csv、text 等格式文件,如果要输出 text 文本文件,则可以采用 write.format("text"),但是,需要注意,只有 select() 中只存在一个列时,才允许保存成文本文件,如果存在两个列,例如 select("name", "age"),就不能保存成文本文件了。

PySpark 支持通过 JDBC 方式连接到其他数据库，并将 DataFrame 存储到数据库中。在下面这个示例中，将分析结果 DataFrame 写到 MySQL 数据库中。

首先，在 MySQL 中新建一个测试 Spark 程序的数据库，数据库的名称是 spark，数据表的名称是 student。登录 MySQL 数据库，执行以下 SQL 语句：

```
mysql> create database spark;
mysql> use spark;
mysql> create table student (id int(4), name char(20), gender char(4), age int(4));
mysql> insert into student values(1,'张三','F',23);
mysql> insert into student values(2,'李四','M',18);
mysql> select * from student;
```

然后，编写 Spark 应用程序，构造一个 DataFrame，包含两行学生信息，代码如下：

```
#第4章/dataframe_demo05.ipynb
from pyspark.sql import SparkSession
from pyspark.sql.types import *
from pyspark.sql.functions import col

#构建 SparkSession 实例
spark = SparkSession.builder \
    .master("spark://xueai8:7077") \
    .appName("pyspark sql demo") \
    .getOrCreate()

#设置两条数据表示两个学生信息
students = [
    (3, "王老五", "F", 44),
    (4, "赵小虎", "M", 27)
]

#指定一个 Schema(模式)
fields = [
    StructField("id", IntegerType(), True),
    StructField("name", StringType(), True),
    StructField("gender", StringType(), True),
    StructField("age", IntegerType(), True)
]
schema = StructType(fields)

#将元组转换为 DataFrame
stusDF = spark.createDataFrame(students, schema=schema)

stusDF.printSchema()
stusDF.show()
```

执行以上代码，输出内容如下：

```
root
 |-- id: integer (nullable = true)
 |-- name: string (nullable = true)
 |-- gender: string (nullable = true)
 |-- age: integer (nullable = true)


+---+------+------+---+
| id|  name|gender|age|
+---+------+------+---+
|  3| 王老五|     F| 44|
|  4| 赵小虎|     M| 27|
+---+------+------+---+
```

接下来，编写 PySpark 代码，连接 MySQL 数据库并且将上面的 studentDF 写入 MySQL 保存。在本例中向 spark.student 表中插入两条记录，代码如下：

```
#下面创建一个prop变量用来保存JDBC连接参数
props = {
    "driver": "org.mariadb.jdbc.Driver",
    "user": "root",
    "password": "admin"
}

#下面就可以连接数据库了，采用append模式，表示追加记录到数据库spark的student表中
url = "jdbc:mysql://localhost:3306/spark?useSSL=false"
stuDF.write \
   .mode("append") \
   .jdbc(url=url, table="spark.students", properties=props)
```

执行上以代码，然后到 MySQL 中去查询 spark.student 表：

```
mysql> select * from spark.student;
```

可以看到以下的查询结果，证明数据已经被正确地写入了数据库中：

```
+------+----------+--------+------+
| id   | name     | gender | age  |
+------+----------+--------+------+
|    1 | 张三     | F      |   23 |
|    2 | 李四     | M      |   18 |
|    3 | 王老五   | 男     |   46 |
|    4 | 赵小花   | 女     |   27 |
+------+----------+--------+------+
```

4.7.2 存储模式

DataFrameWriter 类中的一个重要选项是 save mode，它表示存储模式。在将 DataFrame

中的数据写到存储系统上时，默认行为是创建一个新表。如果指定的输出目录或同名表已经存在，则抛出错误消息。可以使用 PySpark SQL 的 SaveMode 特性来更改此行为。

PySpark 所支持的各种存储模式见表 4-8。

表 4-8　Spark SQL 写出 DataFrame 时支持的存储格式

mode	说　明
"append"	当将一个 DataFrame 保存到数据源时，如果数据/表已经存在，则该 DataFrame 的内容将被追加到已经存储的数据后
"overwrite"	当将一个 DataFrame 保存到数据源时，如果数据/表已经存在，则已经存在的数据将被该 DataFrame 的内容覆盖
"error"或"errorIfExists"(默认)	当将一个 DataFrame 保存到数据源时，如果数据已经存在，则将抛出一个异常
"ignore"	当将一个 DataFrame 保存到数据源时，如果数据/表已经存在，则 save 操作不会保存该 DataFrame 的内容，并且不会改变已经存在的数据。这类似于 SQL 中的 create table if not exists

这些保存模式不使用任何锁定，也不是原子性的。此外，在执行 overwrite 写入时，将在写入新数据之前删除原数据。

例如，将数据以 CSV 格式写出，但使用"#"作为分隔符，代码如下：

```
movies.write \
    .format("csv") \
    .option("sep", "#") \
    .save("/tmp/output/csv")
```

如果使用 overwrite 模式写出数据，并同时写出标题行，则代码如下：

```
movies.write \
    .format("csv") \
    .mode("overwrite") \
    .option("sep", "#") \
    .option("header", "true") \
    .save("/tmp/output/csv")
```

4.7.3　控制 DataFrame 的输出文件数量

PySpark SQL DataFrame API 在转换操作执行 shuffling 时会增加分区。诸如 join()、union()这些 DataFrame 操作及所有聚合函数操作会触发数据 shuffle。

例如，查看一个 DataFrame 拥有的分区数量，代码如下：

```
#第 4 章/dataframe_demo06.ipynb
```

```python
from pyspark.sql import SparkSession
#构建 SparkSession 实例
spark = SparkSession.builder \
    .master("spark://localhost:7077") \
    .appName("pyspark sql demo") \
    .getOrCreate()

simpleData = [
    ("张三","销售部","北京",90000,34,10000),
    ("李四","销售部","北京",86000,56,20000),
    ("王老五","销售部","上海",81000,30,23000),
    ("赵老六","财务部","上海",90000,24,23000),
    ("钱小七","财务部","上海",99000,40,24000),
    ("周扒皮","财务部","北京",83000,36,19000),
    ("孙悟空","财务部","北京",79000,53,15000),
    ("朱八戒","市场部","上海",80000,25,18000),
    ("沙悟净","市场部","北京",91000,50,21000)
]
schema = ["employee_name","department","city","salary","age","bonus"]
df = spark.createDataFrame(simpleData, schema=schema)

print("shuffle 前的分区数: ",df.rdd.getNumPartitions())

#groupBy 操作会触发数据 shuffle
df2 = df.groupBy("city").count()

print("shuffle 后的分区数: ",df2.rdd.getNumPartitions())
```

执行以上代码，输出内容如下：

```
shuffle 前的分区数: 2
shuffle 后的分区数: 200
```

从输出结果可以看出，经过 groupBy() 聚合操作，数据跨分区 shuffling 后，DataFrame 的默认分区数变为 200。当 PySpark 操作执行数据 shuffling（join()、union()、aggregation 函数）时，DataFrame 会自动将分区数增加到 200。这个默认的 shuffle 分区数来自 PySpark SQL 配置 spark.sql.shuffle.partitions，默认设置为 200。

可以使用 SparkSession 对象的 conf() 方法或使用 Spark Submit 命令来修改这个默认的 shuffle 分区数，代码如下：

```
spark.conf.set("spark.sql.shuffle.partitions",100)
println(df.groupBy("city").count().rdd.getNumPartitions())   #100
```

保存到指定输出目录的文件数量与 DataFrame 拥有的分区数量是相对应的。例如

DataFrame 有 100 个分区，那么将其存储到 HDFS 中时就会生成 100 个存储文件，每个分区对应一个存储文件。在下面的示例中，将一部电影数据集加载到 DataFrame 中，并将其重分区为 4 个分区，然后保存到指定的位置，代码如下：

```
//将电影数据集文件加载到 DataFrame 中
val file = "/data/spark/movies/movies.csv"
val movies = spark.read
    .option("header","true")
    .option("inferSchema","true")
    .csv(file)

val movies2 = movies.repartition(4)    //将 DataFrame 重分区为 4 个分区
println("movies2 的分区数是: " + movies2.rdd.getNumPartitions)

//保存 movies2
movies2.write.csv("/data/spark/movies-out-partitions")

#将电影数据集文件加载到 DataFrame 中
file = "/data/spark/movies2/movies.csv"

movies = spark.read \
    .option("header","true") \
    .option("inferSchema","true") \
    .csv(file)

movies2 = movies.repartition(4)    #将 DataFrame 重分区为 4 个分区

#查看分区数
print("movies2 的分区数是: ", movies2.rdd.getNumPartitions())

#保存 movies2
movies2.write.csv("/data/spark/movies-out-partitions")
```

执行以上代码，输出内容如下：

```
movies2 的分区数是：4
```

打开浏览器，访问 http://localhost:9870，查看保存的 HDFS 文件目录，如图 4-35 所示。

在某些情况下，DataFrame 的内容并不大，并且需要写入单个文件中。这时可以使用 coalesce()转换函数，将 DataFrame 的分区数量减少到 1，然后把它写出。coalesce()函数的定义如下：

```
DataFrame.coalesce(numPartitions: int) → pyspark.sql.dataframe.DataFrame
```

这种方法返回一个新的 RDD，该 RDD 被缩减为 numPartitions 所指定的分区数，其中

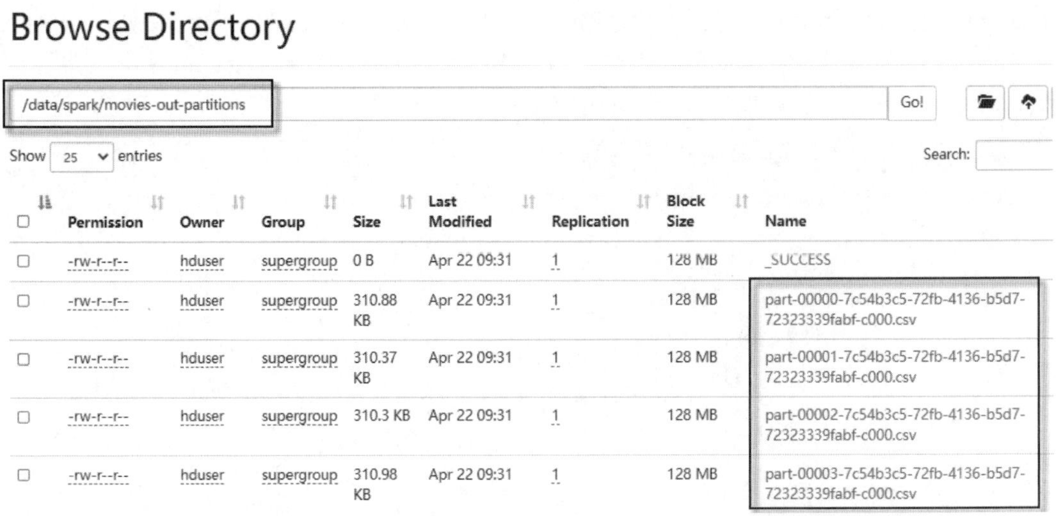

Browse Directory

图 4-35 DataFrame 存储的结果文件数量与分区数量相对应

第 1 个参数 numPartitions 用于指定要被缩减到的目标分区数,第 2 个参数用于指定允不允许数据 shuffle。这导致了一个窄依赖,例如,如果从 1000 个分区缩减到 100 个分区,则不会产生数据 shuffle,而是每 100 个新分区将占用 10 个当前分区。如果请求比当前分区数更多的分区,则将保持当前的分区数。

假设现在想将前面代码中的 movies2 写入一个文件中存储,可将在上面的示例代码中的最后一行修改如下:

```
#保存movies2 到一个存储文件中
movies2.coalesce(1) \
    .write \
    .mode("overwrite") \
    .csv("/data/spark/movies-out-partitions")
```

1. 分区多少对性能的影响

根据数据集大小、CPU 核数量和内存大小,PySpark shuffling 可能对作业有利也可能有害。当处理较少的数据量时,通常应该减少 shuffle 分区,否则将最终得到许多分区文件,每个分区中的记录数量较少。这将导致运行许多需要处理的数据较少的任务。另一方面,当有太多的数据和较少的分区数时同样会导致长时间运行少量任务,这时也可能得到内存错误,所以分区过少或过多都可能是不利的。

(1) 分区过少:将无法充分利用群集中的所有可用 CPU 核。

(2) 分区过多:产生非常多的小任务,从而会产生过多的开销。

在这两者之间,第 1 个对性能的影响相对比较大。对于小于 1000 个分区数的情况而言，调度太多的小任务所产生的影响相对较小，但是，如果有成千上万个分区，则 PySpark 的执行速度会变得非常慢。

获得正确的 shuffle 分区大小总是很棘手的，需要多次尝试不同的值来获得最优的分区数。当在 PySpark 作业上遇到性能问题时，这是需要查找的关键属性之一。

2. 如何设置分区数量

假设要对一个大数据集进行操作，该数据集的分区数也比较大，那么当进行一些操作之后，例如 filter() 过滤操作、sample() 抽样操作，这些操作可能会使结果数据集的数据量大幅减少，但是 PySpark 却不会对其分区进行调整，由此会造成大量的分区没有数据，并且向 HDFS 读取和写入大量的空文件，效率会很低，这种情况就需要我们重新调整分区数量，以此来提升效率。

通常情况下，当结果集的数据量减少时，其对应的分区数也应当相应地减少。那么该如何确定具体的分区数呢？

PySpark 中的 shuffle 分区数是静态的。它不会随着不同的数据大小而变化。上文提到，默认情况下，控制 shuffle 分区数的参数 spark.sql.shuffle.partitions 的值为 200，这将导致以下问题：

（1）对于较小的数据，200 分区过多了，由于调度开销，通常会导致处理速度变慢。

（2）对于大数据，200 分区则太少了，无法有效地使用群集中的所有资源。

一般情况下，可以通过将集群中的 CPU 数量乘以 2、3 或 4 来确定分区的数量。如果要将数据写到文件系统中，则可以选择一个分区大小，以创建合理大小的文件。

3. 如何控制输出文件的大小

如果写入时产生的小文件数量过多，则这时会产生大量的元数据开销。PySpark 和 HDFS 一样，都不能很好地处理这个问题，这被称为"小文件问题"。同时数据文件也不能过大，否则在查询时会有不必要的性能开销，因此要把文件大小控制在一个合理的范围内。

在 PySpark 中，有两种方法可以控制输出文件的大小：

（1）可以通过分区数量来控制生成文件的数量，从而间接地控制文件的大小。

（2）Spark 2.2 引入了一个新的 maxRecordsPerFile 参数，以更自动化的方式控制文件大小。通过它可以控制写入文件的记录数来控制文件的大小。

例如，可以在写出 DataFrame 结果时，指定如下的选项，PySpark 将确保每个输出文件最多包含 5000 条记录，代码如下：

```
df.write.option("maxRecordsPerFile", 5000)
```

4.7.4 控制 DataFrame 实现分区存储

在 PySpark SQL 中写出数据时也可以使用分区技术。DataFrameWriter 有一个 partitionBy()

方法，它指定数据在写入磁盘前如何进行分区。

例如，在 movies 这个 DataFrame 中，year 这个列是最适合用来进行分区的候选列。假设要写出 movies 这个 DataFrame，用 year 这个列来分区，DataFrameWriter 将把所有具有相同年份的电影都写在同一个目录中。输出文件夹中的目录数量将对应于 movies 这个 DataFrame 中的年份的数量，代码如下：

```
#第4章/dataframe_demo07.ipynb

from pyspark.sql import SparkSession

#构建 SparkSession 实例
spark = SparkSession.builder \
    .master("spark://xueai8:7077") \
    .appName("pyspark sql demo") \
    .getOrCreate()

#将电影数据集文件加载到 DataFrame 中
file = "/data/spark/movies2/movies.csv"

movies = spark.read \
    .option("header","true") \
    .option("inferSchema","true") \
    .csv(file)

#查看分区数
print("movies 的分区数是：", movies.rdd.getNumPartitions())

#指定输出路径
outpath = "/data/spark/movies-out-partitions"

#输出 movies DataFrame，使用 Parquet 格式并按 year 列进行分区
movies.repartition("year") \
    .write \
    .mode("overwrite") \
    .partitionBy("year") \
    .save(outpath)
```

注意，在上面的代码中，先用 repartition()方法在内存中按 year 列进行分区，然后调用 repartitionBy()方法按相同的 year 列进行物理分区。这是通常的做法，有利于优化物理分区速度。

执行以上代码，然后在 HDFS 的 Web UI（http://localhost:9870）查看存储目录，可以看到分区存储结果，如图 4-36 所示。

可以看到，/data/spark/movies-out-partitions 目录下包含了多个子目录，从 year=1961 到 year=2012。图中只截取了其中部分内容。

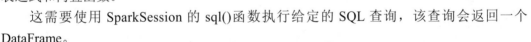

图 4-36　使用 partitionBy 指定分区存储

4.8　临时视图与 SQL 查询

PySpark SQL 支持直接应用标准 SQL 语句进行查询。当在 PySpark SQL 中编写 SQL 命令时，它们会被翻译为 DataFrame 上的关系操作。在 SQL 语句内，可以访问所有 SQL 表达式和内置函数。

这需要使用 SparkSession 的 sql() 函数执行给定的 SQL 查询，该查询会返回一个 DataFrame。

4.8.1　在 PySpark 程序中执行 SQL 语句

PySpark 提供了几种在 PySpark 中运行 SQL 的方法，包括以下几种：

（1）Spark SQL CLI（./bin/spark-sql）。

（2）JDBC/ODBC 服务器。

（3）在 PySpark 应用程序中以编程方式运行 SQL。

前两种选择提供了与 Apache Hive 的集成，以利用 Hive 的元数据。PySpark SQL 支持使用基本 SQL 语法或 HiveQL 编写的 SQL 查询。

在 PySpark Shell 或 Zeppelin Notebook 中，会自动导入 spark.sql，所以可以直接使用该函数编写 SQL 命令，代码如下：

```
spark.sql("select current_date() as today , 1 + 100 as value").show()
```

SparkSession 的 sql()函数可执行给定的 SQL 查询,该查询会返回一个 DataFrame。

本节只讨论最后一个选项,即在 PySpark 应用程序中以编程方式运行 SQL。例如,执行一个不带注册视图的 SQL 语句,代码如下:

```
#第4章/dataframe_demo08.ipynb

from pyspark.sql import SparkSession

#构建 SparkSession 实例
spark = SparkSession.builder \
    .master("spark://localhost:7077") \
    .appName("pyspark sql demo") \
    .getOrCreate()

infoDF = spark.sql("select current_date() as today , 1 + 100 as value")
infoDF.show()
```

执行过程和结果如图 4-37 所示。

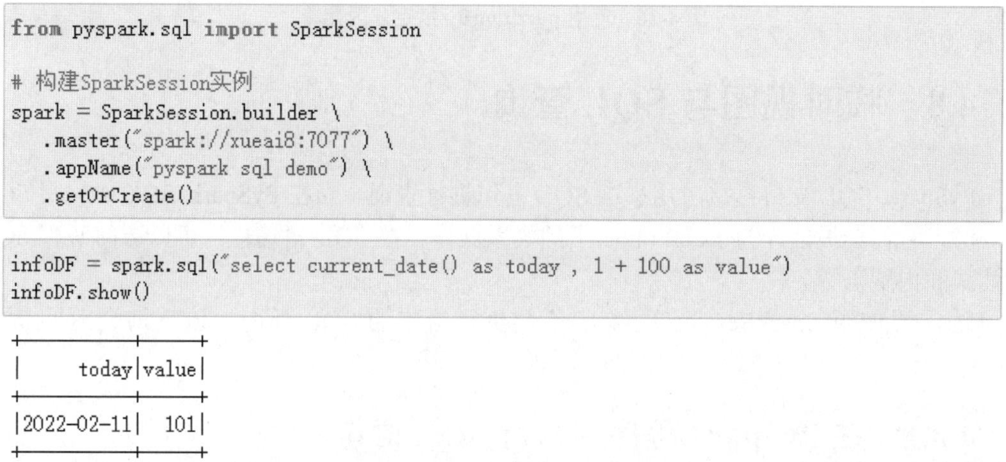

```
from pyspark.sql import SparkSession

# 构建SparkSession实例
spark = SparkSession.builder \
    .master("spark://xueai8:7077") \
    .appName("pyspark sql demo") \
    .getOrCreate()
```

```
infoDF = spark.sql("select current_date() as today , 1 + 100 as value")
infoDF.show()
```

```
+----------+-----+
|     today|value|
+----------+-----+
|2022-02-11|  101|
+----------+-----+
```

图 4-37 编程执行 SQL 查询语句

除了可以使用 PySpark 读 API 将文件加载到 DataFrame 并对其进行查询外,PySpark 也可以使用 SQL 语句直接查询该数据文件,代码如下:

```
sqlDF = spark.sql("""
    SELECT *
    FROM parquet.`/data/spark/resources/users.parquet`
""")
sqlDF.show()
```

执行过程和结果如图 4-38 所示。

```
sqlDF = spark.sql("SELECT * FROM parquet.`/data/spark/resources/users.parquet`")
sqlDF.show()
```

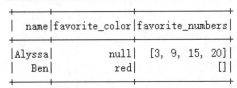

图 4-38　使用 SQL 语句直接查询数据文件

注意：这种方式必须使用 HDFS 文件路径。

4.8.2　注册临时视图并执行 SQL 查询

DataFrame 本质上就像数据库中的表一样，可以通过 SQL 语句来查询它们。不过，在发出 SQL 查询来操作它们之前，需要将它们注册为一个临时视图，然后，就可以使用 SQL 查询从临时表中查询数据了。每个临时视图都有一个名字，通过视图的名字来引用该 DataFrame，该名字在 select 子句中用作表名。

例如，查询电影数据集 movies.parquet 文件，代码如下：

```
#第4章/dataframe_demo09.ipynb

from pyspark.sql import SparkSession

#构建 SparkSession 实例
spark = SparkSession.builder \
    .master("spark://localhost:7077") \
    .appName("pyspark sql demo") \
    .getOrCreate()

#定义文件路径
parquetFile = "/data/spark/movies2/movies.parquet"

#读取 DataFrame 中数据
movies = spark.read.parquet(parquetFile)

#现在将 movies DataFrame 注册为一个临时视图
movies.createOrReplaceTempView("movies")

#从视图 view 查询
sql = """
    select *
    from movies
```

```
        where actor_name like '%Jolie%' and produced_year > 2009
"""
spark.sql(sql).show()
```

执行以上代码，输出内容如下：

```
+--------------+---------------+-------------+
|    actor_name|    movie_title|produced_year|
+--------------+---------------+-------------+
|Jolie, Angelina|           Salt|         2010|
|Jolie, Angelina|Kung Fu Panda 2|         2011|
|Jolie, Angelina|    The Tourist|         2010|
```

也可以在 sql()函数中混合使用 SQL 语句和 DataFrame 转换 API。例如，查询电影数据集，找出参演影片超过 30 部的演员，代码如下：

```
#第 4 章/dataframe_demo10.ipynb

from pyspark.sql import SparkSession
from pyspark.sql.functions import col

#构建 SparkSession 实例
spark = SparkSession.builder \
    .master("spark://localhost:7077") \
    .appName("pyspark sql demo") \
    .getOrCreate()

#定义文件路径
parquetFile = "/data/spark/movies2/movies.parquet"

#读取 DataFrame 中数据
movies = spark.read.parquet(parquetFile)

#现在将 movies DataFrame 注册为一个临时视图
movies.createOrReplaceTempView("movies")

#从视图 view 查询
spark.sql("select actor_name, count(*) as count from movies group by actor_name") \
    .where('count > 30') \
    .orderBy(col("count").desc()) \
    .show()
```

执行以上代码，输出内容如下：

```
+-----------------+-----+
|       actor_name|count|
+-----------------+-----+
```

```
|    Tatasciore, Fred|  38|
|       Welker, Frank|  38|
|Jackson, Samuel L.|  32|
|       Harnell, Jess|  31|
+--------------------+----+
```

当 SQL 语句较长时,可以利用"""(三个双引号)来格式化多行 SQL 语句。例如,查询电影数据集,使用子查询来计算每年上映的电影数量,代码如下:

```
#第4章/dataframe_demo11.ipynb

from pyspark.sql import SparkSession
from pyspark.sql.functions import col

#构建 SparkSession 实例
spark = SparkSession.builder \
   .master("spark://localhost:7077") \
   .appName("pyspark sql demo") \
   .getOrCreate()

#定义文件路径
parquetFile = "/data/spark/movies2/movies.parquet"

#读取 DataFrame 中数据
movies = spark.read.parquet(parquetFile)

#现在将 movies DataFrame 注册为一个临时视图
movies.createOrReplaceTempView("movies")

#使用子查询来计算每年上映的电影数量(利用"""来格式化多行 SQL 语句)
spark.sql("""select produced_year, count(*) as count
         from (select distinct movie_title, produced_year from movies)
         group by produced_year
    """) \
   .orderBy(col("count").desc()) \
   .show(5)
```

执行以上代码,输出内容如下:

```
+-------------+-----+
|produced_year|count|
+-------------+-----+
|         2011|   86|
|         2004|   86|
|         2006|   86|
|         2005|   85|
```

```
|          2008|    82|
+--------------+------+
only showing top 5 rows
```

注意：PySpark 实现了 ANSI SQL:2003 修订版（最流行的 RDBMS 服务器支持）的一个子集。此外，PySpark 2.0 通过包含一个新的 ANSI SQL 解析器扩展了 PySpark SQL 功能，支持子查询和 SQL:2003 标准。更具体地说，子查询支持现在包括相关/不相关的子查询，以及 WHERE / HAVING 子句中的 IN / NOT IN 和 EXISTS / NOT EXISTS 谓词。

4.8.3 使用全局临时视图

PySpark 为临时视图提供了两个级别的会话范围。一个是 PySpark 会话级别，当 DataFrame 在这个级别上注册时，只有在同一个会话中发出的查询才能引用该 DataFrame。当 PySpark 会话关闭时，会话范围的级别将消失。第 2 个作用域级别是全局级别，这意味着可以在所有 PySpark 会话中将这些视图用于 SQL 语句。

PySpark SQL 中的临时视图是有会话范围的，如果创建它的会话终止，它就会消失。如果希望拥有一个在所有会话之间共享的临时视图，并一直保持活动状态，直到 PySpark 应用程序终止，则可以创建一个全局临时视图。全局临时视图通常被绑定到系统保留的数据库 global_temp，必须使用限定名来引用它，例如"SELECT * FROM global_temp.view1"。

如果要注册全局临时视图，则可使用 createOrReplaceGlobalTempView 方法。例如，将 movies 注册为全局临时视图，叫作 movies_g，代码如下：

```
movies.createOrReplaceGlobalTempView("movies_g")
```

从全局视图中查询，需要使用关键字 global_temp 来作为视图名称的前缀，代码如下：

```
spark.sql("select count(*) as total from global_temp.movies_g").show()
```

在下面这个示例中，演示了如何跨多个会话使用全局临时视图进行查询。首先，将 PySpark 自带的 people.json 文件读到一个 DataFrame 中，代码如下：

```
#第4章/dataframe_demo12.ipynb

from pyspark.sql import SparkSession

#构建SparkSession实例
spark = SparkSession.builder \
    .master("spark://localhost:7077") \
    .appName("pyspark sql demo") \
    .getOrCreate()

#将JSON文件数据加载到DataFrame
json_file = "/data/spark/resources/people.json"
```

```
df = spark.read.json(json_file)

#查看 schema
df.printSchema()

#显示 DataFrame 内容
df.show()
```

执行以上代码,输出内容如下:

```
root
 |-- age: long (nullable = true)
 |-- name: string (nullable = true)

+----+-------+
| age|   name|
+----+-------+
|null|Michael|
|  30|   Andy|
|  19| Justin|
+----+-------+
```

然后,将 df 注册为全局视图(注意,需要启动 hive --service metastore),名称为 people,代码如下:

```
#将 DataFrame 注册为全局临时视图
df.createGlobalTempView("people")
```

将全局临时视图绑定到一个系统保留的数据库 global_temp 上,因此查询时需要在数据表前加上此数据库的前缀,代码如下:

```
spark.sql("SELECT * FROM global_temp.people").show()
```

执行以上代码,输出结果如下:

```
+----+-------+
| age|   name|
+----+-------+
|null|Michael|
|  30|   Andy|
|  19| Justin|
+----+-------+
```

全局临时视图是跨会话的,所以可以在另一个会话中使用此全局临时视图,代码如下:

```
spark.newSession().sql("SELECT * FROM global_temp.people").show()
```

执行以上代码,输出结果如下:

```
+----+-------+
| age|   name|
+----+-------+
|null|Michael|
|  30|   Andy|
|  19| Justin|
+----+-------+
```

4.8.4 直接使用数据源注册临时视图

在前面的示例中都是先将数据加载到一个 DataFrame 中，然后将该 DataFrame 注册为临时视图或全局视图。除此之外，也可以使用 SparkSession 的 sql()方法从注册的数据源直接加载数据，以便注册临时视图。

例如，注册一个 Parquet 文件并加载它的内容，代码如下：

```
#第4章/dataframe_demo13.ipynb

from pyspark.sql import SparkSession

#构建 SparkSession 实例
spark = SparkSession.builder \
    .master("spark://localhost:7077") \
    .appName("pyspark sql demo") \
    .getOrCreate()

#从 Parquet 数据源创建临时视图
spark.sql("create temporary view usersParquet "+
    "using org.apache.spark.sql.parquet "+
    "options(path '/data/spark/resources/users.parquet')")

#查询临时视图
spark.sql("select * from usersParquet").show()
```

执行以上代码，输出结果如下：

```
+------+--------------+----------------+
|  name|favorite_color|favorite_numbers|
+------+--------------+----------------+
|Alyssa|          null|   [3, 9, 15, 20]|
|   Ben|           red|              []|
+------+--------------+----------------+
```

下面是另一个使用内置数据源的例子。从 JDBC 注册一个临时视图，然后使用 SQL 语句查询该临时视图（这里连接的是 MySQL 5 数据库，读者如果使用的是其他版本的 MySQL，则应自行修改为相应版本的连接参数），代码如下：

```
#第4章/dataframe_demo14.ipynb

from pyspark.sql import SparkSession

#构建SparkSession实例
spark = SparkSession.builder \
    .master("spark://localhost:7077") \
    .appName("pyspark sql demo") \
    .getOrCreate()

#从JDBC数据源创建临时视图
spark.sql(
    """
        create temporary view people_jdbc
        using org.apache.spark.sql.jdbc
        options(
          url 'jdbc:mysql://localhost:3306/xueai8',
          dbtable 'people',
          user 'root',
          password 'admin'
        )
    """
)

#在临时视图上执行查询操作
spark.sql("select * from people_jdbc").show()
```

执行上面的代码，输出结果如下：

```
+---+------+---+
| id| name|age|
+---+------+---+
|  1|  张三| 23|
|  2|  李四| 18|
|  3|王老五| 35|
+---+------+---+
```

4.8.5 查看和管理表目录

当将 DataFrame 注册为临时视图时，PySpark 会将该视图的定义存储在表目录（Table Catalog）中。所有已注册的视图都保存在这个元数据目录中，PySpark 提供了一个工具，用于管理这个表目录。这个管理工具是作为 Catalog 类实现的，通过 SparkSession 的 catalog 字段访问。

可以使用该 Catalog 对象来查看当前注册的表有哪些（显示 catalog 目录中的表），如果没有，则返回一个空的 list 列表。例如，查看当前注册的临时视图有哪些，代码如下：

```
#查看当前数据库中注册的表有哪些
spark.catalog.listTables().show()
```

执行上面的代码,输出内容如下:

```
+------------+--------+-----------+---------+-----------+
|        name|database|description|tableType|isTemporary|
+------------+--------+-----------+---------+-----------+
|  moviesjdbc|    null|       null|TEMPORARY|       true|
|usersparquet|    null|       null|TEMPORARY|       true|
+------------+--------+-----------+---------+-----------+
```

另外,还可以使用该 Catalog 对象检查一个指定的视图所具有的列。例如,查看 usersParquet 视图的列信息,代码如下:

```
#返回给定表/视图或临时视图的所有列
spark.catalog.listColumns("usersParquet").show()
```

执行上面的代码,输出内容如下:

```
+----------------+-----------+----------+--------+-----------+--------+
|            name|description|  dataType|nullable|isPartition|isBucket|
+----------------+-----------+----------+--------+-----------+--------+
|            name|       null|    string|    true|      false|   false|
|  favorite_color|       null|    string|    true|      false|   false|
|favorite_numbers|       null|array<int>|    true|      false|   false|
+----------------+-----------+----------+--------+-----------+--------+
```

要获得所有可用的 SQL 函数列表,可使用 Catalog 对象的 listFunctions()方法,代码如下:

```
#返回在当前数据库中注册的函数列表
spark.catalog.listFunctions().show(truncate=False)
```

执行上面的代码,输出内容如下:

```
+--------+--------+-----------+--------------------+-----------+
|    name|database|description|           className|isTemporary|
+--------+--------+-----------+--------------------+-----------+
|       !|    null|       null|org.apache.spark....|       true|
|       %|    null|       null|org.apache.spark....|       true|
|       &|    null|       null|org.apache.spark....|       true|
|       *|    null|       null|org.apache.spark....|       true|
|       +|    null|       null|org.apache.spark....|       true|
|       -|    null|       null|org.apache.spark....|       true|
|       /|    null|       null|org.apache.spark....|       true|
|       <|    null|       null|org.apache.spark....|       true|
|      <=|    null|       null|org.apache.spark....|       true|
```

```
|      <=>|    null|    null|org.apache.spark....|        true|
|        =|    null|    null|org.apache.spark....|        true|
|       ==|    null|    null|org.apache.spark....|        true|
|        >|    null|    null|org.apache.spark....|        true|
|       >=|    null|    null|org.apache.spark....|        true|
|        ^|    null|    null|org.apache.spark....|        true|
|      abs|    null|    null|org.apache.spark....|        true|
|     acos|    null|    null|org.apache.spark....|        true|
|    acosh|    null|    null|org.apache.spark....|        true|
|add_months|   null|    null|org.apache.spark....|        true|
|aggregate|    null|    null|org.apache.spark....|        true|
+----------+--------+--------+--------------------+------------+
only showing top 20 rows
```

还可以使用 cacheTable()、uncacheTable()、isCached()和 clearCache()方法来管理这些元数据，包括缓存临时视图、删除临时视图和清除缓存等。

4.9 缓存 DataFrame

可以在内存中对 DataFrame 进行持久/缓存，就像 RDD 一样。在 DataFrame 类中也可以使用同样熟悉的持久性 API，如 persist() 和 unpersist()，然而，DataFrame 的缓存与 RDD 有很大的不同。PySpark SQL 知道 DataFrame 中数据的模式，因此它以一种列格式组织数据，并应用任何适用的压缩来最小化空间。最终的结果是，当两个都由同一个数据文件支持时，在内存中存储 DataFrame 所需的空间比存储 RDD 所需的空间要小得多。

PySpark 缓存和持久化是用于迭代和交互 PySpark 应用程序的 DataFrame 优化技术，以提高作业的性能。

4.9.1 缓存方法

PySpark SQL 提供了 cache()和 persist()方法，用来缓存 DataFrame。当使用 cache()和 persist()方法时，PySpark 提供了一种优化机制来存储 PySpark DataFrame 的中间计算，以便在后续操作中重用。

当持久化一个数据集时，每个节点将它的分区数据存储在自己的内存中，并在该数据集的其他操作中重用它们。PySpark 在节点上的持久化数据是容错的，这意味着如果数据集的任何分区丢失，则它将自动使用创建它的原始转换重新计算。例如，将电影数据集加载到一个 DataFrame 中，并缓存它，代码如下：

```
#第 4 章/dataframe_demo15.ipynb

from pyspark.sql.functions import col
```

```python
from pyspark.sql import SparkSession
#构建 SparkSession 实例
spark = SparkSession.builder \
    .master("spark://localhost:7077") \
    .appName("pyspark sql demo") \
    .getOrCreate()

#读取 CSV 文件源,创建 DataFrame
file = "/data/spark/movies2/movies.csv"
df = spark.read \
    .options(inferSchema="true",delimiter=",",header="true") \
    .csv(file)

#缓存
df2 = df.where(col("year") == "2012").cache()
df2.show(5, truncate = False)

print(f"2012 年上映的电影数量有{df2.count()}部。")

df3 = df2.where(col("title") == "The Hunger Games")
print("电影饥饿游戏的主演是:",df3.collect()[0][0])
print("电影饥饿游戏的主演是:",df3.first()[0])
```

执行以上代码,输出内容如下:

```
+------------------+------------------+----+
|actor             |title             |year|
+------------------+------------------+----+
|Cassavetes, Frank |Battleship        |2012|
|Danner, Blythe    |The Lucky One     |2012|
|Manji, Rizwan     |The Dictator      |2012|
|Harrelson, Woody  |The Hunger Games  |2012|
|Hardy, Tom (I)    |This Means War    |2012|
+------------------+------------------+----+
only showing top 5 rows

2012 年上映的电影数量有 601 部
电影饥饿游戏的主演是: Harrelson, Woody
```

DataFrame 的 cache()方法内部调用了 persist()方法,默认情况下采用的存储级别是 MEMORY_AND_DISK,因为重新计算底层表的内存列表示非常昂贵。注意,这与 RDD.cache()的默认缓存级别 MEMORY_ONLY 不同。

不管是 cache()还是 persist()方法,都是延迟计算的(只有在执行 action 时才进行计算)。实际上,cache()是 persist(StorageLevel.MEMORY_AND_DISK)的别名。

4.9.2 缓存策略

存储级别用于指定如何及在何处持久化或缓存 Spark DataFrame。PySpark SQL 支持的所有不同存储级别都可以在 pyspark.StorageLevel 类上获得，包括以下几个级别：

（1）MEMORY_ONLY。这是 RDD cache()方法的默认行为，并将 DataFrame 作为反序列化对象存储到 JVM 内存中。当没有足够的可用内存时，它将不保存某些分区的 DataFrame，并在需要时重新计算这些 DataFrame。这需要更多的内存，但与 RDD 不同的是，这将比 MEMORY_AND_DISK 级别慢，因为它会重新计算未保存的分区，并且重新计算底层表的内存列表是非常昂贵的。

（2）MEMORY_ONLY_SER。这与 MEMORY_ONLY 相同，但不同之处在于它将 RDD/DataFrame 作为序列化对象存储到 JVM 内存中。它比 MEMORY_ONLY 占用更少的内存（节省空间），因为它将对象保存为序列化的，并且为了反序列化多占用了几个 CPU 周期。

（3）MEMORY_ONLY_2。与 MEMORY_ONLY 存储级别相同，但将每个分区复制到两个集群节点。

（4）MEMORY_ONLY_SER_2。与 MEMORY_ONLY_SER 存储级别相同，但将每个分区复制到两个集群节点。

（5）MEMORY_AND_DISK。这是 DataFrame 或 DataSet 的默认行为。在这个存储级别中，数据帧将作为反序列化对象存储在 JVM 内存中。当所需的存储大于可用内存时，它将一些多余的分区存储到磁盘中，并在需要时从磁盘读取数据。当涉及 I/O 操作时，它会较慢。

（6）MEMORY_AND_DISK_SER。这与 MEMORY_AND_DISK 存储级别相同，不同之处在于，当空间不可用时，它会在内存和磁盘上序列化数据帧对象。

（7）MEMORY_AND_DISK_2。与 MEMORY_AND_DISK 存储级别相同，但将每个分区复制到两个集群节点。

（8）MEMORY_AND_DISK_SER_2。与 MEMORY_AND_DISK_SER 存储级别相同，但将每个分区复制到两个集群节点。

（9）DISK_ONLY。在这个存储级别中，DataFrame 仅存储在磁盘上，CPU 计算时间长，因为涉及 I/O 操作。

（10）DISK_ONLY_2。与 DISK_ONLY 存储级别相同，但将每个分区复制到两个集群节点。

PySpark 会自动监控每个 persist()和 cache()的调用，并检查每个节点上的使用情况，如果没有使用或使用最近最少使用（LRU 算法），则删除持久化数据。也可以使用 unpersist() 方法手动删除。unpersist()将 DataSet/DataFrame 标记为非持久的，并从内存和磁盘中删除它的所有块。在 DataFrame 上调用 unpersist()方法的代码如下：

```
dfPersist = dfPersist.unpersist()
```

4.9.3 缓存表

也可以使用 PySpark 的 catalog 来缓存 DataFrame，以易读的名字持久化表。在下面的示例中，要将一个 DataFrame 持久化到表中，代码如下：

```
#第4章/dataframe_demo16.ipynb

from pyspark.sql import SparkSession

#构建 SparkSession 实例
spark = SparkSession.builder \
   .master("spark://localhost:7077") \
   .appName("pyspark sql demo") \
   .getOrCreate()

#以创建一个 DataFrame
numDF = spark.range(1000).toDF("id")

#注册为一个视图
numDF.createOrReplaceTempView("num_df")

#使用 Spark Catalog 缓存该 numDF，使用名字"num_df"
spark.catalog.cacheTable("num_df")

#通过 count action 操作来强制持久化
numDF.count()              #1000
```

需要注意，cache()和 persist()持久化是延迟执行的，而 cacheTable()是立即执行的。

4.10 PySpark SQL 可视化

PySpark 还没有任何绘图功能。如果想将数据绘制成图，则可以将数据从 SparkContext 中取出并放入"本地"Python 会话中，在那里可以使用 Python 的任意一个绘图库来处理它。

对于 PySpark SQL 中的 DataFrame，可以先将它转换成 Pandas 的 DataFrame，再应用 Python 绘图库进行绘制。

4.10.1 PySpark DataFrame 转换到 Pandas

在 PySpark 中，很容易通过一行代码将 PySpark DataFrame 转换为 Pandas DataFrame，代码如下：

```
df_pd = df.toPandas()
```

在下面的示例中，将演示如何将 PySpark DataFrame Row 对象列表转换为 Pandas DataFrame，代码如下：

```
#第4章/dataframe_demo17.ipynb

from pyspark.sql import SparkSession
from pyspark.sql.functions import collect_list,struct
from pyspark.sql.types import *

from decimal import Decimal
import pandas as pd

#构建SparkSession 实例
spark = SparkSession.builder \
    .master("spark://xueai8:7077") \
    .appName("pyspark rdd demo") \
    .getOrCreate()

#List
data = [ ('Category A', 1, Decimal(12.40)),
        ('Category B', 2, Decimal(30.10)),
        ('Category C', 3, Decimal(100.01)),
        ('Category A', 4, Decimal(110.01)),
        ('Category B', 5, Decimal(70.85))
       ]

#创建 schema
schema = StructType([
    StructField('Category', StringType(), False),
    StructField('ItemID', IntegerType(), False),
    StructField('Amount', DecimalType(scale=2), True)
])

#将 List 转换为 RDD
rdd = spark.sparkContext.parallelize(data)

#创建 DataFrame
df = spark.createDataFrame(rdd, schema)
df.printSchema()
df.show()

#将 PySpark DataFrame 转换为 Pandas DataFrame
df_pd = df.toPandas()
df_pd.info()

#查看 Pandas DataFrame
print(df_pd)
```

执行以上代码,PySpark DataFrame 的输出内容如图 4-39 所示。

```
root
 |-- Category: string (nullable = false)
 |-- ItemID: integer (nullable = false)
 |-- Amount: decimal(10,2) (nullable = true)
```

```
+----------+------+------+
|  Category|ItemID|Amount|
+----------+------+------+
|Category A|     1| 12.40|
|Category B|     2| 30.10|
|Category C|     3|100.01|
|Category A|     4|110.01|
|Category B|     5| 70.85|
+----------+------+------+
```

图 4-39　PySpark DataFrame 的内容

Pandas DataFrame 的输出信息如图 4-40 所示。

```
<class 'pandas.core.frame.DataFrame'>
RangeIndex: 5 entries, 0 to 4
Data columns (total 3 columns):
 #   Column    Non-Null Count  Dtype
---  ------    --------------  -----
 0   Category  5 non-null      object
 1   ItemID    5 non-null      int32
 2   Amount    5 non-null      object
dtypes: int32(1), object(2)
memory usage: 228.0+ bytes
```

图 4-40　Pandas DataFrame 的信息

Pandas DataFrame 的输出内容如图 4-41 所示。

```
     Category  ItemID  Amount
0  Category A       1   12.40
1  Category B       2   30.10
2  Category C       3  100.01
3  Category A       4  110.01
4  Category B       5   70.85
```

图 4-41　Pandas DataFrame 的内容

如果 PySpark SQL DataFrame 具有嵌套结构,该如何转换呢?

例如,在 PySpark 中进行聚合是很常见的操作。按照 Category 属性对上面的 PySpark DataFrame 进行分组,代码如下:

```python
#聚合但仍然保留所有原始属性
df_agg = df \
   .groupby("Category") \
   .agg(collect_list(struct("*")).alias('Items'))

df_agg.printSchema()
df_agg.show(truncate=False)
```

执行以上代码,输出内容如图 4-42 所示。

```
root
 |-- Category: string (nullable = false)
 |-- Items: array (nullable = false)
 |    |-- element: struct (containsNull = false)
 |    |    |-- Category: string (nullable = false)
 |    |    |-- ItemID: integer (nullable = false)
 |    |    |-- Amount: decimal(10,2) (nullable = true)

+----------+---------------------------------------------+
|Category  |Items                                        |
+----------+---------------------------------------------+
|Category C|[{Category C, 3, 100.01}]                    |
|Category B|[{Category B, 2, 30.10}, {Category B, 5, 70.85}] |
|Category A|[{Category A, 1, 12.40}, {Category A, 4, 110.01}]|
+----------+---------------------------------------------+
```

图 4-42　PySpark DataFrame 分组聚合操作

新的 PySpark DataFrame 的 schema 有两个属性:Category 和 Items。这个 Items 属性是 pyspark.sql.Row 对象的数组或列表。要将这个 pyspark.sql.Row 列表转换为 Pandas DataFrame,用户可以使用 foreach 函数来转换 Items 属性,代码如下:

```python
def to_pandas(row):
    print(f'为类别{row["Category"]}创建一个 Pandas DataFrame')
    items = [item.asDict() for item in row["Items"]]
    df_pd_items = pd.DataFrame(items)
    print(df_pd_items)

#将每个类别的 Items 转换为 Pandas DataFrame
for row in df_agg.collect():
    print(to_pandas(row))
```

在上面的代码片段中,首先将 Row list 转换为字典列表,然后使用 pd.DataFrame 函数将该列表转换为 Pandas 的 DataFrame。因为 list 元素是字典对象,它有 key,所以不需要为 pd.DataFrame 函数指定 columns 参数。

执行以上代码,输出结果如图 4-43 所示。

```
为类别Category C创建一个Pandas DataFrame
     Category  ItemID  Amount
0  Category C       3  100.01
None
为类别Category B创建一个Pandas DataFrame
     Category  ItemID  Amount
0  Category B       2   30.10
1  Category B       5   70.85
None
为类别Category A创建一个Pandas DataFrame
     Category  ItemID  Amount
0  Category A       1   12.40
1  Category A       4  110.01
None
```

图 4-43 将 pyspark.sql.Row 列表转换为 Pandas DataFrame

4.10.2 PySpark SQL DataFrame 可视化

PySpark 还没有任何绘图功能。如果想将数据绘制成图，则可以将数据从 Spark 上下文中取出并放入"本地" Python 会话中，在那里可以使用 Python 的任意一个绘图库来处理它。

对于 PySpark SQL 的 DataFrame 绘图，用户可以先将它转换成 Pandas 的 DataFrame，再绘制。在下面的这个示例中，创建了一个包含 3 个正态分布数字列和一种颜色类别列的 DataFrame，然后将其转换为 Pandas 的 DataFrame，并进行可视化绘图展示，代码如下：

```python
#第4章/dataframe_demo18.ipynb

from pyspark.sql import SparkSession
import random

#构建SparkSession实例
spark = SparkSession.builder \
    .master("spark://xueai8:7077") \
    .appName("pyspark sql demo") \
    .getOrCreate()

#创建一个包含3个数字列和一个类别(颜色)的Spark DataFrame
A = [random.normalvariate(0,1) for i in range(100)]
B = [random.normalvariate(1,2) for i in range(100)]
C = [random.normalvariate(-1,0.5) for i in range(100)]
col = [random.choice(['#e41a1c', '#377eb8', '#4eae4b']) for i in range(100)]

df = spark.createDataFrame(zip(A,B,C,col), ["A","B","C","col"])
df.printSchema()
df.show(5)

from pandas.plotting import scatter_matrix
```

```
#转换为Pandas并绘图
pdf = df.toPandas()

stuff = scatter_matrix(pdf, alpha=0.7, figsize=(6, 6), diagonal='kde',
color=pdf.col)
```

执行以上代码,输出的 PySpark SQL DataFrame 内容如图 4-44 所示。

```
root
 |-- A: double (nullable = true)
 |-- B: double (nullable = true)
 |-- C: double (nullable = true)
 |-- col: string (nullable = true)

+-------------------+--------------------+--------------------+------+
|                  A|                   B|                   C|   col|
+-------------------+--------------------+--------------------+------+
|  1.384946838507259|   1.9723595337779216|  -1.7433008332887872|#4eae4b|
| -0.9812217969597337|   1.7046046382241848|  -1.113379241508142|#377eb8|
|  0.575954311201969| -0.10590577165060955|  -1.7451690616923745|#377eb8|
| -1.0115365029459253|  0.30510247214884667| -0.5342195326699025|#377eb8|
| -3.0459506903597946| -0.8299767662273825| -0.12795069365362288|#377eb8|
+-------------------+--------------------+--------------------+------+
only showing top 5 rows
```

图 4-44 构造的 PySpark DataFrame

转换为 Pandas DataFrame 后,绘制结果如图 4-45 所示。

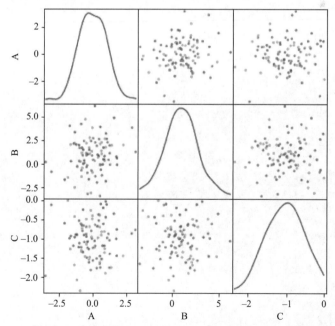

图 4-45 将 PySpark DataFrame 转换为 Pandas DataFrame 后可视化绘图

4.11 PySpark SQL 编程案例

本节通过对几个案例的学习,掌握使用 PySpark SQL 关系型 API 和 SQL 进行大数据分析的方法。

4.11.1 实现单词计数

到目前为止,已经了解到,在 PySpark 中对数据进行处理有以下 3 种方案:
(1) 使用 RDD。
(2) 使用 DataFrame 关系型 API。
(3) 使用 PySpark SQL 语句。

强烈建议不要直接使用 RDD,而是使用 DataFrame 的关系型 API 或 SQL 来对数据进行分析计算。只有这样,才能充分利用 PySpark SQL 的 Catalyst 优化器来对分析过程自动进行优化。

在下面这个示例中,使用 DataFrame 和 SQL 两种方式实现单词计数功能。

【例 4-3】统计某个英文文本中的词频,找出出现频次最高的 3 个单词。

实现过程和代码如下。

(1) 准备数据文件。自行创建一个纯文本文件 word.txt,并编辑内容如下:

```
good good study
day day up
```

将该文件上传到 HDFS 的/data/spark/目录下。

(2) 方法一:使用关系型 API 实现单词计数功能,代码如下:

```
#第 4 章/dataframe_demo19.ipynb

from pyspark.sql import SparkSession

#构建 SparkSession 实例
spark = SparkSession.builder \
    .master("spark://localhost:7077") \
    .appName("pyspark sql demo") \
    .getOrCreate()

#定义文件路径
filePath = "/data/spark/words.txt"
df = spark.read.text(filePath)

#df.printSchema()
#df.show()
```

```
#对DataFrame进行一系列处理,生成一个包含最终结果的DataFrame
wordDF = wordDF.rdd \
    .flatMap(lambda line: line.value.split(" ")) \
    .map(lambda word: (word, 1)) \
    .toDF(["word", "one"])

#wordDF.show()

#获得前3个出现频率最高的词
from pyspark.sql.functions import col
top3 = wordDF \
    .groupBy("word") \
    .count() \
    .orderBy(col("count").desc()) \
    .limit(3)

#输出结果
top3.show()
```

执行以上代码,输出结果如下:

```
+----+-----+
|word|count|
+----+-----+
| day|    2|
|good|    2|
|  up|    1|
+----+-----+
```

(3)方法二:使用 SQL 语句,代码如下:

```
#第4章/dataframe_demo20.ipynb

from pyspark.sql import SparkSession

#构建SparkSession实例
spark = SparkSession.builder \
    .master("spark://localhost:7077") \
    .appName("pyspark sql demo") \
    .getOrCreate()

#定义文件路径
filePath = "/data/spark/words.txt"
df = spark.read.text(filePath)

#df.printSchema()
#df.show()
```

```
#对DataFrame进行一系列处理,生成一个包含最终结果的DataFrame
wordDF = wordDF.rdd \
    .flatMap(lambda line: line.value.split(" ")) \
    .map(lambda word: (word, 1)) \
    .toDF(["word", "one"])

#wordDF.show()

#注册为临时view
wordDF.createOrReplaceTempView("wc_tb")

#执行SQL查询,分析结果
sql = """
        select word,count(1) as count
        from wc_tb
        group by word
        order by count desc
    """
resultDF = spark.sql(sql)
resultDF.limit(3).show()
```

执行以上代码,输出结果如下:

```
+----+-----+
|word|count|
+----+-----+
| day|    2|
|good|    2|
|  up|    1|
+----+-----+
```

4.11.2 用户数据集分析

PySpark SQL 支持通过编程接口自己构造模式,然后将其应用于现有的 RDD。虽然此方法更冗长,但它允许用户在列及其类型尚未知时(直到运行时才能知道)就可以构造数据集。

可以通过以下 3 个步骤以编程方式创建 DataFrame:

(1)从原始 RDD 创建一个 Row RDD。

(2)创建由 StructType 表示的模式,该模式与上一步创建的 RDD 中的 Row 结构相匹配。

(3)通过 SparkSession 提供的 createDataFrame 方法将模式应用到 Row RDD。

【例 4-4】下面这个示例用于分析 PySpark 安装包自带的 people.txt 文件的内容,但是以编程方式指定模式。首先将 people.txt 加载到一个 RDD 中,代码如下:

```
#第4章/dataframe_demo21.ipynb

from pyspark.sql import SparkSession
from pyspark.sql.types import *
from pyspark.sql.functions import *

#构建 SparkSession 实例
spark = SparkSession.builder \
   .master("spark://localhost:7077") \
   .appName("pyspark sql demo") \
   .getOrCreate()

#构造一个 RDD
file = "/data/spark/resources/people.txt"

#创建一个 RDD
peopleRDD = spark.sparkContext.textFile(file)

#指定一个 Schema(模式)
fields = [
    StructField("name", StringType(), True),
    StructField("age", IntegerType(), True)
]
schema = StructType(fields)

#将 RDD[String]转换为 RDD[(String, Int)]
rowRDD = peopleRDD \
     .map(lambda line: line.split(",")) \
     .map(lambda arr:(arr[0], int(arr[1])))

#将这个 schema 应用到该 RDD
peopleDF = spark.createDataFrame(rowRDD, schema)

#输出 schema 和内容
peopleDF.printSchema()
peopleDF.show()
```

执行以上代码，输出结果如下：

```
root
 |-- name: string (nullable = true)
 |-- age: integer (nullable = true)

+-------+---+
|   name|age|
+-------+---+
|Michael| 29|
```

```
|  Andy| 30|
|Justin| 19|
+------+---+
```

然后，注册一个临时视图，使用 SQL 对该视图进行查询，代码如下：

```
#使用该 DataFrame 创建一个临时视图
peopleDF.createOrReplaceTempView("people")

#SQL 可以在使用 DataFrame 创建的临时视图上运行
sqlStr = "SELECT name, age FROM people WHERE age BETWEEN 13 AND 19"
results = spark.sql(sqlStr)

results.show()
```

执行以上代码，输出内容如下：

```
+------------+
|       value|
+------------+
|Name: Justin|
+------------+
```

4.11.3 航空公司航班数据集分析

本案例使用美国交通部的一些航班信息，探索最易导致航班延误的航班属性。使用 PySpark DataFrame，将探索这些航班数据来回答以下问题（只考虑航班延误超过 40min 的情况）：

（1）哪家航空公司的航班延误次数最多？
（2）每周哪几天的航班延误次数最多？
（3）哪些始发机场的航班延误次数最多？
（4）每天什么时候的航班延误次数最多？

航班数据保存在 JSON 格式的 flights20170102.json 文件中，每个航班记录包含的信息见表 4-9。

表 4-9 航班记录各字段含义说明

属　性	含　义
id	ID，由承运人、日期、出发地、目的地、航班号组成
dofW	星期几（1 = Monday 星期一，7 = Sunday 星期日）
carrier	承运人代码
origin	起始机场代码
dest	目的地机场代码

属 性	含 义
crsdephour	规定起飞时间 hour（scheduled departure hour）
crsdeptime	规定起飞时间 time（scheduled departure time）
depdelay	起飞延误分钟数（departure delay in minutes）
crsarrtime	预定到达时间（scheduled arrival time）
arrdelay	到达延误分钟数（arrival delay minutes）
crselapsedtime	飞行时间
dist	距离（distance）

每条航班信息的格式如下：

```
{
    "id": "AA_2017-01-01_ATL_LGA_1678",
    "dofW": 7,
    "carrier": "AA",
    "origin": "ATL",
    "dest": "LGA",
    "crsdephour": 17,
    "crsdeptime": 1700,
    "depdelay": 0.0,
    "crsarrtime": 1912,
    "arrdelay": 0.0,
    "crselapsedtime": 132.0,
    "dist": 762.0
}
```

将 flights20170102.json 文件上传到 HDFS 的/data/spark/flightdelay/目录下，然后按以下步骤实现。

（1）首先导入本案例所需要的依赖包，并创建 SparkSession 实例，代码如下：

```
#第4章/dataframe_demo22.ipynb

from pyspark.sql import SparkSession

#构建 SparkSession 实例
spark = SparkSession.builder \
    .master("spark://localhost:7077") \
    .appName("pyspark sql demo") \
    .getOrCreate()
```

（2）定义 Schema，用于创建 DataFrame，代码如下：

```
#schema
schema = StructType([
```

```
        StructField("_id", StringType(), True),
        StructField("dofW", IntegerType(), True),
        StructField("carrier", StringType(), True),
        StructField("origin", StringType(), True),
        StructField("dest", StringType(), True),
        StructField("crsdephour", IntegerType(), True),
        StructField("crsdeptime", DoubleType(), True),
        StructField("depdelay", DoubleType(), True),
        StructField("crsarrtime", DoubleType(), True),
        StructField("arrdelay", DoubleType(), True),
        StructField("crselapsedtime", DoubleType(), True),
        StructField("dist", DoubleType(), True)
])
```

(3) 将数据集加载到 DataSet 中,应用上一步定义的 Schema,并进行简单探索,代码如下:

```
file = "/data/flightdelay/flights20170102.json"
df = spark.read \
    .option("inferSchema", "false") \
    .schema(schema) \
    .json(file)

print(f"数据集中包含{df.count()}条记录")
#查看前 5 条记录
df.show(5)
```

执行以上代码,输出内容如下:

```
数据集中包含 41348 条记录
+--------------------+----+-------+------+----+----------+----------+--------+----------+--------+--------------+-----+
|                 _id|dofW|carrier|origin|dest|crsdephour|crsdeptime|depdelay|crsarrtime|arrdelay|crselapsedtime| dist|
+--------------------+----+-------+------+----+----------+----------+--------+----------+--------+--------------+-----+
|ATL_BOS_2017-01-0...|   7|     DL|   ATL| BOS|        11|    1141.0|     0.0|    1409.0|     0.0|         148.0|946.0|
|ATL_BOS_2017-01-0...|   7|     WN|   ATL| BOS|        13|    1335.0|     0.0|    1600.0|     0.0|         145.0|946.0|
|ATL_BOS_2017-01-0...|   7|     DL|   ATL| BOS|        14|    1416.0|     0.0|    1644.0|     0.0|         148.0|946.0|
|ATL_BOS_2017-01-0...|   7|     DL|   ATL| BOS|        16|    1616.0|    15.0|    1849.0|     0.0|         153.0|946.0|
|ATL_BOS_2017-01-0...|   7|     WN|   ATL| BOS|        18|    1845.0|     0.0|    2110.0|     0.0|         145.0|946.0|
```

```
+--------------------+----+--------+------+----+----------+----------+-
--------+---------+--------+---------------+-----+
only showing top 5 rows
```

（4）查看上午 10 点起飞的航班有哪些，这里只取前三条，代码如下：

```
print("过滤上午 10 点起飞的航班。take 3")
df.where("crsdephour = 10").take(3)
```

执行以上代码，输出内容如下：

```
过滤上午 10 点起飞的航班。take 3
Flight(ATL_BOS_2017-01-02_10_DL_1202,1,DL,ATL,BOS,10,1004.0,10.0,1236.0,
3.0,152.0,946.0)
Flight(ATL_BOS_2017-01-03_10_DL_1202,2,DL,ATL,BOS,10,1004.0,0.0,1236.0,
0.0,152.0,946.0)
Flight(ATL_BOS_2017-01-03_10_WN_235,2,WN,ATL,BOS,10,1000.0,12.0,1230.0,
6.0,150.0,946.0)
```

（5）按承运人(carrier)分组统计数量，代码如下：

```
print("按承运人(carrier)分组统计：")
df.groupBy("carrier").count().show()
```

执行以上代码，输出内容如下：

```
按承运人(carrier)分组统计：
+-------+-----+
|carrier|count|
+-------+-----+
|     UA|18873|
|     AA|10031|
|     DL|10055|
|     WN| 2389|
+-------+-----+
```

（6）按目的地统计起飞延误超过 40min 的航班信息，并按照延误时间倒序排序，显示延误时间最长的前三趟航班，代码如下：

```
print("按目的地统计超过 40min 的起飞延误，并按延误时间倒序排序：")
df.filter("depdelay > 40") \
  .groupBy("dest") \
  .count() \
  .orderBy(col("count").desc()) \
  .limit(3) \
  .show()
```

执行以上代码，输出内容如下：

```
按目的地统计超过 40min 的起飞延误，并按延误时间倒序排序：
```

```
+----+-----+
|dest|count|
+----+-----+
| SFO|  711|
| EWR|  620|
| ORD|  593|
+----+-----+
```

(7) 接下来执行数据探索。首先以列格式在内存中缓存 DataFrame，并创建临时视图或缓存临时视图，代码如下：

```
#以列格式在内存中缓存 DataFrame
df.cache()

#创建临时表视图
df.createOrReplaceTempView("flights")

#以列格式在内存中缓存表
spark.catalog.cacheTable("flights")
```

(8) 回答问题：显示前 5 个延误时间最长的航班信息。

使用 DataFrame API，代码如下：

```
#print("最长的延误")

#使用 DataFrame Transformation
df.select("carrier", "origin", "dest", "depdelay", "crsdephour", "dist", "dofW") \
  .where("depdelay > 40") \
  .orderBy(col("depdelay").desc()) \
  .show(5)
```

也可以使用 SQL 语句，代码如下：

```
#使用 SQL
spark.sql(
    """
    select carrier,origin, dest, depdelay,crsdephour, dist, dofW
    from flights
    where depdelay > 40
    order by depdelay desc
    limit 5
    """).show()
```

两种方式的执行结果相同，输出内容如下：

```
+-------+------+----+--------+----------+
|carrier|origin|dest|depdelay|crsdephour|
```

```
+-------+----+----+------+---+
|     AA| SFO| ORD|1440.0|  8|
|     DL| BOS| ATL|1185.0| 17|
|     UA| DEN| EWR|1138.0| 12|
|     DL| ORD| ATL|1087.0| 19|
|     UA| MIA| EWR|1072.0| 20|
+-------+----+----+------+---+
```

（9）回答问题：显示承运人的平均起飞延误时间，代码如下：

```
from pyspark.sql.functions import avg

print("承运人的平均起飞延误")
df.groupBy("carrier").agg(avg("depdelay")).show()
```

执行以上代码，输出内容如下：

```
承运人的平均起飞延误
+-------+------------------+
|carrier|      avg(depdelay)|
+-------+------------------+
|     UA|17.477878450696764|
|     AA| 10.45768118831622|
|     DL|15.316061660865241|
|     WN|13.491000418585182|
+-------+------------------+
```

（10）回答问题：统计一周内每天的平均延误起飞时间，代码如下：

```
print("一周内每天的平均延误起飞时间")
spark.sql(
    """
    SELECT dofW, avg(depdelay) as avgdelay
    FROM flights
    GROUP BY dofW
    ORDER BY dofW, avgdelay desc
    """).show()
```

执行以上代码，输出内容如下：

```
一周内每天的平均延误起飞时间
+----+------------------+
|dofW|          avgdelay|
+----+------------------+
|   7|18.754202401372211|
|   1|17.404697121580984|
|   6|16.829883897021706|
|   4|14.609846568875401|
|   2| 13.50552611657835|
```

```
|   5| 13.101932367149761
|   3| 11.3646816543676 |
+----+------------------+
```

（11）回答问题：按承运人统计起飞延误（延误≥40min）次数。

使用 DataFrame API，代码如下：

```
print("按承运人统计起飞延误")
#使用 Transformation API
df.filter("depdelay > 40") \
    .groupBy("carrier") \
    .count() \
    .orderBy(col("count").desc()) \
    .show(5)
```

也可以使用 SQL 语句，代码如下：

```
#使用 SQL 语句
spark.sql(
    """
    select carrier, count(depdelay) as count
    from flights
    where depdelay > 40
    group by carrier
    order by count desc
    """).show(5)
```

两种方式的执行结果相同，输出内容如下：

```
按承运人统计起飞延误
+-------+-----+
|carrier|count|
+-------+-----+
|     UA| 2420|
|     DL| 1043|
|     AA|  757|
|     WN|  244|
+-------+-----+
```

（12）回答问题：按出发机场统计延误次数，代码如下：

```
print("如果按出发地机场延误超过 40min 进行统计，则延误起飞的次数是多少？")
spark.sql(
    """
    select origin, count(depdelay)
    from flights
    where depdelay > 40
    group by origin
```

```
        order by count(depdelay) desc
    """).show()
```

执行以上代码,输出内容如下:

```
如果按出发地机场延误超过 40min 进行统计,则延误起飞的次数是多少?
+------+-----+
|origin|count|
+------+-----+
|   ORD|  679|
|   ATL|  637|
|   SFO|  542|
|   EWR|  518|
|   DEN|  484|
|   IAH|  447|
|   LGA|  432|
|   MIA|  429|
|   BOS|  296|
+------+-----+
```

(13) 回答问题:按每周天数统计延误起飞次数。

使用 DataFrame API,代码如下:

```
print("按每周天数统计延误起飞次数,延误以超过 40min 进行统计")

#使用 transformation API
df.filter("depdelay > 40") \
  .groupBy("dofW") \
  .count() \
  .orderBy("dofW") \
  .show()
```

也可以使用 SQL 语句,代码如下:

```
#使用 SQL
spark.sql(
    """
        select dofW, count(depdelay) as count
        from flights
        where depdelay > 40
        group by dofW
        order by dofW
    """).show()
```

执行以上代码,输出结果相同,如下所示。

```
按每周天数统计延误起飞次数,延误超过 40min 进行统计
+----+-----+
|dofW|count|
```

```
+----+-----+
|   1|  940|
|   2|  712|
|   3|  482|
|   4|  626|
|   5|  579|
|   6|  424|
|   7|  701|
+----+-----+
```

（14）回答问题：按一天中的不同时段统计起飞延误次数，代码如下：

```
print("按小时统计起飞延误次数")
spark.sql(
    """
    select crsdephour, count(depdelay)
    from flights
    where depdelay > 40
    group by crsdephour
    order by crsdephour
    """).show(24)
```

执行以上代码，输出内容如下：

```
按小时统计起飞延误次数
+----------+-----+
|crsdephour|count|
+----------+-----+
|         0|    9|
|         1|    1|
|         5|   15|
|         6|   68|
|         7|  112|
|         8|  190|
|         9|  175|
|        10|  284|
|        11|  280|
|        12|  227|
|        13|  336|
|        14|  353|
|        15|  331|
|        16|  351|
|        17|  474|
|        18|  396|
|        19|  371|
|        20|  230|
|        21|  160|
|        22|   65|
```

```
|          23|   27|
|          24|    9|
+------------+-----+
```

（15）回答问题：按航线统计延误次数，代码如下：

```
print("按航线统计延误次数")
spark.sql(
    """
    select origin,dest,count(depdelay)
    from flights
    where depdelay > 40
    group by origin,dest
    order by count(depdelay) desc
    """).show()
```

执行以上代码，输出内容如下：

```
按航线统计延误次数
+------+----+-----+
|origin|dest|count|
+------+----+-----+
|   DEN| SFO|  172|
|   ORD| SFO|  168|
|   ATL| LGA|  155|
|   ATL| EWR|  141|
|   SFO| DEN|  134|
|   LGA| ATL|  130|
|   ORD| EWR|  122|
|   SFO| ORD|  118|
|   EWR| ORD|  115|
|   ORD| LGA|  100|
|   IAH| SFO|   98|
|   IAH| EWR|   94|
|   MIA| LGA|   92|
|   ORD| ATL|   88|
|   LGA| ORD|   88|
|   ATL| SFO|   87|
|   LGA| MIA|   81|
|   SFO| EWR|   79|
|   EWR| SFO|   77|
|   EWR| ATL|   76|
+------+----+-----+
only showing top 20 rows
```

第 5 章 PySpark SQL（高级）

CHAPTER 5

为了帮助执行复杂的分析，PySpark SQL 提供了一组强大而灵活的聚合函数、连接多个数据集的函数、一组内置的高性能函数和一组高级分析函数。

5.1 PySpark SQL 函数

PySpark DataFrame API 设计的目的是在数据集中操作或转换单行数据，如过滤或分组。如果想要转换一个数据集中某列的值，例如将字符串从大写字母转换成驼峰命名形式，就需要使用一个函数实现。PySpark SQL 内置了一组常用的函数，同时也提供了用户自定义新函数的简单方法。

为了有效地使用 PySpark SQL 执行分布式数据操作，必须熟练掌握 PySpark SQL 函数。PySpark SQL 提供了超过 200 个内置函数，它们被分组到不同的类别中。

按类别来分，SQL 函数可分为以下四类。

（1）标量函数：每行返回单个的值。

（2）聚合函数：每组行返回单个的值。

（3）窗口函数：每组行返回多个值。

（4）用户自定义函数（UDF）：包括自定义的标量函数或聚合函数。

标量函数和聚合函数都位于 pyspark.sql.functions 包内。在使用前，需要先导入它，代码如下：

```
import org.apache.spark.sql.functions._
```

如果使用 PySpark Shell 或 Zeppelin Notebook 进行交互式分析，则会自动导入该包。

3min

5.2 内置标量函数

PySpark SQL 提供了大量的标量函数，主要功能如下。

（1）数学计算：例如 abs()、hypot()、log()、cbrt()等。

（2）字符串操作：例如 length()、trim()、concat()等。

（3）日期操作：例如 year()、date_add()等。

下面详细来了解这些函数及其用法。

5.2.1 日期时间函数

PySpark 内置的日期时间函数大致可分为以下三个类别：

（1）执行日期时间格式转换的函数。

（2）执行日期时间计算的函数。

（3）从日期时间戳中提取特定值（如年、月、日等）的函数。

日期和时间转换函数有助于将字符串转换为日期、时间戳或 UNIX 时间戳，反之亦然。在内部，它使用 Java 日期格式模式语法。这些函数使用的默认的日期格式是 yyyy-mm-dd HH:mm:ss，因此，如果日期或时间戳列的日期格式不同，则需要向这些转换函数传入指定的模式。

1. 将字符串转换为日期或时间戳

例如，将字符串类型的日期和时间戳转换为 PySpark SQL 的 date 和 timestamp 类型，代码如下：

```
#第5章/dataframe2_demo01.ipynb

from pyspark.sql import SparkSession
from pyspark.sql.functions import *

spark = SparkSession.builder \
   .master("spark://xueai8:7077") \
   .appName("pyspark demo") \
   .getOrCreate()

#日期和时间转换函数：这些函数使用的默认的日期格式是 yyyy-mm-dd HH:mm:ss
#构造一个简单的 DataFrame，注意最后两列不遵循默认日期格式
testDate = [(1, "2019-01-01", "2019-01-01 15:04:58", "01-01-2019",
"12-05-2018 45:50")]
testDateTSDF = spark.createDataFrame(testDate,schema=["id", "date",
"timestamp", "date_str", "ts_str"])

#testDateTSDF.printSchema()
#testDateTSDF.show()

#将这些字符串转换为 date、timestamp 和 UNIX timestamp，并指定一个自定义的 date 和
#timestamp 格式

testDateResultDF = testDateTSDF.select(
    to_date('date').alias("date1"),
```

```
        to_timestamp('timestamp').alias("ts1"),
        to_date('date_str',"MM-dd-yyyy").alias("date2"),
        to_timestamp('ts_str',"MM-dd-yyyy mm:ss").alias("ts2"),
        unix_timestamp('timestamp').alias("unix_ts")
    )

testDateResultDF.printSchema()
testDateResultDF.show(truncate=False)
```

执行以上代码，输出结果如下：

```
root
 |-- date1: date (nullable = true)
 |-- ts1: timestamp (nullable = true)
 |-- date2: date (nullable = true)
 |-- ts2: timestamp (nullable = true)
 |-- unix_ts: long (nullable = true)


+----------+-------------------+----------+-------------------+----------+
|date1     |ts1                |date2     |ts2                |unix_ts   |
+----------+-------------------+----------+-------------------+----------+
|2019-01-01|2019-01-01 15:04:58|2019-01-01|2018-12-05 00:45:50|1546326298|
+----------+-------------------+----------+-------------------+----------+
```

2. 将日期或时间戳转换为字符串

将日期或时间戳转换为时间字符串是很容易的，方法是使用 date_format()函数和定制日期格式，或者使用 from_unixtime()函数将 UNIX 时间戳（以秒为单位）转换成字符串。参看日期和时间戳转换为格式字符串的转换例子，代码如下：

```
from pyspark.sql.functions import *

testDateResultDF.select(
    date_format('date1', "dd-MM-yyyy").alias("date_str"),
    date_format('ts1', "dd-MM-yyyy HH:mm:ss").alias("ts_str"),
    from_unixtime('unix_ts',"dd-MM-yyyy HH:mm:ss").alias("unix_ts_str")
).show()
```

执行以上代码，输出结果如下：

```
+----------+-------------------+-------------------+
| date_str |             ts_str|        unix_ts_str|
+----------+-------------------+-------------------+
|01-01-2019|01-01-2019 15:04:58|01-01-2019 15:04:58|
+----------+-------------------+-------------------+
```

3. 日期计算函数

日期-时间计算函数有助于计算两个日期或时间戳的相隔时间,以及执行日期或时间算术运算。关于日期-时间计算的示例，代码如下：

```
from pyspark.sql.functions import *

# 日期-时间（date-time）计算函数
data = [("黄渤", "2016-01-01", "2017-10-15"),
        ("王宝强", "2017-02-06", "2017-12-25")]
employeeData = spark.createDataFrame(data, schema=["name", "join_date", "leave_date"])

employeeData.show()
```

执行以上代码，输出结果如下：

```
+------+----------+----------+
|  name| join_date|leave_date|
+------+----------+----------+
|  黄渤|2016-01-01|2017-10-15|
|王宝强|2017-02-06|2017-12-25|
+------+----------+----------+
```

执行 date()和 month()计算，代码如下：

```
from pyspark.sql.functions import *

employeeData.select(
    'name',
    datediff('leave_date', 'join_date').alias("days"),
    months_between('leave_date', 'join_date').alias("months"),
    last_day('leave_date').alias("last_day_of_mon")
).show()
```

执行以上代码，输出结果如下：

```
+------+----+-----------+---------------+
|  name|days|     months|last_day_of_mon|
+------+----+-----------+---------------+
|  黄渤| 653| 21.4516129|     2017-10-31|
|王宝强| 322|10.61290323|     2017-12-31|
+------+----+-----------+---------------+
```

执行日期加、减计算，代码如下：

```
from pyspark.sql.functions import *

oneDate = spark.createDataFrame([("2019-01-01",)],schema=["new_year"])
oneDate.select(
    date_add('new_year', 14).alias("mid_month"),
    date_sub('new_year', 1).alias("new_year_eve"),
```

```
            next_day('new_year', "Mon").alias("next_mon")
).show()
```

执行上面的代码，输出结果如下：

```
+----------+------------+----------+
|mid_month |new_year_eve| next_mon|
+----------+------------+----------+
|2019-01-15|  2018-12-31|2019-01-07|
+----------+------------+----------+
```

4. 转换不规范的日期

如果采集到的数据是不受控制的，得到的日期可能是不规范的，这就需要将这些不规范的日期转换为规范的日期。对不规范日期的转换，代码如下：

```
from pyspark.sql.functions import *

#转换不规范的日期
data = [("Nov 05, 2018 02:46:47 AM",),("Nov 5, 2018 02:46:47 PM",)]
df = spark.createDataFrame(data,schema=["times"])

df.withColumn(
    "times2",
    from_unixtime(
      unix_timestamp("times", "MMM d, yyyy hh:mm:ss a"),
      "yyyy-MM-dd HH:mm:ss.SSSSSS"
    )
).show(truncate=False)
```

执行以上代码，输出结果如下：

```
+------------------------+--------------------------+
|times                   |times2                    |
+------------------------+--------------------------+
|Nov 05, 2018 02:46:47 AM|2018-11-05 02:46:47.000000|
|Nov 5, 2018 02:46:47 PM |2018-11-05 14:46:47.000000|
+------------------------+--------------------------+
```

5. 处理时间序列数据

在处理时间序列数据（Time-Series Data）时，经常需要提取日期或时间戳值的特定字段（如年、月、日、时、分和秒）。例如，当需要按季、月或周对所有股票交易进行分组时，就可以从交易日期提取该信息，并按这些值分组。从日期或时间戳中提取字段，代码如下：

```
from pyspark.sql.functions import *

#提取日期或时间戳值的特定字段（如年、月、日、时、分和秒）
```

```
#从一个日期值中提取指定的日期字段
valentimeDateDF = spark \
    .createDataFrame([("2019-02-14 13:14:52",)],["date"])

valentimeDateDF.select(
    year('date').alias("year"),              #年
    quarter('date').alias("quarter"),        #季
    month('date').alias("month"),            #月
    weekofyear('date').alias("woy"),         #周
    dayofmonth('date').alias("dom"),         #日
    dayofyear('date').alias("doy"),          #天
    hour('date').alias("hour"),              #时
    minute('date').alias("minute"),          #分
    second('date').alias("second")           #秒
).show()
```

执行以上代码，输出结果如下：

```
+----+-------+-----+---+---+---+----+------+------+
|year|quarter|month|woy|dom|doy|hour|minute|second|
+----+-------+-----+---+---+---+----+------+------+
|2019|      1|    2|  7| 14| 45|  13|    14|    52|
+----+-------+-----+---+---+---+----+------+------+
```

5.2.2 字符串函数

毫无疑问，数据集的大多数列是字符串类型的。PySpark SQL 内置的字符串函数提供了操作字符串类型列的通用和强大的方法。一般来讲，这些函数分为以下两类。

（1）执行字符串转换的函数。

（2）执行字符串提取（或替换）的函数，使用正则表达式。

1. 字符串去空格

最常见的字符串转换包括去空格、填充、大写转换、小写转换和字符串连接等。接下来演示使用各种内置字符串函数转换字符串的方法，代码如下：

```
#第5章/dataframe2_demo02.ipynb

from pyspark.sql import SparkSession
from pyspark.sql.functions import *

spark = SparkSession.builder \
    .master("spark://xueai8:7077") \
    .appName("pyspark demo") \
    .getOrCreate()

#使用各种内置字符串函数转换字符串的方法
```

```
sparkDF = spark.createDataFrame([(" Spark ",)], ["name"])

#去空格
sparkDF.select(
    trim("name").alias("trim"),       #去掉"name"列两侧的空格
    ltrim("name").alias("ltrim"),     #去掉"name"列左侧的空格
    rtrim("name").alias("rtrim")      #去掉"name"列右侧的空格
    ).show()
```

执行上面的代码，输出结果如下：

```
+-----+------+------+
| trim| ltrim| rtrim|
+-----+------+------+
|Spark|Spark | Spark|
+-----+------+------+
```

2. 字符串填充

用给定的填充字符串将字符串填充到指定长度，代码如下：

```
from pyspark.sql.functions import *

#首先去掉"Spark"前后的空格，然后填充到 8 个字符
sparkDF \
    .select(trim("name").alias("trim")) \
    .select(
        lpad("trim", 8, "-").alias("lpad"),    #宽度为 8，若不够，则左侧填充"-"
        rpad("trim", 8, "=").alias("rpad")     #宽度为 8，若不够，则右侧填充"="
    ) \
    .show()
```

执行以上代码，输出结果如下：

```
+--------+--------+
|    lpad|    rpad|
+--------+--------+
|---Spark|Spark===|
+--------+--------+
```

3. 字符串转换

字符串之间可以使用转换函数进行多种转换，代码如下：

```
from pyspark.sql import SparkSession
from pyspark.sql.functions import *

spark = SparkSession.builder \
    .master("spark://xueai8:7077") \
    .appName("pyspark demo") \
```

```
    .getOrCreate()
#使用concatenation、uppercase、lowercase 和 reverse 转换一个字符串
sentenceDF  =  spark.createDataFrame([("Spark", "is", "excellent")],
["subject", "verb", "adj"])
sentenceDF \
    .select(concat_ws(" ","subject", "verb", "adj").alias("sentence")) \
    .select(
      lower("sentence").alias("lower"),            #转小写
      upper("sentence").alias("upper"),            #转大写
      initcap("sentence").alias("initcap"),        #首字母转大写
      reverse("sentence").alias("reverse")         #反转
    ) \
    .show()

#从一个字符转换到另一个字符
sentenceDF \
    .select("subject", translate("subject", "pr", "oc").alias("translate")) \
    .show()
```

执行以上代码，输出结果如下：

```
+-----------------+-----------------+-----------------+-----------------+
|            lower|            upper|          initcap|          reverse|
+-----------------+-----------------+-----------------+-----------------+
|spark is excellent|SPARK IS EXCELLENT|Spark Is Excellent|tnellecxe si krapS|
+-----------------+-----------------+-----------------+-----------------+

+-------+---------+
|subject|translate|
+-------+---------+
|  Spark|    Soack|
+-------+---------+
```

4. 字符串提取

如果想从字符串列值中替换或提取子字符串，则可以使用 regexp_extract()和 regexp_replace()函数。这两个函数通过传入的正则表达式模式实现替换或提取。

如果要提取字符串的某一部分（子字符串），则可使用 regexp_extract()函数。regexp_extract()函数的输入参数是字符串列、匹配的模式和组索引。在字符串中可能会有多个匹配模式，因此，需要组索引（从 0 开始）来确定是哪一个。如果没有指定模式的匹配，则该函数返回空字符串。使用 regexp_extract 函数()，代码如下：

```
from pyspark.sql.functions import *

#使用 regexp_extract 字符串函数来提取"fox"，使用一个正则模式进行匹配
strDF = spark.createDataFrame([("A fox saw a crow sitting on a tree singing
```

```
\"Caw! Caw! Caw!\"",)], ["comment"])
#使用一个正则模式进行匹配
strDF.select(regexp_extract("comment",
"[a-z]*o[xw]",0).alias("substring")).show()
```

执行以上代码,输出结果如下:

```
+---------+
|substring|
+---------+
|      fox|
+---------+
```

5. 字符串替换

如果要替换字符串的某一部分（子字符串），则可使用 regexp_replace()函数。regexp_replace()函数的输入参数是字符串列、匹配的模式及替换的值。使用 regexp_replace()函数，代码如下:

```
from pyspark.sql.functions import *

#用regexp_replace()函数将fox和Caw替换为animal
strDF = spark.createDataFrame([("A fox saw a crow sitting on a tree singing
\"Caw! Caw! Caw!\"",)], ["comment"])

#下面两行代码会生成相同的输出
strDF.select(regexp_replace("comment","fox|crow","animal").alias("new_
comment")).show(truncate=False)
strDF.select(regexp_replace("comment","[a-z]*o[xw]","animal").alias("n
ew_comment")).show(truncate=False)
```

执行以上代码,输出结果如下:

```
+-------------------------------------------------------------+
|new_comm                                                     |
+-------------------------------------------------------------+
|A animal saw a animal sitting on a tree singing "Caw! Caw! Caw!"|
+-------------------------------------------------------------+

+-------------------------------------------------------------+
|new_comm                                                     |
+-------------------------------------------------------------+
|A animal saw a animal sitting on a tree singing "Caw! Caw! Caw!"|
+-------------------------------------------------------------+
```

例如，使用 regexp_replace()函数从混乱的数据中抽取出手机号，代码如下:

```
telDF = spark.createDataFrame([("135a-123b4-c5678",)], ["tel"])
tclDF.withColumn("phone",regexp_replace('tel',"-|\\D","")).show()
```

执行以上代码，输出结果如下：

```
+---------------+-----------+
|            tel|      phone|
+---------------+-----------+
|135a-123b4-c5678|13512345678|
+---------------+-----------+
```

5.2.3 数学计算函数

PySpark SQL 还提供了许多对数值类型列进行计算的函数，其中最常用的是 round() 函数，它对传入的列值采用四舍五入法进行计算。这个函数的定义如下：

```
pyspark.sql.functions.round(col, scale=0)
```

使用 round() 函数对不同格式的数值采用四舍五入法进行计算，代码如下：

```
#第5章/dataframe2_demo03.ipynb

from pyspark.sql import SparkSession
from pyspark.sql.functions import *

spark = SparkSession.builder \
    .master("spark://localhost:7077") \
    .appName("pyspark demo") \
    .getOrCreate()

numberDF = spark.createDataFrame([(3.14159, -3.14159)], ["pie", "-pie"])
numberDF \
    .select(
        "pie",
        round("pie").alias("pie0"),              #整数四舍五入
        round("pie", 2).alias("pie1"),           #四舍五入，保留小数点后 2 位
        round("pie", 4).alias("pie2"),           #四舍五入，保留小数点后 4 位
        "-pie",
        round("-pie").alias("-pie0"),            #整数四舍五入
        round("-pie", 2).alias("-pie1"),         #四舍五入，保留小数点后 2 位
        round("-pie", 4).alias("-pie2")          #四舍五入，保留小数点后 4 位
    ) \
    .show()
```

执行以上代码，输出结果如下：

```
+-------+----+----+------+--------+-----+-----+-------+
|    pie|pie0|pie1|   pie2|    -pie|-pie0|-pie1|  -pie2|
+-------+----+----+------+--------+-----+-----+-------+
|3.14159| 3.0|3.14|3.1416|-3.14159| -3.0|-3.14|-3.1416|
+-------+----+----+------+--------+-----+-----+-------+
```

还有两个数学函数也经常用到，分别是 ceil()和 floor()，它们的方法定义如下：

```
pyspark.sql.functions.ceil(col)
pyspark.sql.functions.floor(col)
```

关于这两个函数的使用，代码如下：

```
from pyspark.sql.functions import *

numberDF = spark.createDataFrame([(3.14159, -3.14159)], ["v1", "v2"])
numberDF \
    .select(
      "v1",
      ceil("v1"),                    #向上取整
      floor("v1"),                   #向下取整
      "v2",
      ceil("v2"),                    #向上取整
      floor("v2")                    #向下取整
    ) \
    .show()
```

执行以上代码，输出结果如下：

```
+-------+-------+--------+--------+--------+--------+
|     v1|CEIL(v1)|FLOOR(v1)|      v2|CEIL(v2)|FLOOR(v2)|
+-------+-------+--------+--------+--------+--------+
|3.14159|      4|       3|-3.14159|      -3|      -4|
+-------+-------+--------+--------+--------+--------+
```

5.2.4　集合元素处理函数

集合被设计用来处理复杂的数据类型，如 arrays、maps 和 struts。本节将介绍两种特定类型的集合函数。第一种使用 array 数据类型，第二种将数据处理为 JSON 数据格式。

1. 数组处理函数

PySpark DataFrame 支持复杂数据类型，也就是列值可以是一个集合。可以使用与数组相关的集合函数来轻松获取数组大小、检查值的存在或者对数组进行排序。下面的示例包含了处理各种数组的相关函数的用法，代码如下：

```
#第5章/dataframe2_demo04.ipynb

from pyspark.sql import SparkSession

spark = SparkSession.builder \
    .master("spark://localhost:7077") \
```

```
    .appName("pyspark demo") \
    .getOrCreate()

#创建一个任务集DataFrame
tasksDF = spark.createDataFrame([("星期天", ["打牌", "下棋", "去游泳"])],
["day", "tasks"])

#tasksDF的schema
tasksDF.printSchema()
tasksDF.show()
```

执行以上代码，输出结果如下：

```
root
 |-- day: string (nullable = true)
 |-- tasks: array (nullable = true)
 |    |-- element: string (containsNull = true)

+------+--------------------+
|day   |        tasks       |
+------+--------------------+
|星期天 |[打牌, 下棋, 去游泳]  |
+------+--------------------+
```

接下来获得该数组的大小，对其进行排序，并检查在该数组中是否存在一个指定的值，代码如下：

```
tasksDF \
    .select(
      "day",
      size("tasks").alias("size"),                              #数组大小
      sort_array("tasks").alias("sorted_tasks"),                #对数组排序
      array_contains("tasks", "去游泳").alias("是否去游泳")      #是否包含
    ) \
    .show(truncate=False)
```

执行以上代码，输出结果如下：

```
+------+----+--------------------+----------+
|day   |size|sorted_tasks        |是否去游泳 |
+------+----+--------------------+----------+
|星期天 |3   |[去游泳, 下棋, 打牌] |true      |
+------+----+--------------------+----------+
```

使用explode()表函数将为数组中的每个元素创建一个新行，代码如下：

```
tasksDF.select("day", explode("tasks").alias("task")).show()
```

执行以上代码，输出结果如下：

```
+------+------+
|   day|  task|
+------+------+
| 星期天|   打牌|
| 星期天|   下棋|
| 星期天| 去游泳|
+------+------+
```

2. JSON 处理函数

许多非结构化数据集都是以 JSON 的形式存在的。对于 JSON 数据类型的列，使用相关的集合函数将 JSON 字符串转换成 struct（结构体）数据类型。主要的函数是 from_json()、get_json_object()和 to_json()。一旦 JSON 字符串被转换为 PySpark struct 数据类型，就可以轻松地提取这些值。下面的代码演示了 from_json()和 to_json()函数的用法。

首先构造一个带有 JSON 字符串内容的 DataFrame，代码如下：

```
#第5章/dataframe2_demo05.ipynb

from pyspark.sql import SparkSession

spark = SparkSession.builder \
    .master("spark://localhost:7077") \
    .appName("pyspark demo") \
    .getOrCreate()

#创建一个字符串，它包含JSON格式的字符串内容
todos = """{"day": "星期天","tasks": ["打牌", "下棋", "去游泳"]}"""
todoStrDF = spark.createDataFrame([(todos,)], ["todos_str"])

#查看Schema和内容
todoStrDF.printSchema()
todoStrDF.show(truncate=False)
```

执行以上代码，输出结果如下：

```
root
 |-- todos_str: string (nullable = true)

+----------------------------------------------------+
|todos_str                                           |
+----------------------------------------------------+
|{"day": "星期天","tasks": ["打牌", "下棋", "去游泳"]}|
+----------------------------------------------------+
```

为了将一个 JSON 字符串转换为一个 PySpark 结构体数据类型，需要将其结构描述给 PySpark，为此需要定义一个 Schema 模式，并在 from_json()函数中应用，代码如下：

```
#第5章/dataframe2_demo06.ipynb
from pyspark.sql.types import *

todoSchema = StructType([
    StructField("day", StringType(), True),
    StructField("tasks", ArrayType(StringType()), True)
])

#使用 from_json 来转换 JSON 字符串
todosDF = todoStrDF \
    .select(from_json("todos_str",todoSchema).alias("todos"))

#todos 是一个 struct 数据类型,包含两个字段: day 和 tasks
todosDF.printSchema()
todosDF.show()
```

执行以上代码,输出结果如下:

```
root
 |-- todos: struct (nullable = true)
 |    |-- day: string (nullable = true)
 |    |-- tasks: array (nullable = true)
 |    |    |-- element: string (containsNull = true)


+-----------------------------+
|                        todos|
+-----------------------------+
|   {星期天, [打牌, 下棋, 去游泳]}|
+-----------------------------+
```

可以使用 Column 类的 getItem()函数检索出结构体数据类型的值,代码如下:

```
todosDF \
    .select(
      col("todos").getItem("day"),
      col("todos").getItem("tasks"),
      col("todos").getItem("tasks")[0].alias("first_task")
    ) \
    .show(truncate=False)
```

执行以上代码,输出结果如下:

```
+---------+---------------------+----------+
|todos.day|todos.tasks          |first_task|
+---------+---------------------+----------+
|星期天    |[打牌, 下棋, 去游泳]    |打牌       |
+---------+---------------------+----------+
```

也可以使用 to_json() 函数将一个 PySpark 结构体数据类型转换为 JSON 格式字符串，代码如下：

```
todosDF.select(to_json("todos")).show(truncate=False)
```

执行以上代码，输出结果如下：

```
+-------------------------------------------------+
|to_json(todos)                                   |
+-------------------------------------------------+
|{"day":"星期天","tasks":["打牌","下棋","去游泳"]}|
+-------------------------------------------------+
```

5.2.5 其他函数

除了前面几节介绍的函数外，还有一些函数在特定的场景下非常有用。本节将介绍以下几个函数：monotonically_increasing_id()、when()、coalesce()和 lit()。

1. monotonically_increasing_id() 函数

有时需要为数据集中的每行生成单调递增的唯一但不一定是连续的 id。例如，如果一个 DataFrame 有 2 亿行，并且是分区存储的，则如何确保这些 id 值是唯一的并且同时增加呢？PySpark SQL 提供了一个 monotonically_increasing_id()函数，它生成 64 位整数作为 id 值。使用 monotonically_increasing_id()函数，代码如下：

```
#第5章/dataframe2_demo07.ipynb

from pyspark.sql import SparkSession
from pyspark.sql.functions import *

spark = SparkSession.builder \
    .master("spark://localhost:7077") \
    .appName("pyspark demo") \
    .getOrCreate()

#首先构造一个 DataFrame，将它的值分散到5个分区中
numDF = spark.range(1,11,1,5)

#验证的确有5个分区
print("分区数为", numDF.rdd.getNumPartitions())

#现在生成单调递增的值，并查看所在的分区
numDF.select(
    "id",
    monotonically_increasing_id().alias("m_ii"),
    spark_partition_id().alias("partition")
).show()
```

执行上面的代码,输出结果如下:

```
分区数为5
+---+-----------+---------+
| id|       m_ii|partition|
+---+-----------+---------+
|  1|          0|        0|
|  2|          1|        0|
|  3| 8589934592|        1|
|  4| 8589934593|        1|
|  5|17179869184|        2|
|  6|17179869185|        2|
|  7|25769803776|        3|
|  8|25769803777|        3|
|  9|34359738368|        4|
| 10|34359738369|        4|
+---+-----------+---------+
```

2. when() 函数

在 DataFrame 中,如果需要根据条件列表来评估一个值并返回一个值,可以使用 when() 函数。例如,使用 when() 函数将数值转换成字符串,代码如下:

```
#第5章/dataframe2_demo08.ipynb

from pyspark.sql import SparkSession
from pyspark.sql.functions import *

spark = SparkSession.builder \
    .master("spark://localhost:7077") \
    .appName("pyspark demo") \
    .getOrCreate()

#创建一个具有从1到7的值的DataFrame表示一周中的每一天
dayOfWeekDF = spark.range(1,8)

#将每个数值转换为字符串
dayOfWeekDF.select(
     "id",
     when(col("id") == 1, "星期一")
       .when(col("id") == 2, "星期二")
       .when(col("id") == 3, "星期三")
       .when(col("id") == 4, "星期四")
       .when(col("id") == 5, "星期五")
       .when(col("id") == 6, "星期六")
       .when(col("id") == 7, "星期日")
       .alias("星期")
).show()
```

执行以上代码,输出结果如下:

```
+---+------+
| id|  星期|
+---+------+
|  1|星期一|
|  2|星期二|
|  3|星期三|
|  4|星期四|
|  5|星期五|
|  6|星期六|
|  7|星期七|
+---+------+
```

处理默认情况时,可以使用 Column 类的 otherwise()函数,代码如下:

```
#第5章/dataframe2_demo09.ipynb

from pyspark.sql import SparkSession
from pyspark.sql.functions import *

spark = SparkSession.builder \
    .master("spark://localhost:7077") \
    .appName("pyspark demo") \
    .getOrCreate()

#创建一个具有从1到7的值的DataFrame 表示一周中的每一天
dayOfWeekDF = spark.range(1,8)

#将每个数值转换为字符串
dayOfWeekDF.select(
    "id",
    when(col("id") == 6, "周末")
      .when(col("id") == 7, "周末")
      .otherwise("工作日")
      .alias("day_type")
).show()
```

执行以上代码,输出结果如下:

```
+---+--------+
| id|day_type|
+---+--------+
|  1|  工作日|
|  2|  工作日|
|  3|  工作日|
|  4|  工作日|
|  5|  工作日|
```

```
|  6|    周末|
|  7|    周末|
+---+--------+
```

3. coalesce() 函数和 lit() 函数

在处理数据时，正确处理 null 值（Python 中表示为 None）是很重要的。PySpark SQL 提供了一个名为 coalesce()的函数，该函数接受一个或多个列值，并返回第 1 个非空值，而 coalesce()函数中的每个参数都必须是 Column 类型，所以如果想传入字面量值，则需要使用 lit()函数，将字面量值包装为 Column 类的实例。例如，构造一个包含 None 值的 DataFrame，使用 coalesce()函数进行处理，并使用 lit()函数将字符串值转换为 Column 类型，代码如下：

```
#第5章/dataframe2_demo10.ipynb

from pyspark.sql import SparkSession
from pyspark.sql.functions import *

spark = SparkSession.builder \
    .master("spark://localhost:7077") \
    .appName("pyspark demo") \
    .getOrCreate()

#构造一个 DataFrame，带有 null 值
data = [(None, None, 2018),("黄渤", "一出好戏", 2018)]
badMoviesDF = spark.createDataFrame(data,
schema=["actor_name","movie_title","produced_year"])

badMoviesDF.show()
#使用 coalese()函数来处理 null 值
badMoviesDF \
    .select(
      coalesce("actor_name", lit("路人甲")).alias("演员"),
      coalesce("movie_title", lit("烂片")).alias("电影"),
      coalesce("produced_year", lit("烂片")).alias("年份")
    ) \
    .show()
```

执行以上代码，输出结果如下：

```
+----------+-----------+-------------+
|actor_name|movie_title|produced_year|
+----------+-----------+-------------+
|      null|       null|         2018|
|      黄渤|   一出好戏|         2018|
+----------+-----------+-------------+
```

```
+------+--------+----+
|  演员 |   电影  |年份 |
+------+--------+----+
| 路人甲 |   烂片  |2018|
| 黄渤  | 一出好戏 |2018|
+------+--------+----+
```

5.2.6 函数应用示例

前面几节了解了 PySpark SQL 常用的一些内置函数。本节通过一个示例演示如何使用其中一些函数实现 PySpark DataFrame 二次排序。

【例 5-1】应用 PySpark DataFrame API 实现数据的二次排序。

假设在 HDFS 的/data/spark/路径下有一个文件 data.txt，内容如下：

```
2018,5,22
2019,1,24
2018,2,128
2019,3,56
2019,1,3
2019,2,-43
2019,4,5
2019,3,46
2018,2,64
2019,1,4
2019,1,21
2019,2,35
2019,2,0
```

其中每行由逗号分隔的分别是年、月和总数。现在想要对这些数据排序，期望的输出结果是先按年、月进行排序，年、月相同的情况下，数值列表按大小排序，排序后的列表如下：

```
2018-2    64,128
2018-5    22
2019-1    3,4,21,24
2019-2    -43,0,35
2019-3    46,56
2019-4    5
```

建议按以下步骤实现。

（1）加载数据集，代码如下：

```
#第 5 章/dataframe2_demo11.ipynb

from pyspark.sql import SparkSession
from pyspark.sql.functions import *
```

```
spark = SparkSession.builder \
    .master("spark://localhost:7077") \
    .appName("pyspark demo") \
    .getOrCreate()

#加载数据集
inputPath = "/data/spark/data.txt"
inputDF = spark.read \
            .option("inferSchema","true") \
            .option("header","false") \
            .csv(inputPath) \
            .toDF("year","month","cnt")

inputDF.show()
```

执行以上代码,输出结果如下:

```
+----+-----+---+
|year|month|cnt|
+----+-----+---+
|2018|    5| 22|
|2019|    1| 24|
|2018|    2|128|
|2019|    3| 56|
|2019|    1|  3|
|2019|    2|-43|
|2019|    4|  5|
|2019|    3| 46|
|2018|    2| 64|
|2019|    1|  4|
|2019|    1| 21|
|2019|    2| 35|
|2019|    2|  0|
+----+-----+---+
```

(2) 将 year 和 month 组合为一列,并取别名为 ym,代码如下:

```
from pyspark.sql.functions import *

df2 = inputDF.select(concat_ws("-","year","month").alias("ym"), "cnt")
df2.printSchema()
df2.show()
```

执行以上代码,输出结果如下:

```
root
 |-- ym: string (nullable = false)
 |-- cnt: integer (nullable = true)
```

```
+------+---+
|    ym|cnt|
+------+---+
|2018-5| 22|
|2019-1| 24|
|2018-2|128|
|2019-3| 56|
|2019-1|  3|
|2019-2|-43|
|2019-4|  5|
|2019-3| 46|
|2018-2| 64|
|2019-1|  4|
|2019-1| 21|
|2019-2| 35|
|2019-2|  0|
+------+---+
```

(3) 先按 ym 进行分组聚合，然后对每组的 cnt 列进行排序，并输出，代码如下：

```
df2.groupBy("ym") \
   .agg(sort_array(collect_list("cnt")).alias("cnt")) \
   .orderBy("ym") \
   .show()
```

执行以上代码，输出结果如下：

```
+------+---------------+
|    ym|            cnt|
+------+---------------+
|2018-2|      [64, 128]|
|2018-5|           [22]|
|2019-1|  [3, 4, 21, 24]|
|2019-2|    [-43, 0, 35]|
|2019-3|       [46, 56]|
|2019-4|            [5]|
+------+---------------+
```

(4) 最后，可以把上面的代码写到一个 ETL 处理逻辑中，代码如下：

```
from pyspark.sql.functions import *

inputPath = "/data/spark/data.txt"

spark.read \
   .option("inferSchema","true") \
   .option("header","false") \
   .csv(inputPath) \
```

```
    .toDF("year","month","cnt") \
    .select(concat_ws("-","year","month").alias("ym"),"cnt") \
    .groupBy("ym") \
    .agg(sort_array(collect_list("cnt")).alias("cnt")) \
    .orderBy("ym") \
    .show()
```

5.2.7　PySpark 3 数组函数

PySpark 3 增加了一些新的数组函数，其中的 transform()和 aggregate()数组函数是功能特别强大的通用函数。它们提供的功能相当于 Python 中的 map()和 fold()，使 ArrayType 列更容易处理。

1）pyspark.sql.functions.exists(col, f)

如果函数对数组中的任何值有返回值 true，则 exists()函数就返回值 true。下面创建一个 DataFrame，然后运行 pyspark.sql.functions.exists()函数，用于将一个 even_best_number_exists 列附加到 DataFrame 上。

首先，构造一个 DataFrame，代码如下：

```
#第5章/dataframe2_demo12.ipynb

from pyspark.sql import SparkSession
from pyspark.sql.functions import *

spark = SparkSession.builder \
    .master("spark://localhost:7077") \
    .appName("pyspark demo") \
    .getOrCreate()

#数据集
data = [
    ("a", [3, 4, 5]),
    ("b", [8, 12]),
    ("c", [7, 13]),
    ("d", None),
]

#列名
columns = ("person_id","best_numbers")

#创建 DataFrame
df = spark.createDataFrame(data, columns)

#查看
df.printSchema()
df.show()
```

执行以上代码，输出结果如下：

```
root
 |-- person_id: string (nullable = true)
 |-- best_numbers: array (nullable = true)
 |    |-- element: integer (containsNull = false)

+---------+------------+
|person_id|best_numbers|
+---------+------------+
|        a|   [3, 4, 5]|
|        b|     [8, 12]|
|        c|     [7, 13]|
|        d|        null|
+---------+------------+
```

然后，判断 best_numbers 列中包含偶数元素的记录，代码如下：

```
#增加一个新列，判断'best_numbers'列是否包含偶数值
resDF2 = df.withColumn("even_exists", exists("best_numbers", lambda x: x % 2 == 0))
resDF2.show()
```

执行以上代码，输出结果如下：

```
+---------+------------+-----------+
|person_id|best_numbers|even_exists|
+---------+------------+-----------+
|        a|   [3, 4, 5]|       true|
|        b|     [8, 12]|       true|
|        c|     [7, 13]|      false|
|        d|        null|       null|
+---------+------------+-----------+
```

2）pyspark.sql.functions.forall(col, f)

该函数用来遍历数组中的每个元素。下面的代码演示了该函数的用法。首先构造一个 DataFrame，代码如下：

```
#第5章/dataframe2_demo13.ipynb

from pyspark.sql import SparkSession
from pyspark.sql.functions import *

spark = SparkSession.builder \
    .master("spark://localhost:7077") \
    .appName("pyspark demo") \
    .getOrCreate()

#数据集
```

```
data = [
    (["ants", "are", "animals"],),
    (["italy", "as", "interesting"],),
    (["brazilians", "love", "soccer"],),
    (None,)
]

#创建 DataFrame
df = spark.createDataFrame(data, schema=["words"])

df.printSchema()
df.show(truncate=False)
```

执行以上代码，输出结果如下：

```
root
 |-- words: array (nullable = true)
 |    |-- element: string (containsNull = true)

+--------------------------+
|words                     |
+--------------------------+
|[ants, are, animals]      |
|[italy, is, interesting]  |
|[brazilians, love, soccer]|
|null                      |
+--------------------------+
```

然后用 forall() 来标识所有单词以字母 a 开头的数组，代码如下：

```
resDF = df.withColumn("start_with_a", forall("words", lambda x: x.startswith("a") ))
resDF.show(truncate=False)
```

执行以上代码，输出结果如下：

```
+--------------------------+------------+
|words                     |start_with_a|
+--------------------------+------------+
|[ants, are, animals]      |true        |
|[italy, is, interesting]  |false       |
|[brazilians, love, soccer]|false       |
|null                      |null        |
+--------------------------+------------+
```

3）pyspark.sql.functions.filter(col, f)

该函数用于过滤数组中的每个元素。下面的代码演示了该函数的用法。首先构造一个 DataFrame，代码如下：

```
#第5章/dataframe2_demo14.ipynb

from pyspark.sql import SparkSession
from pyspark.sql.functions import *

spark = SparkSession.builder \
    .master("spark://localhost:7077") \
    .appName("pyspark demo") \
    .getOrCreate()

#创建 DataFrame
df = spark.createDataFrame(
    [
        (["bad", "bunny", "is", "funny"],),
        (["food", "is", "bad", "tasty"],),
        (None,),
    ],
    ["words"]
)

df.printSchema()
df.show(truncate=False)
```

执行以上代码，输出结果如下：

```
root
 |-- words: array (nullable = true)
 |    |-- element: string (containsNull = true)

+----------------------+
|words                 |
+----------------------+
|[bad, bunny, is, funny]|
|[food, is, bad, tasty] |
|null                  |
+----------------------+
```

然后过滤掉数组中 bad 这个单词，代码如下：

```
resDF = df.withColumn("filtered_words", filter("words", lambda x: x != "bad"))
resDF.show(truncate=False)
```

执行以上代码，输出结果如下：

```
+----------------------+------------------+
|words                 |filtered_words    |
+----------------------+------------------+
|[bad, bunny, is, funny]|[bunny, is, funny]|
```

```
|[food, is, bad, tasty] |[food, is, tasty]  |
|null                   |null               |
+-----------------------+-------------------+
```

在下面这个示例中,只保存月份大于 6 的数组元素值,代码如下:

```
df = spark.createDataFrame(
    [(1, ["2018-09-20", "2019-02-03", "2019-07-01", "2020-06-01"])],
    ("key", "values")
)

def after_second_quarter(x):
    return month(to_date(x)) > 6

df.select(
    filter("values", after_second_quarter).alias("after_second_quarter")
).show(truncate=False)
```

执行以上代码,输出结果如下:

```
+------------------------+
|after_second_quarter    |
+------------------------+
|[2018-09-20, 2019-07-01]|
+------------------------+
```

4) pyspark.sql.functions.transform(col, f)

该函数相当于 map()操作,用来转换数组中的每个元素。下面的代码演示了该函数的用法。首先构造一个 DataFrame,代码如下:

```
#第5章/dataframe2_demo15.ipynb

from pyspark.sql import SparkSession
from pyspark.sql.functions import *

spark = SparkSession.builder \
    .master("spark://localhost:7077") \
    .appName("pyspark demo") \
    .getOrCreate()

#创建 DataFrame
df = spark.createDataFrame(
    [
        (["New York", "Seattle"],),
        (["Barcelona", "Bangalore"],),
        (None,),
    ],
```

```
        ["places"]
)

df.printSchema()
df.show(truncate=False)
```

执行以上代码,输出结果如下:

```
root
 |-- places: array (nullable = true)
 |    |-- element: string (containsNull = true)

+---------------------+
|places               |
+---------------------+
|[New York, Seattle]  |
|[Barcelona, Bangalore]|
|null                 |
+---------------------+
```

然后调用 transform()转换函数,将 places 列中的每个数组元素与另一个字符串进行连接,代码如下:

```
resDF = df.withColumn("fun_places", transform("places", lambda x:
concat(x, lit(" is fun!")) ))
resDF.show(truncate=False)
```

执行以上代码,输出结果如下:

```
+---------------------+----------------------------------------+
|places               |fun_places                              |
+---------------------+----------------------------------------+
|[New York, Seattle]  |[New York is fun!, Seattle is fun!]     |
|[Barcelona, Bangalore]|[Barcelona is fun!, Bangalore is fun!] |
|null                 |null                                    |
+---------------------+----------------------------------------+
```

该方法也支持引用数组元素的索引。例如,将数组中索引为偶数的元素取反,代码如下:

```
df = spark.createDataFrame([(1, [1, 2, 3, 4])], ("key", "values"))

def alternate(x, i):
    return when(i % 2 == 0, x).otherwise(-x)

df.select(transform("values", alternate).alias("alternated")).show()
```

执行以上代码,输出结果如下:

```
+--------------+
|   alternated |
+--------------+
|[1, -2, 3, -4]|
+--------------+
```

5）pyspark.sql.functions.aggregate(col, initialValue, merge, finish=None)

该函数用来执行数组元素的聚合。下面的代码演示了该函数的用法。首先构造一个DataFrame，代码如下：

```
#第5章/dataframe2_demo16.ipynb
from pyspark.sql import SparkSession
from pyspark.sql.functions import *

spark = SparkSession.builder \
    .master("spark://localhost:7077") \
    .appName("pyspark demo") \
    .getOrCreate()

#定义schema
schema = StructType([
    StructField("numbers",ArrayType(IntegerType()))
])

#创建DataFrame
df = spark.createDataFrame(
    [
        ([1, 2, 3, 4],),
        ([5, 6, 7],),
        (None,),
    ],
    schema=schema
)

df.printSchema()
df.show(truncate=False)
```

执行以上代码，输出结果如下：

```
root
 |-- numbers: array (nullable = true)
 |    |-- element: integer (containsNull = false)


+-------------+
```

```
|numbers    |
+-----------+
|[1, 2, 3, 4]|
|[5, 6, 7]  |
|null       |
+-----------+
```

然后调用 aggregate()函数对数组元素执行聚合运算,代码如下:

```
resDF = df.withColumn("numbers_sum", aggregate("numbers", lit(0), lambda acc,x: acc + x))
resDF.show(truncate=False)
```

执行以上代码,输出结果如下:

```
+-----------+-----------+
|    numbers|numbers_sum|
+-----------+-----------+
|[1, 2, 3, 4]|         10|
|   [5, 6, 7]|         18|
|       null|       null|
+-----------+-----------+
```

在下面这个示例中,使用 aggregate()来计算一个数组元素的平均值,代码如下:

```
df = spark.createDataFrame([(1, [20.0, 4.0, 2.0, 6.0, 10.0])], ("id", "values"))

def merge(acc, x):
    count = acc.count + 1
    sum = acc.sum + x
    return struct(count.alias("count"), sum.alias("sum"))

df.select(
    aggregate(
        "values",
        struct(lit(0).alias("count"), lit(0.0).alias("sum")),
        merge,
        lambda acc: acc.sum / acc.count,
    ).alias("mean")
).show()
```

执行以上代码,输出结果如下:

```
+----+
|mean|
+----+
| 8.4|
+----+
```

6）pyspark.sql.functions.zip_with(left, right, f)

该函数执行拉链操作，用来组合两个数组中相同索引位置的元素。下面的代码演示了该函数的用法。首先构造一个 DataFrame，代码如下：

```
#第5章/dataframe2_demo17.ipynb
from pyspark.sql import SparkSession
from pyspark.sql.functions import *

spark = SparkSession.builder \
    .master("spark://localhost:7077") \
    .appName("pyspark demo") \
    .getOrCreate()

#创建 DataFrame
df = spark.createDataFrame(
    [
        (["a", "b"],["c", "d"]),
        (["x", "y"],["p", "o"]),
        (None,["e", "r"]),
    ],
    ("letters1","letters2")
)

df.printSchema()
df.show(truncate=False)
```

执行上面的代码，输出内容如下：

```
root
 |-- letters1: array (nullable = true)
 |    |-- element: string (containsNull = true)
 |-- letters2: array (nullable = true)
 |    |-- element: string (containsNull = true)

+--------+--------+
|letters1|letters2|
+--------+--------+
|[a, b]  |[c, d]  |
|[x, y]  |[p, o]  |
|null    |[e, r]  |
+--------+--------+
```

然后执行拉链操作，将 df 中的 letters1 列和 letters2 列的数组元素一一对应组合在一起，代码如下：

```
resDF = df.withColumn(
   "zipped_letters",
   zip_with("letters1","letters2",lambda x,y: concat_ws("***",x,y)
 )
)

resDF.show(truncate=False)
```

执行以上代码，输出结果如下：

```
+--------+--------+--------------+
|letters1|letters2|zipped_letters|
+--------+--------+--------------+
| [a, b]|  [c, d]|[a***c, b***d]|
| [x, y]|  [p, o]|[x***p, y***o]|
|   null|  [e, r]|          null|
+--------+--------+--------------+
```

使用 zip_with() 函数来合并两个数组列的元素值以计算新的列，代码如下：

```
df = spark.createDataFrame([(1, [1, 3, 5, 8], [0, 2, 4, 6])], ("id", "xs", "ys"))
df.select(zip_with("xs", "ys", lambda x, y: x ** y).alias("powers")).show(truncate=False)
```

执行以上代码，输出结果如下：

```
+---------------------------+
|powers                     |
+---------------------------+
|[1.0, 9.0, 625.0, 262144.0]|
+---------------------------+
```

5.3 聚合与透视函数

对大数据进行分析通常需要对数据进行聚合操作。聚合通常需要某种形式的分组，要么在整个数据集上，要么在一列或多列上，然后对它们应用聚合函数，例如对每组进行求和、计数或求平均值等。PySpark SQL 提供了许多常用的聚合函数。

5.3.1 聚合函数

在 PySpark 中，所有的聚合都是通过函数完成的。聚合函数被设计用来在一组行上执行聚合，不管那组行是由 DataFrame 中的所有行还是一组子行组成的。PySpark 中常见的聚合函数见表 5-1。

表 5-1 常见聚合函数

聚合函数	描述
count(col)	返回每组中可能有重复值,例如1,2,3,3,4成员数量
countDistinct(col)	返回每组中可能有重复值,例如1,2,3,3,4成员数量,则为5
approx_count_distinct(col)	返回每组中可能有重复值,例如1,2,3,3,4成员唯一数量,则为4
min(col)	返回每组中给定列的最小值
max(col)	返回每组中给定列的最大值
sum(col)	返回每组中给定列的值的和
sumDistinct(col)	返回每组中给定列的唯一值的和
avg(col)	返回每组中给定列的值的平均值
skewness(col)	返回每组中给定列的值的分布的偏斜度
kurtosis(col)	返回每组中给定列的值的分布的峰度
variance(col)	返回每组中给定列的值的无偏方差
stddev(col)	返回每组中给定列的值的标准差
collect_list(col)	返回每组中给定列的值的集合。返回的集合可能包含重复的值
collect_set(col)	返回每组中给定列的唯一值的集合

为了演示这些函数的用法,在下面的示例中将使用"2018年11月14日深圳市价格定期监测信息"数据集。这个数据集包含一些主要副食品的监测信息,以 CSV 格式存储在文件中。需要将该数据集上传到 HDFS 分布式文件系统的/data/spark 目录下。

首先读取价格监测信息数据集,并创建 DataFrame,代码如下:

```
#第5章/dataframe2_demo18.ipynb

from pyspark.sql import SparkSession
from pyspark.sql.functions import *

spark = SparkSession.builder \
   .master("spark://localhost:7077") \
   .appName("pyspark demo") \
   .getOrCreate()

#读取数据源文件,创建 DataFrame
filePath = "/data/spark/2018年11月14日深圳市价格定期监测信息.csv"
priceDF = spark \
      .read \
      .option("header","true") \
      .option("inferSchema","true") \
      .csv(filePath)
```

```
priceDF.printSchema()
priceDF.show(5)
```

执行以上代码,输出结果如下:

```
root
 |-- RECORDID: string (nullable = true)
 |-- JCLB: string (nullable = true)
 |-- JCMC: string (nullable = true)
 |-- BQ: double (nullable = true)
 |-- SQ: double (nullable = true)
 |-- TB: double (nullable = true)
 |-- HB: double (nullable = true)

+--------------------+----+--------+-----+-----+------+------+
|            RECORDID|JCLB|    JCMC|   BQ|   SQ|    TB|    HB|
+--------------------+----+--------+-----+-----+------+------+
|537B9A6E0C836F36E...|null|    椰菜| 2.27|2.261|-0.067| 0.004|
|537B9A6E0C846F36E...|null|  东北米|2.863|2.843|-0.067| 0.007|
|537B9A6E0C856F36E...|null|　早籼米| 3.08|3.044| 0.219| 0.012|
|537B9A6E0C866F36E...|null|  晚籼米|3.217| 3.22| 0.081|-0.001|
|537B9A6E0C876F36E...|null|泰国香米| 9.48| 9.48| 0.047|   0.0|
+--------------------+----+--------+-----+-----+------+------+
only showing top 5 rows
```

数据集中每行代表一条商品的价格监测信息,其中各个字段的含义如下。

(1) JCLB:监测类别。

(2) JCMC:监测名称。

(3) BQ:本期价格。

(4) SQ:上期价格。

(5) TB:同比价格变化。

(6) HB:环比价格变化。

对数据集进行简单探索。首先,找出这个数据集总共有多少行,代码如下:

```
print(f"监测的商品数量有{priceDF.count()}")
```

执行以上代码,输出结果如下:

```
监测的商品数量有331
```

下面使用一些常用的聚合函数进行统计。

1) count(col)

该函数统计指定列的数量。统计数据集中监测名称(JCMC)的数量和监测类别(JCLB)的数量,代码如下:

```
#监测名称的数量
priceDF.select(count("JCMC").alias("监测名称")).show()

#当统计一列中的项目数量时,count(col)函数不包括计数中的null值
priceDF.select(count("JCLB").alias("监测类别")).show()

#判断"JCLB"列为null值的有多少
nullJclb = priceDF.where(col("JCLB").isNull()).count()
print("\"JCLB\"列为null值的有", nullJclb)
```

执行以上代码,输出结果如下:

```
+--------+
|监测名称 |
+--------+
|    331 |
+--------+


+--------+
|监测类别 |
+--------+
|    298 |
+--------+

"JCLB"列为null值的有 33
```

2) countDistinct(col)

该函数只计算每个组的唯一项。例如,统计总共有多少个商品类别和多少种商品,代码如下:

```
priceDF.select(
    countDistinct("JCLB").alias("监测类别"),
    countDistinct("JCMC").alias("监测名称"),
    count("*").alias("总数量")
).show()
```

执行以上代码,输出结果如下:

```
+--------+--------+------+
|监测类别|监测名称|总数量|
+--------+--------+------+
|       6|      35|   331|
+--------+--------+------+
```

注意,CountDistinct()和distinct()不同。distinct()函数用来按行去重,包括null值。例如,计算检测的商品类别有多少,代码如下:

```
priceDF.select("JCLB").distinct().count()          #7
priceDF.select("JCLB").distinct().show()
```

执行以上代码，输出结果如下：

```
7
+------+
| JCLB|
+------+
|   粮食|
|  食用油|
|  null|
|  水产品|
|  肉奶蛋|
|   蔬菜|
|  肉蛋奶|
+------+
```

3）approx_count_distinct (col, max_estimated_error=0.05)

唯一数量。在一个大数据集里计算每个组中唯一项的确切数量是一个成本很高且很耗时的操作。在某些用例中，有一个近似唯一的计数就足够了。例如，在线广告业务中，每小时有数亿个广告曝光并且需要生成一份报告来显示每个特定类型的成员段的独立访问者的数量。PySpark 实现了 approx_count_distinct()函数，用来统计近似唯一计数。因为唯一计数是一个近似值，所以会有一定的误差。这个函数允许指定一个可接受估算误差的值。

使用 approx_count_distinct()函数，代码如下：

```
#统计price DataFrame 的"JCMC"列。默认估算误差是0.05 (5%)
priceDF.select(
    count("JCMC"),
    countDistinct("JCMC"),
    approx_count_distinct("JCMC", 0.05)
).show()
```

执行以上代码，输出结果如下：

```
+-----------+--------------------+---------------------------+
|count(JCMC)|count(DISTINCT JCMC)|approx_count_distinct(JCMC)|
+-----------+--------------------+---------------------------+
|        331|                  35|                         33|
+-----------+--------------------+---------------------------+
```

4）min(col)和 max(col)

该函数用于获取 col 列的最小值和最大值。例如，统计本期价格的最大值和最小值，代码如下：

```
#统计本期价格(BQ)的最大值和最小值
priceDF.select(
    min("BQ").alias("最便宜的"),
```

```
        max("BQ").alias("最贵的")
).show()
```

执行以上代码,输出结果如下:

```
+--------+------+
|最便宜的|最贵的|
+--------+------+
|     1.7|43.515|
+--------+------+
```

5) sum(col)

这个函数用于计算一个数列中的值的总和。例如,计算本期价格之和,代码如下:

```
priceDF.select(sum("BQ")).show()
```

执行以上代码,输出结果如下:

```
+-----------------+
|          sum(BQ)|
+-----------------+
|2904.477000000003|
+-----------------+
```

6) sumDistinct(col)

该函数只汇总了一个数列的不同值。例如,计算本期价格(唯一值)之和,代码如下:

```
priceDF.select(sumDistinct("BQ")).show()
```

执行以上代码,输出结果如下:

```
+-----------------+
| sum(DISTINCT BQ)|
+-----------------+
|2544.9589999999994|
+-----------------+
```

7) avg(col)

这个函数用于计算一个数列的平均值。这个函数简单地取总并除以项目的数量。例如,计算本期的平均价格,代码如下:

```
priceDF.select(avg("BQ"), sum("BQ") / count("BQ")).show()
```

执行以上代码,输出结果如下:

```
+----------------+--------------------+
|         avg(BQ)|(sum(BQ) / count(BQ))|
+----------------+--------------------+
|8.774854984894262|    8.774854984894262|
+----------------+--------------------+
```

8）skewness(col)和kurtosis(col)

在统计领域中，峰度（Kurtosis）与偏态（Skewness）是测量数据正态分布特性的两个指标。了解偏态和峰度这两个统计量的含义很重要，在对数据进行正态转换时，需要将其作为参考，选择合适的转换方法。

偏态是一种度量数据集的值分布对称性的度量：

（1）当偏态≈0时，可认为分布是对称的，服从正态分布。

（2）当偏态>0时，分布为右偏，即拖尾在右边，峰尖在左边，也称为正偏态。

（3）当偏态<0时，分布为左偏，即拖尾在左边，峰尖在右边，也称为负偏态。

数据分布的左偏或右偏，指的是数值拖尾的方向，而不是峰的位置，如图5-1所示。

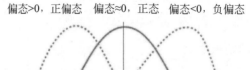
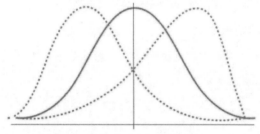

图5-1 数据分布的偏态特征

峰度是对分布曲线性状的度量，用来衡量数据分布的平坦度（曲线有可能是正常的、平坦的，还有可能是尖的）。正的Kurtosis表示曲线是细而尖的，负的Kurtosis表示曲线是宽而平的。正态分布的峰度值为3。不同大小的峰度值的含义如下：

（1）当峰度≈0时，可认为分布的峰度合适，服从正态分布（不胖不瘦）。

（2）当峰度>0时，分布的峰度陡峭（高尖）。

（3）当峰度<0时，分布的峰度平缓（矮胖）。

这几种峰度特征分布如图5-2所示。

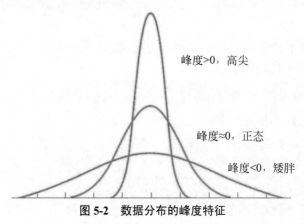

图5-2 数据分布的峰度特征

计算 BQ 列（表示本期价格列）的偏态和峰度，代码如下：

```
priceDF.select(skewness("BQ"), kurtosis("BQ")).show()
```

执行以上代码，输出结果如下：

```
+------------------+------------------+
|      skewness(BQ)|      kurtosis(BQ)|
+------------------+------------------+
|2.2954037816302346|4.8142280508421510|
+------------------+------------------+
```

结果表明，BQ 列的分布是不对称的，呈正偏态分布（右边的尾巴比左边的尾巴长或宽）。峰度值表明分布曲线是高而尖的。

9）variance(col)和 stddev(col)

在统计学中，方差（Variance）和标准差（Stddev）用于测量数据的分散性或分布。换句话说，它们被用来说明 values 到平均值的平均距离。当方差值较低时，意味着该值接近均值。方差和标准差是相关的，后者是前者的平方根。

variance()和 stddev()函数分别用于计算方差和标准差。PySpark 提供了这些函数的两种不同实现：一种是利用抽样来加速度计算，另一种使用全样数据。例如，计算 priceDF DataFrame 中的 BQ 列的方差和标准差，代码如下：

```
#使用方差和标准差的两种变化
priceDF \
    .select(variance("BQ"),var_pop("BQ"),stddev("BQ"),stddev_pop("BQ")) \
    .show()
```

执行以上代码，输出结果如下：

```
+------------------+------------------+-----------------+-----------------+
|      var_samp(BQ)|       var_pop(BQ)|  stddev_samp(BQ)|   stddev_pop(BQ)|
+------------------+------------------+-----------------+-----------------+
| 85.96212834860381| 85.70242403335124|9.271576368051110|9.257560371574751|
+------------------+------------------+-----------------+-----------------+
```

结果表明，priceDF DataFrame 中的 BQ 值很分散。

5.3.2 分组聚合

分组聚合不会在 DataFrame 中对全局组执行聚合，而是在 DataFrame 中的每个子组中执行聚合。通常分组执行聚合的过程分通过 groupBy（col1、col2、…）转换来执行分组，也就是指定要按哪些列分组。与其他返回 DataFrame 的转换不同，这个 groupBy()转换返回 RelationalGroupedDataset 类的一个实例。类 RelationalGroupedDataset 提供了一组标准的聚合函数，可以将它们应用到每个子组中。这些聚合函数有 avg(cols)、count()、mean(cols)、min(cols)、max(cols)和 sum(cols)。除了 count()函数之外，其余函数都在数列上运行。

例如，按 JCLB（监测类别）列分组并执行一个 count()聚合（groupBy()列将自动包含在输出中），代码如下：

```
#按监测类别分组统计
priceDF.groupBy("JCLB").count().show(truncate=False)
```

执行以上代码，输出结果如下：

```
+------+-----+
|JCLB  |count|
+------+-----+
|粮食   |55   |
|食用油 |37   |
|水产品 |27   |
|null  |33   |
|肉奶蛋 |7    |
|蔬菜   |117  |
|肉蛋奶 |55   |
+------+-----+
```

按 JCLB 和 JCMC 分组之后，执行 count()聚合，并按统计数量降序排序，代码如下：

```
#按监测类别和商品小类分组统计
priceDF \
    .groupBy("JCLB", "JCMC") \
    .count() \
    .orderBy(col("count").desc()) \
    .show(truncate=False)

priceDF \
    .groupBy("JCLB", "JCMC") \
    .count() \
    .where("JCMC=='花生油'") \
    .orderBy(col("count").desc()) \
    .show(truncate=False)
```

执行以上代码，输出结果如下：

```
+------+----------+-----+
|JCLB  |JCMC      |count|
+------+----------+-----+
|粮食   |东北米     |10   |
|食用油 |花生油     |10   |
|蔬菜   |蔬菜均价   |9    |
|蔬菜   |大白菜     |9    |
|粮食   |散装面粉   |9    |
```

```
|食用油 |菜籽油   |9   |
|蔬菜   |萝卜     |9   |
|粮食   |袋装面粉 |9   |
|蔬菜   |菠菜     |9   |
|蔬菜   |芹菜     |9   |
|蔬菜   |青椒     |9   |
|蔬菜   |西红柿   |9   |
|蔬菜   |黄瓜     |9   |
|蔬菜   |茄子     |9   |
|水产品 |大头鱼   |9   |
|食用油 |调和油   |9   |
|粮食   |泰国香米 |9   |
|粮食   |早籼米   |9   |
|水产品 |草鱼     |9   |
|粮食   |晚籼米   |9   |
+------+---------+----+
only showing top 20 rows


+------+------+-----+
|JCLB  |JCMC  |count|
+------+------+-----+
|食用油|花生油|10   |
|null  |花生油|1    |
+------+------+-----+
```

有时需要在同一时间对每组执行多个聚合。例如，除了计数之外，还想知道最小值和最大值。RelationalGroupedDataset 类提供了一个名为 agg() 的功能强大的函数，它接受一个或多个列表达式，这意味着可以使用任何聚合函数。这些聚合函数会返回 Column 类的一个实例，这样就可以使用所提供的函数来应用任何列表达式了。一个常见的需求是在聚合完成后重命名列，使之更短、更可读、更易于引用。

例如，按 JCLB 分组之后，执行多个聚合，代码如下：

```
#同时对每个组执行多个聚合
from pyspark.sql.functions import *

priceDF.na.drop() \
    .groupBy("JCLB") \
    .agg(
        count("BQ").alias("本期数量"),
        min("BQ").alias("本期最低价格"),
        max("BQ").alias("本期最高价格"),
        avg("BQ").alias("本期平均价格")
    ).show()
```

执行以上代码，输出结果如下：

```
+------+---------+------------+------------+--------------------+
| JCLB |本期数量 |本期最低价格|本期最高价格| 本期平均价格        |
+------+---------+------------+------------+--------------------+
| 粮食 |     55 |      2.844 |       9.48 |  4.258527272727273 |
|食用油|     37 |        4.5 |      11.43 |  7.179567567567568 |
|水产品|     27 |       8.33 |      37.26 | 17.51562962962963  |
|肉奶蛋|      7 |       2.99 |      42.93 | 19.195714285714285 |
| 蔬菜 |    117 |        1.7 |      8.765 |  4.372042735042735 |
|肉蛋奶|     55 |        2.8 |     43.515 | 18.250181818181815 |
+------+---------+------------+------------+--------------------+
```

函数 collect_list(col)和 collect_set(col)用于在应用分组后收集特定组的所有值。一旦每组的值被收集到一个集合中，就可以自由地以任何选择的方式对其进行操作。这两个函数的返回集合之间有一个小的区别，那就是唯一性。collect_list()函数返回一个可能包含重复值的集合，collect_set()函数返回一个只包含唯一值的集合。

例如，使用 collection_list()函数来收集每个监测类别下的商品名称，代码如下：

```
from pyspark.sql.functions import *

priceDF.na.drop() \
    .groupBy(col('JCLB').alias("监测类别")) \
    .agg(collect_set("JCMC").alias("监测的商品")) \
    .withColumn("监测商品数量",size('监测的商品')) \
    .show(truncate=False)
```

执行以上代码，输出结果如下：

```
+--------+------------------------------+------------+
|监测类别|          监测的商品          |监测商品数量|
+--------+------------------------------+------------+
|  粮食  |[早籼米, 散装面粉, 泰国香米,...]|     6     |
| 食用油 |[花生油, 菜籽油, 豆油, 调和油] |     4     |
| 水产品 |     [带鱼, 草鱼, 大头鱼]      |     3     |
| 肉奶蛋 |[其中:精瘦肉, 鸡蛋, 牛肉,...]  |     7     |
|  蔬菜  |[蔬菜均价, 西红柿, 大白菜,...] |    14     |
| 肉蛋奶 |[其中:精瘦肉, 鸡蛋, 牛肉,...]  |     8     |
+--------+------------------------------+------------+
```

5.3.3 数据透视

数据透视是一种通过聚合和旋转把数据行转换成数据列的技术，它是一种将行转换成列同时应用一个或多个聚合的方法。这样一来，分类值就会从行转到单独的列中。这种技术通常用于数据分析或报告。

数据透视过程从一列或多列的分组开始，然后在一列上旋转，最后在一列或多列上应用一个或多个聚合，因此，当透视数据时，需要确定 3 个要素：①要在行（分组元素）中看到的元素；②要在列（扩展元素）上看到的元素；③要在数据部分看到的元素（聚合元素）。

【例 5-2】有一个包含学生信息的数据集，每行包含学生的姓名、性别、体重、毕业年份。现在想要知道每个毕业年份每个性别学生的平均体重。

首先，创建一个 DataFrame，代码如下：

```
#第5章/dataframe2_demo19.ipynb

from pyspark.sql import SparkSession
from pyspark.sql.functions import *

spark = SparkSession.builder \
    .master("spark://localhost:7077") \
    .appName("pyspark demo") \
    .getOrCreate()

#创建 DataFrame
studentsDF = spark.createDataFrame([
    ("刘宏明", "男", 180, 2015),
    ("赵薇", "女", 110, 2015),
    ("黄海波", "男", 200, 2015),
    ("杨幂", "女", 109, 2015),
    ("楼一萱", "女", 105, 2015),
    ("龙梅子", "女", 115, 2016),
    ("陈知远", "男", 195, 2016)
], ["name","gender","weight","graduation_year"])

studentsDF.show()
```

执行以上代码，输出结果如下：

```
+------+------+------+---------------+
|  name|gender|weight|graduation_year|
+------+------+------+---------------+
|刘宏明 |    男 |   180|           2015|
| 赵薇 |    女 |   110|           2015|
|黄海波 |    男 |   200|           2015|
|  杨幂 |    女 |   109|           2015|
|楼一萱 |    女 |   105|           2015|
|龙梅子 |    女 |   115|           2016|
|陈知远 |    男 |   195|           2016|
+------+------+------+---------------+
```

然后调用 pivot() 函数在 gender 列上旋转，统计不同性别的平均体重，代码如下：

```
#计算每个毕业年份每个性别学生的平均体重
studentsDF \
    .groupBy("graduation_year") \
    .pivot("gender") \
    .avg("weight") \
    .show()
```

执行以上代码，输出结果如下：

```
+---------------+-----+-----+
|graduation_year|   女|   男|
+---------------+-----+-----+
|           2015|108.0|190.0|
|           2016|115.0|195.0|
+---------------+-----+-----+
```

可以利用 agg() 函数来执行多个聚合，这会在结果中创建更多的列，代码如下：

```
studentsDF \
    .groupBy("graduation_year") \
    .pivot("gender") \
    .agg(
        min("weight").alias("min"),
        max("weight").alias("max"),
        avg("weight").alias("avg")
    ).show()
```

执行以上代码，输出结果如下：

```
+---------------+------+------+------+------+------+------+
|graduation_year|女_min|女_max|女_avg|男_min|男_max|男_avg|
+---------------+------+------+------+------+------+------+
|           2015|   105|   110| 108.0|   180|   200| 190.0|
|           2016|   115|   115| 115.0|   195|   195| 195.0|
+---------------+------+------+------+------+------+------+
```

如果 pivot 列有许多不同的值，可以选择性地选择生成聚合的值，代码如下：

```
studentsDF \
    .groupBy("graduation_year") \
    .pivot("gender", ["男"]) \
    .agg(
        min("weight").alias("min"),
        max("weight").alias("max"),
        avg("weight").alias("avg")
    ).show()
```

执行以上代码，输出结果如下：

```
+---------------+------+------+------+
|graduation_year|男_min|男_max|男_avg|
+---------------+------+------+------+
|           2015|   180|   200| 190.0|
|           2016|   195|   195| 195.0|
+---------------+------+------+------+
```

为 pivot 列指定一个 distinct 值的列表实际上会加速旋转过程。

5.4 高级分析函数

PySpark SQL 提供了许多高级分析函数，如多维聚合函数、时间窗口聚合函数和窗口分析函数等。本节就介绍这些高级分析函数的使用。

5.4.1 使用多维聚合函数

在高级分析函数中，第 1 个是关于多维聚合的，它对于涉及分层数据分析的用例非常有用，在这种情况下，通常需要在一组分组列中计算子总数和总数。常用的多维聚合函数包括 rollup() 和 cube()，它们基本上是在多列上进行分组的高级版本，通常用于在这些列的组合和排列中生成子总数和总数。

1) rollup()

当使用分层数据时，例如不同部门和分部的销售收入数据等，rollup() 可以很容易地计算出它们的子总数和总数。rollup() 按给定的列集的层次结构，并且总是在层次结构中的第 1 列启动 rolling up 过程。使用 rollup() 函数的代码如下：

```
#第5章/dataframe2_demo20.ipynb
from pyspark.sql import SparkSession
from pyspark.sql.functions import *

spark = SparkSession.builder \
    .master("spark://localhost:7077") \
    .appName("pyspark demo") \
    .getOrCreate()

#读取超市订单汇总数据
filePath = "/data/spark/超市订单.csv"
ordersDF = spark \
    .read \
    .option("header", "true") \
    .option("inferSchema","true") \
    .csv(filePath)
```

```
print(f"订单数量: {ordersDF.count()}")
ordersDF.printSchema()
```

执行以上代码，输出结果如下：

```
订单数量: 10000
root
 |-- 行 ID: integer (nullable = true)
 |-- 订单 ID: string (nullable = true)
 |-- 订单日期: string (nullable = true)
 |-- 发货日期: string (nullable = true)
 |-- 邮寄方式: string (nullable = true)
 |-- 客户 ID: string (nullable = true)
 |-- 客户名称: string (nullable = true)
 |-- 细分: string (nullable = true)
 |-- 城市: string (nullable = true)
 |-- 省/自治区/直辖市: string (nullable = true)
 |-- 国家: string (nullable = true)
 |-- 地区: string (nullable = true)
 |-- 产品 ID: string (nullable = true)
 |-- 类别: string (nullable = true)
 |-- 子类别: string (nullable = true)
 |-- 产品名称: string (nullable = true)
 |-- 销售额: double (nullable = true)
 |-- 数量: integer (nullable = true)
 |-- 折扣: double (nullable = true)
 |-- 利润: double (nullable = true)
```

查看前 10 条数据，代码如下：

```
ordersDF.show(10)
```

执行以上代码，输出结果如图 5-3 所示。

行 ID	订单 ID	订单日期	发货日期	邮寄方式	客户 ID	客户名称	细分	城市	省/自治区/直辖市	国家	地区	产品 ID	类别	子类别	产品名称	销售额	数量	折扣	利润
1	US-2017-1357144	2017/4/27	2017/4/29	二级	曾伍-14485	曾伍	公司	杭州	浙江	中国	华东	办公用-用品-10002717	办公用品	用品	Fiskars 剪刀,蓝色	129.696	2	0.4	-60.704
2	CN-2017-1973789	2017/6/15	2017/6/19	标准级	许安-10165	许安	消费者	内江	四川	中国	西南	办公用-信封-10004832	办公用品	信封	GlobeWeis 搭扣信封,红色	125.44	2	0.0	42.56
3	CN-2017-1973789	2017/6/15	2017/6/19	标准级	许安-10165	许安	消费者	内江	四川	中国	西南	办公用-装订-10001505	办公用品	装订机	Cardinal 孔加固材料,回收	31.92	2	0.4	4.2
4	CN-2017-3017568	2017/12/9	2017/12/13	标准级	宋良-17170	宋良	公司	镇江	江苏	中国	华东	办公用-装订-10003746	办公用品	用品	Kleencut 开信刀,工业	321.216	4	0.4	-27.104
5	CN-2016-2975416	2016/5/31	2016/6/2	二级	万兰-15730	万兰	消费者	汕头	广东	中国	中南	办公用-器具-10003452	办公用品	器具	KitchenAid 搅拌机,黑色	1375.92	3	0.0	550.2
6	CN-2015-4497736	2015/10/27	2015/10/31	标准级	俞明-18325	俞明	消费者	景德镇	江西	中国	华东	技术-设备-10001640	技术	设备	柯尼卡 打印机,红色	11129.58	5	0.0	3783.78
7	CN-2015-4497736	2015/10/27	2015/10/31	标准级	俞明-18325	俞明	消费者	景德镇	江西	中国	华东	办公用-装订-10001029	办公用品	装订机	Ibico 订书机,实惠	479.92	1	0.0	172.76
8	CN-2015-4497736	2015/10/27	2015/10/31	标准级	俞明-18325	俞明	消费者	景德镇	江西	中国	华东	家具-椅子-10000578	家具	椅子	SAFCO 扶手椅,可调	8659.84	4	0.0	2684.08
9	CN-2015-4497736	2015/10/27	2015/10/31	标准级	俞明-18325	俞明	消费者	景德镇	江西	中国	华东	办公用-纸张-10001629	办公用品	纸张	Green Bar 计划信息表,多色	588.0	5	0.0	46.9
10	CN-2015-4497736	2015/10/27	2015/10/31	标准级	俞明-18325	俞明	消费者	景德镇	江西	中国	华东	办公用-系固-10004801	办公用品	系固件	Stockwell 橡皮筋,整包	154.28	4	0.0	33.88

only showing top 10 rows

图 5-3 订单数据

对数据进行过滤，以便更容易地看到 rollup()的结果，代码如下：

```
twoSummary = ordersDF.select("地区","省/自治区/直辖市","订单 ID").where("`地
区`=='华东' or `地区`=='华北'")

#让我们看一看数据是什么样子的
```

```
print("数据量: ", twoSummary.count())
twoSummary.show()
```

执行以上代码,输出结果如下:

```
数据量: 4327
+----+--------------+---------------+
|地区|省/自治区/直辖市|        订单 ID|
+----+--------------+---------------+
|华东|          浙江|US-2017-1357144|
|华东|          江苏|US-2017-3017568|
|华东|          江西|CN-2015-4497736|
|华东|          江西|CN-2015-4497736|
|华东|          江西|CN-2015-4497736|
|华东|          江西|CN-2015-4497736|
|华东|          江西|CN-2015-4497736|
|华东|          山东|CN-2015-2752724|
|华东|          山东|CN-2015-2752724|
|华东|          山东|CN-2015-2752724|
|华东|          江苏|US-2016-2511714|
|华东|          江苏|US-2016-2511714|
|华东|          上海|CN-2017-5631342|
|华东|          上海|CN-2017-5631342|
|华东|          上海|CN-2017-5631342|
|华东|          上海|CN-2017-5631342|
|华东|          上海|CN-2017-5631342|
|华东|          上海|CN-2017-5631342|
|华东|          上海|CN-2017-5631342|
|华东|          浙江|US-2016-4150614|
+----+--------------+---------------+
only showing top 20 rows
```

接下来按地区、省/自治区/直辖市执行 rollup()操作,然后计算计数的总和,最后按 null 排序,代码如下:

```
from pyspark.sql.functions import *

twoSummary.rollup("地区", "省/自治区/直辖市") \
        .agg(count("订单 ID").alias("total")) \
        .orderBy(col("地区").asc_nulls_last(),
                col("省/自治区/直辖市").asc_nulls_last()) \
        .show()
```

执行以上代码,输出结果如下:

```
+----+--------------+-----+
|地区|省/自治区/直辖市|total|
+----+--------------+-----+
```

```
|华东|            上海|  292|
|华东|            安徽|  347|
|华东|            山东|  914|
|华东|            江苏|  583|
|华东|            江西|  139|
|华东|            浙江|  424|
|华东|            福建|  259|
|华东|          null| 2958|
|华北|          内蒙古|  224|
|华北|            北京|  252|
|华北|            天津|  304|
|华北|            山西|  201|
|华北|            河北|  388|
|华北|          null| 1369|
|null|          null| 4327|
+----+--------------+-----+
```

这个输出结果显示了华东地区和华北地区的每个城市的子总数，而总数显示在最后一行，并带有在"地区"和"省/自治区/直辖市"的列上的 null 值。注意带有 asc_nulls_last 选项进行排序，因此 PySpark SQL 会将 null 值排序到最后位置。

2）cube()

一个 cube()函数可以看作 rollup()函数的更高级版本。它在分组列的所有组合中执行聚合操作，因此，结果包括 rollup()提供的及其他组合所提供的。在上面的"地区"和"省/自治区/直辖市"的例子中，结果将包括每个"省/自治区/直辖市"的聚合。使用 cube()函数的方法类似于使用 rollup()函数，代码如下：

```
twoSummary \
    .cube("地区", "省/自治区/直辖市") \
    .agg(count("订单 ID").alias("total")) \
    .orderBy(col("地区").asc_nulls_last(),
             col("省/自治区/直辖市").asc_nulls_last()) \
    .show(30)
```

执行以上代码，输出结果如下：

```
+----+--------------+-----+
|地区|省/自治区/直辖市|total|
+----+--------------+-----+
|华东|            上海|  292|
|华东|            安徽|  347|
|华东|            山东|  914|
|华东|            江苏|  583|
|华东|            江西|  139|
|华东|            浙江|  424|
|华东|            福建|  259|
```

```
|华东|          null| 2958|
|华北|          内蒙古|  224|
|华北|          北京|  252|
|华北|          天津|  304|
|华北|          山西|  201|
|华北|          河北|  388|
|华北|          null| 1369|
|null|          上海|  292|
|null|          内蒙古|  224|
|null|          北京|  252|
|null|          天津|  304|
|null|          安徽|  347|
|null|          山东|  914|
|null|          山西|  201|
|null|          江苏|  583|
|null|          江西|  139|
|null|          河北|  388|
|null|          浙江|  424|
|null|          福建|  259|
|null|          null| 4327|
+----+---------------+-----+
```

在结果表格中,在"地区"列中有 null 值的行表示一个地区中所有城市的聚合,因此,一个 cube() 的计算结果总会比 rollup() 的结果有更多的行。

5.4.2 使用时间窗口聚合

在高级分析函数中,第 2 个功能是基于时间窗口执行聚合,这在处理来自物联网设备的事务或传感器值等时间序列数据时非常有用。

这些时序数据由一系列的时间顺序数据点组成。这种数据集在金融或电信等行业很常见。例如,股票市场交易数据集有交易日期、开盘价、收盘价、交易量和每个股票代码的其他信息,以京东股票的历史数据为例,如图 5-4 所示。

在 PySpark 2.0 中引入了时间窗口的聚合,使其能够轻松地处理时间序列数据。可以使用时间窗口聚合分析时序数据,例如京东股票的周平均收盘价,或者京东股票跨每周的月移动平均收盘价。

时间窗口函数有几个版本,但是它们都需要一个时间戳类型列和一个窗口长度,该窗口长度可以指定为几秒、几分、几小时、几天或几周。窗口长度代表一个时间窗口,它有一个开始时间和结束时间,它被用来确定一个特定的时间序列数据应该属于哪个桶。

有两种类型的时间窗口:滚动窗口和滑动窗口。与滚动窗口(也叫固定窗口)相比,滑动窗口需要提供额外的输入参数,用来说明在计算下一个桶时,一个时间窗口应该滑动多少。

JD历史数据

时间范围: 每日

下载数据 2022/04/05 - 2022/05/05

日期	收盘	开盘	高	低	交易量	涨跌幅
2022年5月4日	63.18	60.23	63.33	59.11	9.32M	1.62%
2022年5月3日	62.17	62.80	63.96	61.65	9.10M	-1.89%
2022年5月2日	63.37	61.64	63.55	60.97	10.17M	2.77%
2022年4月29日	61.66	65.08	65.29	61.56	19.76M	6.66%
2022年4月28日	57.81	57.82	58.80	56.15	8.07M	0.68%
2022年4月27日	57.42	55.49	58.89	55.35	14.69M	7.91%
2022年4月26日	53.21	54.13	55.49	52.92	10.16M	-0.95%
2022年4月25日	53.72	50.83	54.02	50.59	9.09M	3.23%
2022年4月22日	52.04	52.48	54.27	51.33	12.52M	2.64%
2022年4月21日	50.70	52.89	53.68	50.25	10.45M	-5.67%
2022年4月20日	53.75	56.60	56.60	53.03	9.30M	-5.52%
2022年4月19日	56.89	56.35	56.92	55.07	8.44M	-1.06%
2022年4月18日	57.50	55.44	58.13	55.34	9.22M	1.66%
2022年4月14日	56.56	57.34	57.97	56.27	7.35M	-2.95%
2022年4月13日	58.28	57.74	59.70	56.99	11.72M	3.19%
2022年4月12日	56.48	57.48	58.31	56.44	7.96M	-0.62%
2022年4月11日	56.83	55.78	58.23	54.85	8.77M	0.51%
2022年4月8日	56.54	56.75	57.85	55.82	10.47M	-0.98%
2022年4月7日	57.10	57.85	57.85	56.21	10.82M	-3.34%
2022年4月6日	59.07	59.50	59.75	57.59	8.98M	-3.04%
2022年4月5日	60.92	62.32	62.40	60.37	6.88M	-3.78%

最高: 65.29   最低: 50.25   差价: 15.04   平均: 57.39   涨跌幅: -0.21

图 5-4   股票历史价格时序数据

【例 5-3】编写 PySpark SQL 批处理程序，分析京东股票历史交易数据，统计京东股票的周平均价格，以及京东股票的月平均收盘价（每周计算一次）。

计算京东股票的周平均价格，代码如下：

```
#第 5 章/dataframe2_demo21.ipynb

from pyspark.sql import SparkSession
from pyspark.sql.functions import *

spark = SparkSession.builder \
    .master("spark://localhost:7077") \
    .appName("pyspark demo") \
    .getOrCreate()
```

```
#加载京东股票历史交易数据
csvPath = "/data/spark/jd/jd-formated.csv"
jdDF = spark \
    .read \
    .option("header", "true") \
    .option("inferSchema","true") \
    .csv(csvPath)

#显示该schema，第1列是交易日期
jdDF.printSchema()
jdDF.show(10)
```

执行以上代码，输出结果如下：

```
root
 |-- Date: string (nullable = true)
 |-- Close: double (nullable = true)
 |-- Volume: integer (nullable = true)
 |-- Open: double (nullable = true)
 |-- High: double (nullable = true)
 |-- Low: double (nullable = true)

+----------+-----+-------+------+-----+-----+
|      Date|Close| Volume|  Open| High|  Low|
+----------+-----+-------+------+-----+-----+
|2022-02-15|76.13|6766205| 75.35|76.35| 74.8|
|2022-02-14|74.45|5244967| 73.94|74.62|73.01|
|2022-02-11|73.98|6673354| 75.97|76.55|73.55|
|2022-02-10| 76.4|6432184|75.955|78.39|75.24|
|2022-02-09|78.29|7061571| 76.83|78.67|76.61|
|2022-02-08|75.36|7903249| 73.12|76.07|72.05|
|2022-02-07|73.15|6135832| 74.09|74.99|72.81|
|2022-02-04|73.77|6082889| 71.94|74.95|71.86|
|2022-02-03|71.85|7493688| 72.08| 73.3|71.33|
|2022-02-02|73.21|5887066| 75.58|75.71|72.41|
+----------+-----+-------+------+-----+-----+
only showing Top 10 rows
```

从结果可以注意到，其中 Date 字段被自动推断为 string 类型，因此需要对交易日期进行整理（将 Date 字段的字符串类型转换为 Date 类型），代码如下：

```
from pyspark.sql.functions import *

jdStock = jdDF.withColumn("Date",to_date('Date',"yyyy/MM/dd"))
jdStock.printSchema()
jdStock.show(10)
```

执行以上代码，输出结果如下：

```
root
 |-- Date: date (nullable = true)
 |-- Close: double (nullable = true)
 |-- Volume: integer (nullable = true)
 |-- Open: double (nullable = true)
 |-- High: double (nullable = true)
 |-- Low: double (nullable = true)

+----------+-----+-------+------+-----+-----+
|      Date|Close| Volume|  Open| High|  Low|
+----------+-----+-------+------+-----+-----+
|2022-02-15|76.13|6766205| 75.35|76.35| 74.8|
|2022-02-14|74.45|5244967| 73.94|74.62|73.01|
|2022-02-11|73.98|6673354| 75.97|76.55|73.55|
|2022-02-10| 76.4|6432184|75.955|78.39|75.24|
|2022-02-09|78.29|7061571| 76.83|78.67|76.61|
|2022-02-08|75.36|7903249| 73.12|76.07|72.05|
|2022-02-07|73.15|6135832| 74.09|74.99|72.81|
|2022-02-04|73.77|6082889| 71.94|74.95|71.86|
|2022-02-03|71.85|7493688| 72.08| 73.3|71.33|
|2022-02-02|73.21|5887066| 75.58|75.71|72.41|
+----------+-----+-------+------+-----+-----+
only showing Top 10 rows
```

可以看到，Date 字段的数据类型已经被转换为 date 日期类型。接下来，使用时间窗口函数来计算京东股票的周平均收盘价，代码如下：

```
#使用窗口函数计算 groupBy 变换内的周平均价格
#这是一个滚动窗口的例子，也就是固定窗口
jdWeeklyAvg = jdStock \
    .groupBy(window('Date', "1 week")) \
    .agg(avg("Close").alias("weekly_avg"))

#结果模式有窗口启动和结束时间
jdWeeklyAvg.printSchema()

#按开始时间顺序显示结果，并四舍五入到小数点后 2 位
jdWeeklyAvg \
    .orderBy("window.start") \
    .selectExpr("window.start", "window.end", "round(weekly_avg, 2) as weekly_avg") \
    .show(10)
```

执行以上代码，输出结果如下：

```
root
 |-- window: struct (nullable = false)
 |    |-- start: timestamp (nullable = true)
 |    |-- end: timestamp (nullable = true)
 |-- weekly_avg: double (nullable = true)


+-------------------+-------------------+----------+
|              start|                end|weekly_avg|
+-------------------+-------------------+----------+
|2017-02-09 08:00:00|2017-02-16 08:00:00|     30.23|
|2017-02-16 08:00:00|2017-02-23 08:00:00|     30.29|
|2017-02-23 08:00:00|2017-03-02 08:00:00|     30.65|
|2017-03-02 08:00:00|2017-03-09 08:00:00|     30.93|
|2017-03-09 08:00:00|2017-03-16 08:00:00|     31.41|
|2017-03-16 08:00:00|2017-03-23 08:00:00|     31.09|
|2017-03-23 08:00:00|2017-03-30 08:00:00|     31.45|
|2017-03-30 08:00:00|2017-04-06 08:00:00|     31.65|
|2017-04-06 08:00:00|2017-04-13 08:00:00|     32.43|
|2017-04-13 08:00:00|2017-04-20 08:00:00|     33.22|
+-------------------+-------------------+----------+
only showing Top 10 rows
```

上面的例子使用了一个星期的滚动窗口，其中交易数据没有重叠，因此，每个交易只使用一次来计算移动平均值，而下面的例子使用滑动窗口来计算京东股票的月平均收盘价，每周计算一次。这意味着在计算平均每月移动平均值时，一些交易数据将被多次使用。在这个滑动窗口中，窗口的大小是4周，每个窗口一次滑动一周，代码如下：

```
#使用时间窗口函数来计算京东股票的月平均收盘价
#4 周窗口长度，每次滑动 1 周
jdMonthlyAvg = jdStock \
    .groupBy(window('Date', "4 week", "1 week")) \
    .agg(avg("Close").alias("monthly_avg"))

#按开始时间显示结果
jdMonthlyAvg \
    .orderBy("window.start") \
    .selectExpr("window.start", "window.end", "round(monthly_avg, 2) as monthly_avg") \
    .show(10)
```

执行以上代码，输出结果如下：

```
+-------------------+-------------------+-----------+
|              start|                end|monthly_avg|
+-------------------+-------------------+-----------+
|2017-01-19 08:00:00|2017-02-16 08:00:00|      30.23|
|2017-01-26 08:00:00|2017-02-23 08:00:00|      30.28|
```

```
|2017-02-02 08:00:00|2017-03-02 08:00:00|          30.46|
|2017-02-09 08:00:00|2017-03-09 08:00:00|          30.62|
|2017-02-16 08:00:00|2017-03-16 08:00:00|          30.85|
|2017-02-23 08:00:00|2017-03-23 08:00:00|          31.02|
|2017-03-02 08:00:00|2017-03-30 08:00:00|          31.22|
|2017-03-09 08:00:00|2017-04-06 08:00:00|           31.4|
|2017-03-16 08:00:00|2017-04-13 08:00:00|          31.66|
|2017-03-23 08:00:00|2017-04-20 08:00:00|          32.13|
+-------------------+-------------------+---------------+
only showing Top 10 rows
```

由于滑动窗口的间隔是一周,所以这个结果显示了两个连续行的起始时间间隔是一个星期。在连续两行之间,有大约三周的重叠交易,这意味着一个交易被多次使用,以此来计算移动平均值。

5.4.3 使用窗口分析函数

第三类高级分析函数是在逻辑分组中执行聚合的函数,这个逻辑分组被称为窗口,这些函数被称为窗口函数。有时需要对一组数据进行操作,并为每组输入行返回一个值,而窗口函数提供了这种独特的功能,使其易于执行计算,如移动平均、累计和或每行的排名。它们显著地提高了 PySpark 的 SQL 和 DataFrame API 的表达能力。

注意:窗口函数是 SQL2003 标准中定义的一项新特性,并在 SQL2011、SQL2016 中又加以完善,添加了若干拓展。窗口函数不同于用户熟悉的常规函数及聚合函数,它对每行数据进行一次计算,特点是输入多行(一个窗口),但只返回一个值。

使用窗口函数有两个主要步骤,具体步骤如下:

(1)第 1 步是定义一个窗口规范,该规范定义了称为 frame 的行逻辑分组,这是每行被计算的上下文。

(2)第 2 步是应用一个合适的窗口函数。

窗口规范定义了窗口函数将使用的 3 个重要组件,这 3 个组件分别介绍如下:

(1)第 1 个组件被称为 partition by,指定用来对行进行分组的列(一列或多列)。

(2)第 2 个组件称为 order by,它定义了如何根据一列或多列来排序各行,以及顺序是升序还是降序。

(3)最后一个组件称为 frame,它定义了窗口相对于当前行的边界。换句话说,frame 限制了在计算当前行的值时包括哪些行。可以通过行索引或 order by 表达式的实际值来指定在 Window Frame 中包含的一系列行。

最后一个组件 frame 是可选的,有的窗口函数需要,有的窗口函数或场景不需要。窗口规范使用在 pyspark.sql.Window 类中定义的函数构建。rowsBetween 和 rangeBetween 函数分别用来定义行索引和实际值的范围。

窗口函数可分为 3 种类型：排序函数、分析函数和聚合函数。关于排序函数的描述，见表 5-2。

表 5-2　排序函数

函 数 名 称	描　　述
rank	返回一个 frame 内行的排名和排序，基于一些排序规则
dense_rank	类似于 rank，但是在不同的排名之间没有间隔，紧密衔接显示
ntile(n)	在一个有序的窗口分区中返回 ntile 分组 ID。例如，如果 n 是 4，则前 25%行得到的 ID 值为 1，第 2 个 25%行得到的 ID 值为 2，以此类推
row_number	返回一个序列号，每个 frame 从 1 开始

关于分析函数的描述，见表 5-3。

表 5-3　分析函数

函 数 名 称	描　　述
cume_dist	返回一个 frame 的值的累积分布。换句话说，低于当前行的比例
lag(col,offset)	返回当前行之前 offset 行的列值
lead(col,offset)	返回当前行之后 offset 行的列值

对于聚合函数，可以使用前面提到的任何聚合函数作为窗口函数。

【例 5-4】下面通过一个小的样本数据集 MonthlySales.csv 来演示窗口函数的功能。这个小的样本数据集文件包含两个产品（P1 和 P2）的月销售数据，共 24 个观察数据，内容如下：

```
Product,Month,Sales
P1,1,66
P1,2,24
P1,3,54
P1,4,0
P1,5,56
P1,6,34
P1,7,48
P1,8,46
P1,9,76
P1,10,12
P1,11,8
P1,12,24
P2,1,98
P2,2,16
P2,3,78
P2,4,66
P2,5,14
P2,6,76
```

```
P2,7,62
P2,8,92
P2,9,60
P2,10,68
P2,11,10
P2,12,82
```

将该数据文件上传到 HDFS 的/data/spark/目录下。现要求计算过去 3 个月中每个月的平均销售额。

首先创建一个 DataFrame，包含两个产品的月销售数据，代码如下：

```
#第 5 章/dataframe2_demo22.ipynb

from pyspark.sql import SparkSession
from pyspark.sql.functions import *

spark = SparkSession.builder \
   .master("spark://localhost:7077") \
   .appName("pyspark demo") \
   .getOrCreate()

file = "/data/spark/MonthlySales.csv"
monthlySales = spark \
   .read \
   .option("header","true") \
   .option("inferSchema","true") \
   .csv(file)

monthlySales.printSchema()
monthlySales.show()
```

执行以上代码，输出结果如下：

```
root
 |-- Product: string (nullable = true)
 |-- Month: integer (nullable = true)
 |-- Sales: integer (nullable = true)

+-------+-----+-----+
|Product|Month|Sales|
+-------+-----+-----+
|     P1|    1|   66|
|     P1|    2|   24|
|     P1|    3|   54|
|     P1|    4|    0|
|     P1|    5|   56|
|     P1|    6|   34|
```

```
|    P1|    7|   48|
|    P1|    8|   46|
|    P1|    9|   76|
|    P1|   10|   12|
|    P1|   11|    8|
|    P1|   12|   24|
|    P2|    1|   98|
|    P2|    2|   16|
|    P2|    3|   78|
|    P2|    4|   66|
|    P2|    5|   14|
|    P2|    6|   76|
|    P2|    7|   62|
|    P2|    8|   92|
+------+-----+-----+
only showing top 20 rows
```

然后定义一个窗口规范,为每个产品创建一个包含 3 个月的滑动窗口,并在该滑动窗口上定义一个求移动平均值的计算,代码如下:

```
from pyspark.sql import Window

#准备窗口规范,为每个产品创建一个包含 3 个月的滑动窗口
#负数下标表示在当前行之上(前)的行
w = Window.partitionBy("Product").orderBy("Month").rangeBetween(-2,0)

#应用该滑动窗口和计算,检查结果
monthlySales \
    .select("Product",
            "Sales",
            "Month",
            bround(avg("Sales").over(w), 2).alias("MovingAvg")
    ) \
    .orderBy("Product","Month") \
    .show()
```

执行以上代码,输出结果如下:

```
+-------+-----+-----+---------+
|Product|Sales|Month|MovingAvg|
+-------+-----+-----+---------+
|     P1|   66|    1|     66.0|
|     P1|   24|    2|     45.0|
|     P1|   54|    3|     48.0|
|     P1|    0|    4|     26.0|
|     P1|   56|    5|    36.67|
|     P1|   34|    6|     30.0|
```

```
|    P1|   48|    7|     46.0|
|    P1|   46|    8|    42.67|
|    P1|   76|    9|    56.67|
|    P1|   12|   10|    44.67|
|    P1|    8|   11|     32.0|
|    P1|   24|   12|    14.67|
|    P2|   98|    1|     98.0|
|    P2|   16|    2|     57.0|
|    P2|   78|    3|     64.0|
|    P2|   66|    4|    53.33|
|    P2|   14|    5|    52.67|
|    P2|   76|    6|     52.0|
|    P2|   62|    7|    50.67|
|    P2|   92|    8|    76.67|
+------+-----+-----+---------+
only showing top 20 rows
```

【例 5-5】现在假设有两个用户 user01 和 user02，两个用户的购物交易数据如下：

用户 ID	交易日期	交易金额
user01	2018-07-02	13.35
user01	2018-07-06	27.33
user01	2018-07-04	21.72
user02	2018-07-07	69.74
user02	2018-07-01	59.44
user02	2018-07-05	80.14

有了这个购物交易数据，尝试使用窗口函数来回答以下问题：

（1）对于每个用户，最高的交易金额是多少？
（2）每个用户的交易金额和最高交易金额之间的差是多少？
（3）每个用户的交易金额相对于上一次交易的变化是多少？
（4）每个用户的移动平均交易金额是多少？
（5）每个用户的累计交易金额是多少？

首先，构造一个 DataFrame，包含这个小型购物交易数据，代码如下：

```
#第5章/dataframe_demo23.ipynb

from pyspark.sql import SparkSession
from pyspark.sql.functions import *

spark = SparkSession.builder \
    .master("spark://localhost:7077") \
    .appName("pyspark demo") \
    .getOrCreate()

#为两个用户设置的小型购物交易数据集
```

```
txDataDF= spark.createDataFrame([
    ("user01", "2018-07-02", 13.35),
    ("user01", "2018-07-06", 27.33),
    ("user01", "2018-07-04", 21.72),
    ("user02", "2018-07-07", 69.74),
    ("user02", "2018-07-01", 59.44),
    ("user02", "2018-07-05", 80.14)
],["uid", "tx_date", "amount"])

txDataDF.printSchema()
txDataDF.show()
```

执行以上代码，输出内容如下：

```
root
 |-- uid: string (nullable = true)
 |-- tx_date: string (nullable = true)
 |-- amount: double (nullable = false)

+------+----------+------+
|   uid|   tx_date|amount|
+------+----------+------+
|user01|2018-07-02| 13.35|
|user01|2018-07-06| 27.33|
|user01|2018-07-04| 21.72|
|user02|2018-07-07| 69.74|
|user02|2018-07-01| 59.44|
|user02|2018-07-05| 80.14|
+------+----------+------+
```

下面应用窗口函数来回答问题：

（1）为了回答第 1 个问题，可以将 rank()窗口函数应用于一个窗口规范，该规范按用户 ID 对数据进行分区，并按交易金额对其进行降序排序。rank()窗口函数根据每个 frame 中每行的排序顺序给每行分配一个排名，代码如下：

```
#导入 Window 类
from pyspark.sql import Window

#定义窗口规范，按用户 ID 分区，按数量降序排序
w = Window.partitionBy("uid").orderBy(desc("amount"))

#增加一个新列，以包含每行的等级，应用 rank()函数以对每行分级
txDataWithRankDF = txDataDF.withColumn("rank", rank().over(w))
#txDataWithRankDF.show()

#根据等级过滤行，以找到第一名并显示结果
```

```
txDataWithRankDF.where('rank == 1').show()
```

执行以上代码，输出结果如下：

```
+------+----------+------+----+
|   uid|   tx_date|amount|rank|
+------+----------+------+----+
|user02|2018-07-05| 80.14|   1|
|user01|2018-07-06| 27.33|   1|
+------+----------+------+----+
```

可以看出，用户 user01 的最高交易金额是 27.33，用户 user02 的最高交易金额是 80.14。

（2）解决第 2 个问题的方法是在每个分区的所有行的 amount 列上应用 max()函数。除了按用户 ID 分区之外，它还需要定义一个包含每个分区中所有行的 frame 边界。要定义这个 frame，可以使用 Window.rangeBetween()函数，以 Window.unboundedPreceding 作为开始值，以 Window.unboundedFollowing 作为结束值，代码如下：

```
#使用 rangeBetween 来定义 frame 边界，它包含每个 frame 中的所有行
w = Window \
    .partitionBy("uid") \
    .orderBy(desc("amount")) \
    .rangeBetween(Window.unboundedPreceding,Window.unboundedFollowing)

#增加 amount_diff 列，将 max()函数应用于 amount 列，然后计算差值
txDiffWithHighestDF = txDataDF.withColumn(
    "amount_diff", round((max("amount").over(w) - col("amount")), 3)
)

#显示结果
txDiffWithHighestDF.show()
```

执行以上代码，输出结果如下：

```
+------+----------+------+-----------+
|   uid|   tx_date|amount|amount_diff|
+------+----------+------+-----------+
|user02|2018-07-05| 80.14|        0.0|
|user02|2018-07-07| 69.74|       10.4|
|user02|2018-07-01| 59.44|       20.7|
|user01|2018-07-06| 27.33|        0.0|
|user01|2018-07-04| 21.72|       5.61|
|user01|2018-07-02| 13.35|      13.98|
+------+----------+------+-----------+
```

（3）解决第 3 个问题的方法是使用每个分区的当前行的 amount 列减去上一行的 amount 列。获取上一行的指定字段用 lag()函数。除了按用户 ID 分区之外，它还需要定义一个包含每个分区中所有行的 frame 边界。默认 frame 包括所有前面的行和当前行，代码

如下：

```
#定义窗口规范
w = Window.partitionBy("uid").orderBy("tx_date")

#增加 amount_diff 列，计算交易量的变动值
lagDF = txDataDF.withColumn(
    "amount_var", round((col("amount") - lag("amount",1).over(w)), 3)
)

#显示结果
lagDF.show()
```

执行以上代码，输出结果如下：

```
+------+----------+------+----------+
|   uid|   tx_date|amount|amount_var|
+------+----------+------+----------+
|user02|2018-07-01| 59.44|      null|
|user02|2018-07-05| 80.14|      20.7|
|user02|2018-07-07| 69.74|     -10.4|
|user01|2018-07-02| 13.35|      null|
|user01|2018-07-04| 21.72|      8.37|
|user01|2018-07-06| 27.33|      5.61|
+------+----------+------+----------+
```

（4）为了计算每个用户按交易日期顺序移动的平均移动数量，将利用 avg()函数来根据 frame 中的一组行计算每行的平均数量。这里希望每个 frame 都包含 3 行：当前行、前面的一行和后面的一行。与前面的例子类似，窗口规范将按用户 ID 对数据进行分区，但是每个 frame 中的行将按交易日期排序，代码如下：

```
#应用 avg()函数来计算移动平均交易量
#定义窗口规范，一个好的做法是指定相对于 Window.currentRow 的偏移量
w = Window \
    .partitionBy("uid") \
    .orderBy("tx_date") \
    .rowsBetween(Window.currentRow-1, Window.currentRow+1)

#在窗口上将 avg()函数应用到 amount 列，并将移动平均量四舍五入为两位小数
avgDF = txDataDF.withColumn("moving_avg",round(avg("amount").over(w), 2))

#显示结果
avgDF.show()
```

执行以上代码，输出结果如下：

```
+------+----------+------+----------+
|  uid |   tx_date|amount|moving_avg|
+------+----------+------+----------+
|user02|2018-07-01| 59.44|     69.79|
|user02|2018-07-05| 80.14|     69.77|
|user02|2018-07-07| 69.74|     74.94|
|user01|2018-07-02| 13.35|     17.54|
|user01|2018-07-04| 21.72|      20.8|
|user01|2018-07-06| 27.33|     24.53|
+------+----------+------+----------+
```

（5）为了计算每个用户的交易金额的和，将把 sum()函数应用于一个 frame，该 frame 由所有以前的行和当前行组成。其 partitionBy()和 orderBy()方法与移动平均示例相同，实现代码如下：

```
#定义每个 frame 的窗口规范，包括所有以前的行和当前行
w = Window \
    .partitionBy("uid") \
    .orderBy("tx_date") \
    .rowsBetween(Window.unboundedPreceding, Window.currentRow)

#将 sum()函数应用于窗口规范
sumDF = txDataDF.withColumn("culm_sum",round(sum("amount").over(w),2))

#显示结果
sumDF.show()
```

执行以上代码，输出结果如下：

```
+------+----------+------+--------+
|  uid |   tx_date|amount|culm_sum|
+------+----------+------+--------+
|user02|2018-07-01| 59.44|   59.44|
|user02|2018-07-05| 80.14|  139.58|
|user02|2018-07-07| 69.74|  209.32|
|user01|2018-07-02| 13.35|   13.35|
|user01|2018-07-04| 21.72|   35.07|
|user01|2018-07-06| 27.33|    62.4|
+------+----------+------+--------+
```

窗口规范的默认 frame 包括所有前面的行和当前行。对于前面的例子，没有必要指定 frame，所以应该得到相同的结果。

前面的窗口函数示例是使用 DataFrame API 编写的。也可以通过 SQL 实现相同的目标，使用关键字 PARTITION BY、ORDER BY、ROWS BETWEEN 和 RANGE BETWEEN。frame 边界可以使用以下关键字来指定：UNBOUNDED PRECEDING、UNBOUNDED FOLLOWING、CURRENT ROW、<value> PRECEDING 和<value> FOLLOWING。

使用 SQL 的窗口函数的示例，代码如下：

```
#第5章/dataframe2_demo24.ipynb

#将txDataDF注册为一个临时视图，叫作tx_data
txDataDF.createOrReplaceTempView("tx_data")

#使用RANK()函数来找出前两个较高的交易量
spark.sql("""select uid, tx_date, amount, rank from
( select uid, tx_date, amount,RANK() OVER (PARTITION BY uid ORDER BY amount
DESC) as rank
  from tx_data
) where rank=1
""").show()

#与最大交易金额的差额
spark.sql("""select uid, tx_date, amount, round((max_amount - amount),2) as
amount_diff from
( select uid, tx_date, amount,
  MAX(amount) OVER (PARTITION BY uid ORDER BY amount DESC
  RANGE BETWEEN UNBOUNDED PRECEDING AND UNBOUNDED FOLLOWING
) as max_amount from tx_data)
""").show()

#与上一交易金额的差额
spark.sql("""select uid, tx_date, amount, round(amount - lag_amount, 2) as
amount_var from
( select uid, tx_date, amount,
  lag(amount,1) OVER(PARTITION BY uid ORDER BY tx_date) as lag_amount from
tx_data)
""").show()

#移动平均
spark.sql("""select uid, tx_date, amount, round(moving_avg,2) as moving_avg
from
( select uid, tx_date, amount,
  AVG(amount) OVER(
      PARTITION BY uid
      ORDER BY tx_date
      ROWS BETWEEN 1 PRECEDING AND 1 FOLLOWING
  ) as moving_avg from tx_data)
""").show()

#累计和
spark.sql("""select uid, tx_date, amount, round(culm_sum,2) as moving_avg
from
```

```
( select uid, tx_date, amount,
  SUM(amount) OVER(
    PARTITION BY uid
    ORDER BY tx_date
    ROWS BETWEEN UNBOUNDED PRECEDING AND CURRENT ROW
  ) as culm_sum from tx_data)
""").show()
```

执行上面的代码，输出结果应当与使用 DataFrame API 进行窗口操作的结果一样。当使用 SQL 中的窗口函数时，必须在单个语句中指定 partition by、order by 和 frame 窗口。

5.5 用户自定义函数（UDF）

尽管 PySpark SQL 为大多数常见用例提供了大量的内置函数，但总会在一些情况下，这些内置函数都不能提供用户的用例所需要的功能。PySpark SQL 提供了一个相当简单的工具来编写用户自定义的函数，并可在 PySpark 数据处理逻辑或应用程序中使用它们，就像使用内置函数一样。

UDF 是用于扩展框架的函数，并在多个 DataFrame 上重用这些函数。UDF 实际上是用户可以扩展 PySpark 的功能以满足特定需求的一种方式。

5.5.1 内部原理

PySpark 实际上是用 Scala 编写的 Spark Core 的包装器。当在 Python 中启动 SparkSession 时，在后台 PySpark 会使用 Py4J 启动一个 JVM 并创建一个 Java SparkContext。所有的 PySpark 操作，例如对 df.filter()方法调用，都会在后台被转换为对 JVM SparkContext 中各自的 Spark DataFrame 对象的相应调用。这通常是非常快的，只要不数百万次地调用函数，开销可以被忽略，因此，在 df.filter()调用中，DataFrame 操作和筛选条件将被发送到 Java SparkContext，在那里它被编译成一个整体优化的查询计划。一旦执行查询，过滤条件将在 Java 中的分布式 DataFrame 上计算，而不需要回调到 Python。

在这种情况下，在整个查询执行过程中，所有的数据操作都是在 Java Spark Worker 中以分布式的方式执行的，这使 Spark 对于大型数据集的查询非常快。那么，为什么 PySpark RDD filter()方法会慢得多呢?原因是 lambda 函数不能直接应用 JVM 内存中的 DataFrame。在内部实际发生的是，Spark 在集群节点上的 Spark Executor 启动后再启动 Python Worker。在执行时，Spark Workers 将 lambda 函数发送给那些 Python Worker。接下来，Spark Worker 开始序列化它们的 RDD 分区，并通过 Socket 套接字将它们由管道发送到 Python Worker，在那里，lambda 函数将对每行进行计算。对于结果，整个序列化/反序列化过程再次以相反的方向发生，以便使实际的 filter()可以应用于结果集。

在 PySpark 中使用任意 Python 函数时的整个数据流如图 5-5 所示。

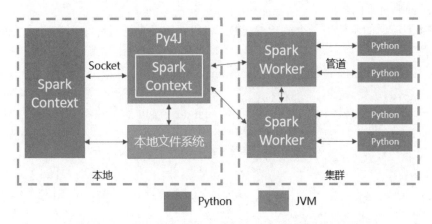

图 5-5　在 PySpark 中使用任意 Python 函数时的整个数据流

当创建 UDF 时，需要非常仔细地设计它们，否则可能会遇到优化和性能问题。UDF 是 PySpark 的黑盒，因此它不能应用优化，用户将失去 PySpark 在 DataFrame 上所做的所有优化。如果可能，则应该使用 PySpark SQL 内置函数，因为这些函数提供了优化。考虑只在现有的内置 SQL 函数不能满足业务需求时才创建 UDF。

从概念上讲，UDF 只是一些常规的函数，它们接受一些输入并提供输出。尽管 UDF 可以用 Scala、Java 或 Python 编写，但是必须注意当 UDF 用 Python 编写时的性能差异。UDF 必须在使用 Spark 之前注册，因此 Spark 知道将它们发送到 executor，以便使用和执行。鉴于 executor 是用 Scala 编写的 JVM 进程，它们可以在同一个进程中本地执行 Scala 或 Java UDF。如果一个 UDF 是用 Python 编写的，则 executor 就不能本地执行它，因此它必须生成一个单独的 Python 进程来执行 Python UDF。除了生成 Python 过程的成本之外，在数据集中的每行中都要对数据进行序列化，这是一个很大的开销。

由上可知，就执行时间而言，在分布式 Java 系统中执行 Python 函数是非常昂贵的，这是因为来回过度地复制数据。

注意：简单总结一下这种低级的变化：只要用户避免所有类型的 Python UDF，一个 PySpark 程序将会和基于 Scala 的 Spark 程序一样快。如果用户不能避免 UDF，则至少应该让它们尽可能高效。

5.5.2　创建和使用 UDF

在 PySpark 中，使用 UDF 涉及以下 3 个步骤：
第 1 步，用 Python 语法创建一个函数并进行测试。
第 2 步，通过将函数名传递给 PySpark SQL 的 udf()函数来注册它。

第 3 步，在 DataFrame 代码或发出 SQL 查询时使用 UDF。在 SQL 查询中使用 UDF 时，注册过程略有不同。

【例 5-6】下面的示例用一个简单的 UDF 将成绩转换为考查等级，它演示了前面提到的 3 个步骤。

首先，创建一个包含学生成绩的 DataFrame，代码如下：

```
#第5章/dataframe2_demo25.ipynb

from pyspark.sql import SparkSession
from pyspark.sql.functions import *

spark = SparkSession.builder \
    .master("spark://localhost:7077") \
    .appName("pyspark sql demo") \
    .getOrCreate()

#创建学生成绩 DataFrame
studentDF = spark.createDataFrame(
    [
        ("张三", 85),
        ("李四", 90),
        ("王老五", 55)
    ],["name","score"]
)

studentDF.printSchema()
studentDF.show()
```

执行以上代码，输出内容如下：

```
root
 |-- name: string (nullable = true)
 |-- score: integer (nullable = false)

+------+-----+
|  name|score|
+------+-----+
|  张三|   85|
|  李四|   90|
|王老五|   55|
+------+-----+
```

将 studentDF 注册到名为 students 的临时视图，代码如下：

```
#注册为视图
studentDF.createOrReplaceTempView("students")
```

```
#spark.sql("select * from students").show()
```

接下来创建一个普通的 Python 函数，用来将成绩转换为考查等级，代码如下：

```
#创建一个函数(普通的 Python 函数)将成绩转换为考查等级
def convertGrade(score):
    if score > 100:
        return "作弊"
    elif score >= 90:
        return "优秀"
    elif score >= 80:
        return "良好"
    elif score >= 70:
        return "中等"
    else:
        return "不及格"

#注册为一个 UDF（在 DataFrame API 中使用时的注册方法）
convertGradeUDF = udf(convertGrade)

#使用该 UDF 将成绩转换为考查等级
studentDF.select("name","score",
convertGradeUDF(col("score")).alias("grade")).show()
```

最后，可以像使用普通 PySpark 内置函数一样使用该 UDF，将成绩转换为考查等级，代码如下：

```
#使用该 UDF 将成绩转换为考查等级
studentDF \
    .select("name","score",convertGradeUDF(col("score")).alias("grade")) \
    .show()
```

执行以上代码，输出结果如下：

```
+------+-----+------+
|  name|score| grade|
+------+-----+------+
|  张三|   85|  良好|
|  李四|   90|  优秀|
|王老五|   55|不及格|
+------+-----+------+
```

当在 SQL 查询中使用 UDF 时，注册过程与上面略有不同，代码如下：

```
#注册为 UDF，在 SQL 中使用
spark.udf.register("convertGrade", convertGrade)
```

```
spark.sql("""
   select name, score, convertGrade(score) as grade
   from students"""
).show()
```

执行以上代码,输出结果如下:

```
+------+-----+------+
| name|score| grade|
+------+-----+------+
|  张三|   85|  良好|
|  李四|   90|  优秀|
|王老五|   55|不及格|
+------+-----+------+
```

【例 5-7】把 DataFrame 中名字字符串中每个单词的第 1 个字母都转换成大写字母。
首先,创建一个 DataFrame,代码如下:

```
#第5章/dataframe2_demo26.ipynb

from pyspark.sql import SparkSession
from pyspark.sql.functions import *

spark = SparkSession.builder \
   .master("spark://localhost:7077") \
   .appName("pyspark sql demo") \
   .getOrCreate()

columns = ["Seqno","Name"]
data = [
    ("1", "john jones"),
    ("2", "tracey smith"),
    ("3", "amy sanders")
]

df = spark.createDataFrame(data=data,schema=columns)

df.show(truncate=False)
```

执行以上代码,输出结果如下:

```
+-----+------------+
|Seqno|Name        |
+-----+------------+
|1    |john jones  |
|2    |tracey smith|
|3    |amy sanders |
+-----+------------+
```

接下来，创建一个普通的 Python 函数，它接受一个字符串参数并将每个单词的第 1 个字母转换为大写字母，代码如下：

```
def convertCase(str):
    resStr=""
    arr = str.split(" ")
    for x in arr:
        resStr= resStr + x[0:1].upper() + x[1:len(x)] + " "
    return resStr
```

然后，将函数传递给 PySpark SQL 的 pyspark.sql.functions.udf()函数，以此将函数 convertCase()注册为 UDF，代码如下：

```
convertUDF = udf(lambda z: convertCase(z), StringType())
```

因为 udf()函数的默认类型是 StringType，因此，也可以编写不带返回类型的语句，代码如下：

```
convertUDF = udf(lambda z: convertCase(z))
```

现在可以在 DataFrame 列上将 convertUDF()作为常规内置函数来使用，代码如下：

```
df.select(col("Seqno"), \
    convertUDF(col("Name")).alias("Name") ) \
    .show(truncate=False)
```

执行以上代码，输出结果如下：

```
+-----+-------------+
|Seqno|Name         |
+-----+-------------+
|1    |John Jones   |
|2    |Tracey Smith |
|3    |Amy Sanders  |
+-----+-------------+
```

也可以在 DataFrame 的 withColumn()函数上使用 udf()函数。下面创建另一个 upperCase()函数，它将输入字符串转换为大写，代码如下：

```
def upperCase(str):
    return str.upper()
```

将 Python 函数 upperCase()转换为 UDF，然后将其与 DataFrame withColumn()一起使用。下面的例子将 Name 列的值转换为大写，并创建一个新列 Cureated Name，代码如下：

```
upperCaseUDF = udf(lambda z:upperCase(z),StringType())

df.withColumn("Cureated Name", upperCaseUDF(col("Name"))) \
  .show(truncate=False)
```

执行以上代码，输出结果如下：

```
+-----+------------+--------------+
|Seqno|Name        | Cureated Name|
+-----+------------+--------------+
|1    |john jones  |JOHN JONES    |
|2    |tracey smith|TRACEY SMITH  |
|3    |amy sanders |AMY SANDERS   |
+-----+------------+--------------+
```

为了在 PySpark SQL 上使用 convertCase()函数，需要使用 spark.udf.register()在 PySpark 上注册这个函数，代码如下：

```
#注册函数
spark.udf.register("convertUDF", convertCase,StringType())

#创建临时视图
df.createOrReplaceTempView("NAME_TABLE")

#执行 SQL 查询，在 SQL 语句中使用自定义函数
spark.sql("select Seqno, convertUDF(Name) as Name from NAME_TABLE") \
    .show(truncate=False)
```

执行以上代码，输出结果如下：

```
+-----+------------+
|Seqno|Name        |
+-----+------------+
|1    |John Jones  |
|2    |Tracey Smith|
|3    |Amy Sanders |
+-----+------------+
```

前面要创建 UDF，需要两步处理：先创建一个 Python 函数，再将该函数注册为 UDF。用户也可以通过注解来创建 UDF，只需一步，代码如下：

```
@udf(returnType=StringType())
def upperCase(str):
    return str.upper()

df.withColumn("Cureated Name", upperCase(col("Name"))) \
  .show(truncate=False)
```

执行以上代码，输出结果如下：

```
+-----+------------+--------------+
|Seqno|Name        |Cureated Name |
+-----+------------+--------------+
|1    |john jones  |JOHN JONES    |
```

```
|2    |tracey smith|TRACEY SMITH |
|3    |amy sanders |AMY SANDERS  |
+-----+------------+-------------+
```

参考更多的 PySpark SQL UDF 示例，代码如下：

```
from pyspark.sql.types import IntegerType

#注册一个匿名的 UDF 函数
slen = udf(lambda s: len(s), IntegerType())

#定义并注册一个 UDF 函数
@udf
def to_upper(s):
    if s is not None:
        return s.upper()

#定义并注册另一个 UDF 函数
@udf(returnType=IntegerType())
def add_one(x):
    if x is not None:
        return x + 1

#构造一个 DataFrame
df = spark.createDataFrame([(1, "John Doe", 21)], ("id", "name", "age"))
df.select(slen("name").alias("slen(name)"),
          to_upper("name"), add_one("age")
  ).show()
```

执行以上代码，输出结果如下：

```
+----------+--------------+------------+
|slen(name)|to_upper(name)|add_one(age)|
+----------+--------------+------------+
|         8|      JOHN DOE|          22|
+----------+--------------+------------+
```

5.5.3 特殊处理

在 PySpark 中应用用户自定义函数（UDF）时，有几个特殊方面需要处理。

1. 执行顺序

需要注意的一点是，PySpark 并不保证子表达式的求值顺序，这意味着表达式不能保证从左到右或以任何其他固定的顺序求值。PySpark 为了查询优化而重新排序执行，因此 AND、OR、WHERE 和 HAVING 表达式会产生副作用，因此，在设计和使用 UDF 时，必须非常小心，特别是对空值的处理，因为这些结果是运行时异常。

例如，在下面的示例中，不能保证 Name is not null 将首先被执行。如果 convertUDF(Name) like '%John%' 先执行，有可能得到一个运行时错误（如果 Name 为空），代码如下：

```
spark.sql("""
    select Seqno, convertUDF(Name) as Name
    from NAME_TABLE
    where Name is not null and convertUDF(Name) like '%John%'
""").show(truncate=False)
```

如果设计不周全，UDF 很容易出错。例如，当某些记录上包含值为 null 的列时，代码如下：

```
columns = ["Seqno","Name"]
data = [
    ("1", "john jones"),
    ("2", "tracey smith"),
    ("3", "amy sanders"),
    ('4', None)
]

df2 = spark.createDataFrame(data=data,schema=columns)
df2.show(truncate=False)
df2.createOrReplaceTempView("NAME_TABLE2")

spark.sql("select convertUDF(Name) from NAME_TABLE2").show(truncate=False)
```

注意，在上面的代码片段中，Seqno 为 4 的记录的 Name 列的值为 None。因为我们没有在 UDF 函数中处理 null，因此使用这个 UDF 时会在 DataFrame 返回错误，错误信息如图 5-6 所示。

```
PythonException:
    An exception was thrown from the Python worker. Please see the stack trace below.
Traceback (most recent call last):
    File "/home/hduser/bigdata/spark-3.1.2/python/lib/pyspark.zip/pyspark/worker.py", line 604, in main
        process()
    File "/home/hduser/bigdata/spark-3.1.2/python/lib/pyspark.zip/pyspark/worker.py", line 596, in process
        serializer.dump_stream(out_iter, outfile)
    File "/home/hduser/bigdata/spark-3.1.2/python/lib/pyspark.zip/pyspark/serializers.py", line 211, in dump_stream
        self.serializer.dump_stream(self._batched(iterator), stream)
    File "/home/hduser/bigdata/spark-3.1.2/python/lib/pyspark.zip/pyspark/serializers.py", line 132, in dump_stream
        for obj in iterator:
    File "/home/hduser/bigdata/spark-3.1.2/python/lib/pyspark.zip/pyspark/serializers.py", line 200, in _batched
        for item in iterator:
    File "/home/hduser/bigdata/spark-3.1.2/python/lib/pyspark.zip/pyspark/worker.py", line 450, in mapper
        result = tuple(f(*[a[o] for o in arg_offsets]) for (arg_offsets, f) in udfs)
    File "/home/hduser/bigdata/spark-3.1.2/python/lib/pyspark.zip/pyspark/worker.py", line 450, in <genexpr>
        result = tuple(f(*[a[o] for o in arg_offsets]) for (arg_offsets, f) in udfs)
    File "/home/hduser/bigdata/spark-3.1.2/python/lib/pyspark.zip/pyspark/worker.py", line 85, in <lambda>
        return lambda *a: f(*a)
    File "/home/hduser/bigdata/spark-3.1.2/python/lib/pyspark.zip/pyspark/util.py", line 73, in wrapper
        return f(*args, **kwargs)
    File "<ipython-input-13-6cc992af9cb0>", line 3, in convertCase
AttributeError: 'NoneType' object has no attribute 'split'
```

图 5-6　在 UDF 中未处理 null 值时的错误信息

注意，在 Python 中 None 被认为是 null。

2. 空值检查

以下几点需要记住：

（1）最好的做法是在 UDF 的内部检查 null，而不是在 UDF 的外部检查 null。

（2）在任何情况下，如果不能在 UDF 中检查 null，则应该至少使用 if 或 case when 来检查 null，并有条件地调用 UDF。

例如，创建一个 null 值安全的 UDF，代码如下：

```
spark.udf.register("_nullsafeUDF", lambda str: convertCase(str) if not
str is None else "" , StringType())

spark.sql("select _nullsafeUDF(Name) from NAME_TABLE2") \
    .show(truncate=False)

spark.sql("""
    select Seqno, _nullsafeUDF(Name) as Name
    from NAME_TABLE2
    where Name is not null and _nullsafeUDF(Name) like '%John%'
""").show(truncate=False)
```

执行以上代码，输出结果如下：

```
+------------------+
|_nullsafeUDF(Name)|
+------------------+
|John Jones        |
|Tracey Smith      |
|Amy Sanders       |
|                  |
+------------------+

+-----+----------+
|Seqno|Name      |
+-----+----------+
|1    |John Jones|
+-----+----------+
```

可以看到，在注册 UDF 时检查 null/None 值，代码将成功地执行，没有错误。

5.6 数据集的 join 连接

PySpark SQL 支持对两个或多个 DataFrame 执行各种类型的 join 连接操作。本节将介绍这些连接类型和使用方法，并介绍在 PySpark SQL 内部执行 join 连接的一些细节。

5.6.1　join 表达式和 join 类型

在高级分析函数中，第 1 个是关于多维聚合的，它对于涉及分层数据分析的用例非常有用，在这种情况下，通常需要在一组分组列中计算子总数和总数。常用的多维聚合函数包括 rollup()和 cube()，它们基本上是在多列上进行分组的高级版本，通常用于在这些列的组合和排列中生成子总数和总数。

执行两个数据集的连接需要指定两个内容。第 1 个是连接表达式，它指定来自每个数据集的哪些列应该用于确定来自两个数据集的哪些行将被包含在连接后的数据集中（确定连接列/等值列）。第 2 种是连接类型，它决定了连接后的数据集中应该包含哪些内容。在 PySpark SQL 所支持的 join 类型见表 5-4。

表 5-4　PySpark SQL 所支持的 join 类型

类　　型	描　　述
内连接（又叫等值连接）	当连接表达式计算结果为 true 时，返回来自两个数据集的行
左外连接	当连接表达式计算结果为 false 时，返回来自左侧数据集的行
右外连接	当连接表达式计算结果为 false 时，返回来自右侧数据集的行
外连接	当连接表达式计算结果为 false 时，返回来自两侧数据集的行
左反连接	当连接表达式计算结果为 false 时，只返回来自左侧数据集的行
左半连接	当连接表达式计算结果为 true 时，只返回来自左侧数据集的行
交叉连接（又名笛卡儿连接）	返回左侧数据集中每行和右侧数据集中每行合并后的行。行的数量将是两个数据集的行数的乘积

左半连接和左反连接是唯一的只有值来自左侧数据集的连接类型。左半连接与过滤左侧数据集中只有在右侧数据集中存在键的行相同。左反连接也只返回来自左侧数据集的数据，但只返回右侧数据集中不存在的记录。

DataFrame 支持自连接，但是在连接结果集中会得到重复的列名，因此当执行自连接时，需要为连接列取一个别名。一旦为每个 DataFrame 设置了别名，在结果中就可以使用 dfName.colName 访问每个 DataFrame 的各列，代码如下：

```
joined = df.alias("a") \
    .join(df.alias("b")) \
    .where(col("a.name") == col("b.name"))
```

在 PySpark SQL 中，通过调用 queryExecution.executedPlan()方法可以看到正在执行的连接类型。如果其中一个表比另一个表小得多，可能需要广播散列连接（Broadcast Hash Join）。可以向 PySpark SQL 提供一个提示：某个 DataFrame 应该在 join 连接之前通过在 DataFrame 上调用 broadcast 对其进行广播，以便进行 join 连接，代码如下：

```
df1.join(broadcast(df2), "key")
```

PySpark 还自动使用 pyspark.sql.conf.autoBroadcastJoinThreshold 来确定是否应该广播表。

5.6.2 执行 join 连接

为了演示如何在 PySpark SQL 中使用 join 连接，需要先准备两个小型的 DataFrame。第 1 个 DataFrame 代表一个员工列表，每行包含员工的姓名和所属部门。第 2 个 DataFrame 包含一个部门列表，每行包含一个部门的 ID 和部门名称。

为此，首先创建两个测试 DataFrame，代码如下：

```
#第5章/dataframe2_demo27.ipynb

from pyspark.sql import SparkSession
from pyspark.sql.functions import *

spark = SparkSession.builder \
   .master("spark://localhost:7077") \
   .appName("pyspark sql demo") \
   .getOrCreate()

#员工
employeeDF = spark.createDataFrame([
    ("刘宏明", 31),
    ("赵薇", 33),
    ("黄海波", 33),
    ("杨幂", 34),
    ("楼一萱", 34),
    ("龙梅子", None)
], ["name","dept_no"])

#部门
deptDF = spark.createDataFrame([
    (31, "销售部"),
    (33, "工程部"),
    (34, "财务部"),
    (35, "市场营销部")
], ["id","name"])
```

将这两个 DataFrame 注册为临时视图，代码如下：

```
employeeDF.createOrReplaceTempView("employees")
deptDF.createOrReplaceTempView("departments")
```

然后就可以使用 SQL 来执行 join 连接测试了。

1. 内连接

这是最常用的连接类型，它使用相等比较的连接表达式，包含来自两个数据集与连接条件相匹配的列。连接的数据集只有当连接表达式结果为真时才包含行。没有匹配列值的行将被排除在连接数据集之外。在 PySpark SQL 中，内连接是默认连接类型。

对 employeeDF 和 deptDF 这两个数据集按 id 执行内连接操作，代码如下：

```
#执行join
employeeDF.join(deptDF, col("dept_no") == col("id"), "inner").show()

#不需要指定join类型，因为"inner"是默认的
#employeeDF.join(deptDF, col("dept_no") == col("id")).show()

#使用SQL
#spark.sql("select * from employees JOIN departments on dept_no == id").show()
```

执行上面的代码，输出结果如下：

```
+--------+-------+---+------+
|emp_name|dept_no| id| name |
+--------+-------+---+------+
|  刘宏明|     31| 31|销售部|
|    赵薇|     33| 33|工程部|
|  黄海波|     33| 33|工程部|
|    杨幂|     34| 34|财务部|
|  楼一萱|     34| 34|财务部|
+--------+-------+---+------+

+--------+-------+---+------+
|emp_name|dept_no| id| name |
+--------+-------+---+------+
|  刘宏明|     31| 31|销售部|
|    赵薇|     33| 33|工程部|
|  黄海波|     33| 33|工程部|
|    杨幂|     34| 34|财务部|
|  楼一萱|     34| 34|财务部|
+--------+-------+---+------+
```

连接表达式可以在 join() 转换中指定，也可以使用 where() 变换。如果列名是唯一的，则可以使用简写引用 join 表达式中的列。如果没有，则需要通过 col() 函数指定特定列来自哪个 DataFrame，代码如下：

```
#join表达式的简写版本
employeeDF.join(deptDF, col("dept_no") == col("id")).show()

#指定特定列来自哪个DataFrame
```

```
employeeDF.join(deptDF, employeeDF.dept_no == deptDF.id).show()

#使用 where transformation 指定 join 表达式
employeeDF.join(deptDF).where('dept_no == id').show()
```

执行上面的代码,输出结果如下:

```
+--------+-------+---+------+
|emp_name|dept_no| id|  name|
+--------+-------+---+------+
|  刘宏明|     31| 31|销售部|
|    赵薇|     33| 33|工程部|
|  黄海波|     33| 33|工程部|
|    杨幂|     34| 34|财务部|
|  楼一萱|     34| 34|财务部|
+--------+-------+---+------+

+--------+-------+---+------+
|emp_name|dept_no| id|  name|
+--------+-------+---+------+
|  刘宏明|     31| 31|销售部|
|    赵薇|     33| 33|工程部|
|  黄海波|     33| 33|工程部|
|    杨幂|     34| 34|财务部|
|  楼一萱|     34| 34|财务部|
+--------+-------+---+------+

+--------+-------+---+------+
|emp_name|dept_no| id|  name|
+--------+-------+---+------+
|  刘宏明|     31| 31|销售部|
|    赵薇|     33| 33|工程部|
|  黄海波|     33| 33|工程部|
|    杨幂|     34| 34|财务部|
|  楼一萱|     34| 34|财务部|
+--------+-------+---+------+
```

2. 左外连接

这个 join 类型的连接后的数据集包括来自内连接的所有行加上来自左侧数据集的连接表达式的计算结果为 False 的所有行。对于那些不匹配的行,它将为右侧数据集的列填充 null 值。例如,对 employeeDF 和 deptDF 执行左外连接,代码如下:

```
#连接类型既可以是"left_outer",也可以是"leftouter"
employeeDF \
    .join(deptDF, col("dept_no") == col("id"), "left_outer") \
    .show()
```

```
#使用SQL
spark.sql("""
    select *
    from employees
        LEFT OUTER JOIN departments
        on dept_no == id
""").show()
```

执行上面的代码，输出结果如下：

```
+--------+-------+----+------+
|emp_name|dept_no|  id|  name|
+--------+-------+----+------+
|  刘宏明|     31|  31|销售部|
|    赵薇|     33|  33|工程部|
|  黄海波|     33|  33|工程部|
|    杨幂|     34|  34|财务部|
|  楼  萱|     34|  34|财务部|
|  龙梅子|      0|null|  null|
+--------+-------+----+------+

+--------+-------+----+------+
|emp_name|dept_no|  id|  name|
+--------+-------+----+------+
|  刘宏明|     31|  31|销售部|
|    赵薇|     33|  33|工程部|
|  黄海波|     33|  33|工程部|
|    杨幂|     34|  34|财务部|
|  楼一萱|     34|  34|财务部|
|  龙梅子|      0|null|  null|
+--------+-------+----+------+
```

3. 右外连接

这种 join 类型的行为类似于左外连接类型的行为，除了将相同的处理应用于右侧数据集之外。换句话说，连接后的数据集包括来自内连接的所有行加上来自右侧数据集的连接表达式的计算结果为 False 的所有行。对于那些不匹配的行，它将为左侧数据集的列填充 null 值。例如，对 employeeDF 和 deptDF 执行右外连接，代码如下：

```
#连接类型既可以是"right_outer"，也可以是"rightouter"
employeeDF \
    .join(deptDF, col("dept_no") == col("id"), "right_outer") \
    .show()

#使用SQL
spark.sql("""
```

```
    select *
    from employees
        RIGHT OUTER JOIN departments
        on dept_no == id
""").show()
```

执行上面的代码,输出结果如下:

```
+--------+-------+----+----------+
|emp_name|dept_no| id|      name|
+--------+-------+----+----------+
|  刘宏明|     31|  31|    销售部|
|  黄海波|     33|  33|    工程部|
|    赵薇|     33|  33|    工程部|
|  楼一萱|     34|  34|    财务部|
|    杨幂|     34|  34|    财务部|
|    null|   null|  35|市场营销部|
+--------+-------+----+----------+

+--------+-------+----+----------+
|emp_name|dept_no| id|      name|
+--------+-------+----+----------+
|  刘宏明|     31|  31|    销售部|
|  黄海波|     33|  33|    工程部|
|    赵薇|     33|  33|    工程部|
|  楼一萱|     34|  34|    财务部|
|    杨幂|     34|  34|    财务部|
|    null|   null|  35|市场营销部|
+--------+-------+----+----------+
```

4. 全外连接

这种 join 类型的行为实际上与将左外连接和右外连接的结果结合起来是一样的。例如,对 employeeDF 和 deptDF 执行全外连接,代码如下:

```
#使用join转换
employeeDF.join(deptDF, col("dept_no") == col("id"), "outer").show()

#使用SQL
spark.sql("""
    select *
    from employees
        FULL OUTER JOIN departments
        on dept_no == id
""").show()
```

执行上面的代码,输出结果如下:

```
+--------+-------+----+----------+
|emp_name|dept_no| id|      name|
+--------+-------+----+----------+
|   龙梅子|      0|null|      null|
|     杨幂|     34|  34|    财务部|
|   楼一萱|     34|  34|    财务部|
|   刘宏明|     31|  31|    销售部|
|     赵薇|     33|  33|    工程部|
|   黄海波|     33|  33|    工程部|
|     null|   null|  35|市场营销部|
+--------+-------+----+----------+


+--------+-------+----+----------+
|emp_name|dept_no| id|      name|
+--------+-------+----+----------+
|   龙梅子|      0|null|      null|
|     杨幂|     34|  34|    财务部|
|   楼一萱|     34|  34|    财务部|
|   刘宏明|     31|  31|    销售部|
|     赵薇|     33|  33|    工程部|
|   黄海波|     33|  33|    工程部|
|     null|   null|  35|市场营销部|
+--------+-------+----+----------+
```

5. 左反连接

这种 join 类型能够发现来自左侧数据集的哪些行在右侧数据集上没有任何匹配的行，而连接后的数据集只包含来自左侧数据集的列。

例如，对 employeeDF 和 deptDF 执行左反连接，代码如下：

```
#使用 join 转换
employeeDF.join(deptDF, col("dept_no") == col("id"), "left_anti").show()

#使用 SQL
spark.sql("""
    select *
    from employees
        LEFT ANTI JOIN departments
        on dept_no == id
""").show()
```

注意：没有右反连接（right anti-join）类型。

执行上面的代码，输出结果如下：

```
+--------+-------+
|emp_name|dept_no|
+--------+-------+
```

```
|emp_name|dept_no|
+--------+-------+
|  龙梅子 |      0|
+--------+-------+


+--------+-------+
|emp_name|dept_no|
+--------+-------+
|  龙梅子 |      0|
+--------+-------+
```

6. 左半连接

这种 join 类型的行为类似于内连接类型，除了连接后的数据集不包括来自右侧数据集的列。可以将这种 join 类型看作与左反连接类型相反，在这里，连接后的数据集只包含匹配的行。

例如，对 employeeDF 和 deptDF 执行左半连接，代码如下：

```
#使用join转换
employeeDF.join(deptDF, col("dept_no") == col("id"), "left_semi").show()

#使用SQL
spark.sql("""
    select *
    from employees
        LEFT SEMI JOIN departments
        on dept_no == id
""").show()
```

执行上面的代码，输出结果如下：

```
+--------+-------+
|emp_name|dept_no|
+--------+-------+
|  刘宏明 |     31|
|   赵薇  |     33|
|  黄海波 |     33|
|   杨幂  |     34|
|  楼一萱 |     34|
+--------+-------+


+--------+-------+
|emp_name|dept_no|
+--------+-------+
|  刘宏明 |     31|
|   赵薇  |     33|
|  黄海波 |     33|
|   杨幂  |     34|
|  楼一萱 |     34|
+--------+-------+
```

7. 交叉连接

交叉连接又称笛卡儿连接。执行交叉连接的示例代码如下：

```
#使用crossJoin transformation 并显示该count
print(employeeDF.crossJoin(deptDF).count())              #24

#使用SQL，并显示前30行以观察连接后的数据集中所有的行
spark.sql("select * from employees CROSS JOIN departments").show(30)
```

执行上面的代码，输出结果如下：

```
24
+--------+-------+---+----------+
|emp_name|dept_no| id|      name|
+--------+-------+---+----------+
|  刘宏明|     31| 31|    销售部|
|  刘宏明|     31| 33|    工程部|
|  刘宏明|     31| 34|    财务部|
|  刘宏明|     31| 35|市场营销部|
|    赵薇|     33| 31|    销售部|
|    赵薇|     33| 33|    工程部|
|    赵薇|     33| 34|    财务部|
|    赵薇|     33| 35|市场营销部|
|  黄海波|     33| 31|    销售部|
|  黄海波|     33| 33|    工程部|
|  黄海波|     33| 34|    财务部|
|  黄海波|     33| 35|市场营销部|
|    杨幂|     34| 31|    销售部|
|    杨幂|     34| 33|    工程部|
|    杨幂|     34| 34|    财务部|
|    杨幂|     34| 35|市场营销部|
|  楼一萱|     34| 31|    销售部|
|  楼一萱|     34| 33|    工程部|
|  楼一萱|     34| 34|    财务部|
|  楼一萱|     34| 35|市场营销部|
|  龙梅子|      0| 31|    销售部|
|  龙梅子|      0| 33|    工程部|
|  龙梅子|      0| 34|    财务部|
|  龙梅子|      0| 35|市场营销部|
+--------+-------+---+----------+
```

5.6.3 处理重复列名

有时，在join两个具有同名列的DataFrame之后，会出现一个意想不到的问题。当这种情况发生时，连接后的DataFrame会有多个同名的列。在这种情况下，在对连接后的DataFrame进行某种转换时，就不太方便引用其中一列。

例如，向 deptDF 增加一个新的列，列名为 dept_no，值来自于 id 列，代码如下：

```
#将 deptDF 的列名 dept_id 修改为 dept_no
deptDF2 = deptDF.withColumn("dept_no", "id")

deptDF2.printSchema()
```

执行上面的代码，输出结果如下：

```
root
 |-- id: long (nullable = false)
 |-- name: string (nullable = true)
 |-- dept_no: long (nullable = false)
```

现在，使用 employeeDF 连接 deptDF2，基于 dept_no 列进行连接，代码如下：

```
dupNameDF = employeeDF.join(deptDF2, employeeDF.dept_no == deptDF2.dept_no)

dupNameDF.printSchema()
```

执行上面的代码，输出结果如下：

```
root
 |-- emp_name: string (nullable = true)
 |-- dept_no: long (nullable = false)
 |-- id: long (nullable = false)
 |-- name: string (nullable = true)
 |-- dept_no: long (nullable = false)
```

注意，dupNameDF 现在有两个名称相同的列，都为 dept_no。当试图在 dupNameDF 中投影 dept_no 列时，PySpark 会抛出一个错误。例如，选择 dept_no 列，代码如下：

```
dupNameDF.select("dept_no")
```

执行上面的代码，会抛出异常信息，内容如下：

```
AnalysisException: Reference 'dept_no' is ambiguous, could be: dept_no, dept_no.
......
```

这个异常信息的意思是，因为 dept_no 列是模糊的（不知道应该引用哪个 DataFrame 中的 dept_no 列），所以无法正确执行，抛出异常。

要解决这个问题，有以下几种方法。

1. 使用原始的 DataFrame

连接后的 DataFrame 记得在连接过程中哪些列来自哪个原始的 DataFrame。为了消除某个特定列来自哪个 DataFrame 的歧义，可以告诉 PySpark 以其原始的 DataFrame 名称作为前缀，代码如下：

```
#解决方法一：明确来自哪个DataFrame
#dupNameDF.select(employeeDF.dept_no).show()
dupNameDF.select(deptDF2.dept_no).show()
```

执行上面的代码，会发现可以正常执行而没有抛出异常。

2. 在 join 之前重命名列

为了避免列名称的模糊性问题，另一种方法是使用 withColumnRenamed 转换来重命名其中一个 DataFrame 中的列，代码如下：

```
#解决方法二：在join之前重命名列，使用withColumnRenamed转换
deptDF3 = deptDF2.withColumnRenamed("dept_no","dept_id")
deptDF3.printSchema()

dupNameDF3 = employeeDF.join(deptDF3, col('dept_no') == col('dept_id'))
dupNameDF3.printSchema()

dupNameDF3.select("dept_no").show()
```

执行上面的代码，会发现可以正常执行而没有抛出异常，因为在 join 时已经不存在重名的列了。

3. 使用一个连接后的列名

在两个 DataFrame 中，当连接的列名相同时，在 join()函数中指定一个连接列名即可，这会自动从连接后的 DataFrame 中删除重复列名，但是，如果这是一个自连接，也就是说连接一个 DataFrame 本身，就没有办法引用其他重复的列名了。在这种情况下，需要使用第二种方法来重命名一个 DataFrame 的列，代码如下：

```
noDupNameDF = employeeDF.join(deptDF2, "dept_no")

noDupNameDF.printSchema()
noDupNameDF.show()
```

执行上面的代码，输出结果如下：

```
root
 |-- dept_no: long (nullable = false)
 |-- emp_name: string (nullable = true)
 |-- id: long (nullable = false)
 |-- name: string (nullable = true)

+-------+--------+---+------+
|dept_no|emp_name| id|  name|
+-------+--------+---+------+
```

```
|    31|    刘宏明|    31|  销售部|
|    33|    赵薇  |    33|  工程部|
|    33|    黄海波|    33|  工程部|
|    34|    杨幂  |    34|  财务部|
|    34|    楼一萱|    34|  财务部|
+------+----------+------+--------+
```

5.6.4　join 连接策略

可以说，join 连接是 PySpark 中最昂贵的操作之一。不过，有两种不同的策略可以用来连接两个数据集，它们是 Shuffle Hash Join 和 Broadcast Hash Join。选择特定策略的主要标准是基于两个数据集的大小。当两个数据集的大小都很大时，应使用 Shuffle Hash Join 策略。当其中一个数据集的大小足够小且可以装入 executors 的内存时，应使用 Broadcast Hash Join 策略。下面分别详细介绍这两种策略。

1. Shuffle Hash Join

Shuffle Hash Join 的具体实现由两个步骤组成。首先计算在每个数据集的每行的连接表达式中的列的哈希值，然后将这些具有相同哈希值的行 Shuffle 到同一分区。为了确定某一行将被移动到哪个分区，Spark 执行一个简单的算术操作，它通过分区的数量来计算哈希值的模。一旦第 1 步完成，第 2 步就将那些具有相同列哈希值的行的列组合起来。在较高的层次上，这两个步骤与 MapReduce 编程模型的步骤很相似。Shuffle Hash Join 中 Shuffling 的过程如图 5-7 所示。

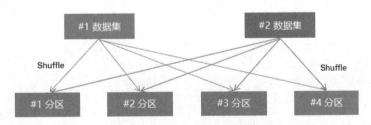

图 5-7　Shuffle Hash Join 中 Shuffling 的过程

正如前面提到的，这是一个昂贵的操作，因为它需要通过网络在多台机器上移动大量数据。当在网络上移动数据时，数据通常会经过数据序列化和反序列化过程。想象一下，在两个大型数据集上执行一个 join 连接，其中每个数据集的大小为 100GB。在这个场景中，它需要移动大约 200GB 的数据。在 join 两个大型数据集时，不可能完全避免 Shuffle Hash Join，但重要的是要注意在可能的情况下减少 join 它们的频率。

2. Broadcast Hash Join

只有当其中一个数据集足够小且小到可以装入内存时，这种 join 策略才适用。从上文已经知道 Shuffle Hash Join 是一项昂贵的操作，Broadcast Hash Join 避免了对两个数据集都进行 Shuffling，而是只对较小的数据进行 Shuffling。与 Shuffle Hash Join 策略类似，这个策略也包含两个步骤。第 1 步是将整个小数据集的副本广播到较大数据集的每个分区上。第 2 步是遍历较大数据集中的每行，并在较小的数据集中按匹配列值查找对应的行。Broadcast Hash Join 连接策略中广播较小数据集的过程如图 5-8 所示。

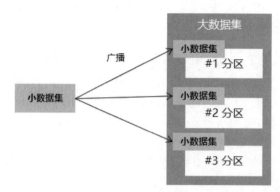

图 5-8　Broadcast Hash Join 连接策略中广播小数据集的过程

很容易理解，在可能的情况下，首选 Broadcast Hash Join。PySpark SQL 在大多数情况下可以根据在读取数据时对数据集的一些统计数据自动判断是否使用 Broadcast Hash Join 或 Shuffle Hash Join，然而，也可以在使用 join 转换时，将一个明确提示提供给 PySpark SQL，以便使用 Broadcast Hash Join，代码如下：

```
import org.pyspark.sql.functions.broadcast

#提供一个提示，使用一个Broadcast Hash Join 来广播deptDF
from pyspark.sql.functions import broadcast
#输出执行计划以验证Broadcast Hash Join 策略被使用了
joined = employeeDF.join(broadcast(deptDF), col("dept_no") == col("id"))

joined.show()

#输出执行计划以验证Broadcast Hash Join 策略被使用了
joined = employeeDF \
    .join(broadcast(deptDF), employeeDF.dept_no == deptDF.id)
```

或者使用 SQL 语句执行连接，代码如下：

```
#使用SQL
joined = spark.sql("""
    select /*+ MAPJOIN(departments) */ *
    from employees
```

```
            JOIN departments on dept_no == id
"""")
joined.show()
```

执行上面的代码,输出结果如下:

```
+--------+-------+---+------+
|emp_name|dept_no| id|  name|
+--------+-------+---+------+
|   刘宏明|     31| 31|销售部|
|     赵薇|     33| 33|工程部|
|   黄海波|     33| 33|工程部|
|     杨幂|     34| 34|财务部|
|   楼一萱|     34| 34|财务部|
+--------+-------+---+------+
```

可以在 DataFrame 上调用 explain()方法输出物理执行计划,代码如下:

```
joined.explain()
```

执行上面的代码,输出的物理执行计划如图 5-9 所示。

```
== Physical Plan ==
*(2) BroadcastHashJoin [dept_no#1L], [dept_id#13L], Inner, BuildRight, false
:- *(2) Filter isnotnull(dept_no#1L)
:  +- *(2) Scan ExistingRDD[name#0,dept_no#1L]
+- BroadcastExchange HashedRelationBroadcastMode(List(input[0, bigint, false]),false), [id=#46]
   +- *(1) Filter isnotnull(dept_id#13L)
      +- *(1) Scan ExistingRDD[dept_id#13L,name#14]
```

图 5-9 输出的物理执行计划

从输出的物理计划可以看到,利用了 Broadcast Hash Join。

5.7 读写 Hive 表

PySpark SQL 还支持读取和写入存储在 Apache Hive 中的数据。PySpark 支持两种 SQL 方言:PySpark 的 SQL 方言和 Hive 查询语言(HQL)。PySpark SQL 支持 HiveQL 语法,同时支持 Hive SerDes 和 UDF,可以访问现有的 Hive 仓库。

用户可以通过 PySpark SQL 访问已经存在的 Hive 表并使用已经存在的社区构建的 Hive UDF。没有部署 Hive 的用户仍然可以启用 Hive 支持。

注意:PySpark 中与每个表相关联的是它的相关元数据,它是关于表及其数据的信息:模式、描述、表名、数据库名、列名、分区、实际数据所在的物理位置等。所有这些都存储在一个中央元存储中。PySpark 默认情况下没有为 PySpark 表提供单独的元存储,而是使用 Apache Hive Metastore，位于/user/hive/warehouse 目录下，用于持久化关于表的所有元数据，但是，可以通过将 PySpark 的配置变量 spark.sql.warehouse.dir 设置为另一个位置来更改默认位置，该位置可以设置为本地或外部分布式存储。

5.7.1 PySpark SQL 的 Hive 配置

要通过 PySpark SQL 访问 Hive 数据表，需要先进行以下配置。

1. 配置 Hive 支持

要配置 Hive 支持，需要将文件 hive-site.xml、core-site.xml（用于安全配置）和 hdfs-site.xml（用于 HDFS 配置）复制到 Spark 安装的 conf 目录下。

注意：如果没有配置 hive-site.xml 文件，则上下文会自动在当前目录中创建 metastore_db，并创建一个由 spark.sql.warehouse.dir 配置的目录，它默认为启动 Spark 应用程序的当前目录下的 spark-warehouse 目录。注意，从 Spark 2.0.0 开始，hive-site.xml 文件中的 hive.metastore.warehouse.dir 属性已被弃用，取而代之的是 spark.sql.warehouse.dir，以指定 warehouse 中数据库的默认位置。用户需要向启动 Spark 应用程序的用户授予写权限。

在 hive-site.xml 文件中，需要包含以下几项：

（1）hive.metastore.warehouse.dir：数据仓库目录，用来存放托管表对应的数据文件。

（2）spark.sql.hive.metastore.version：指定 Hive Metastore 的版本。在 Spark 3.x 中，Hive Metastore 的默认版本为 2.3.7。

（3）spark.sql.hive.metastore.jar：设置为 maven（由 Spark 负责检索 JAR 包）或 Hive jar 存在的系统路径。

（4）hive.metastore.uris：Hive Metastore Server 的 thrift URL。

（5）MySQL 连接相关配置。

关于 PySpark SQL 中的 hive-site.xml 文件的参考配置，内容如下：

```xml
<?xml version="1.0" encoding="UTF-8" standalone="no"?>
<?xml-stylesheet type="text/xsl" href="configuration.xsl"?>
<configuration>
    <property>
            <name>hive.metastore.warehouse.dir</name>
            <value>/user/hive/warehouse</value>
    </property>
    <property>
            <name>spark.sql.hive.metastore.version</name>
            <value>3.1.2</value>
    </property>
    <property>
            <name>spark.sql.hive.metastore.jars</name>
```

```xml
            <value>/home/hduser/bigdata/hive-3.1.2/lib/*</value>
        </property>

        <property>
            <name>hive.metastore.uris</name>
            <value>thrift://xueai8:9083</value>
        </property>

        <property>
            <name>hive.aux.jars.path</name>
            <value>/home/hduser/bigdata/hive-3.1.2/lib/*,/home/hduser/bigdata/hive-3.1.2/jdbc/*
            </value>
        </property>

        <!--配置Hive Metastore:MySQL连接信息 -->
        <property>
            <name>Javax.jdo.option.ConnectionURL</name>
            <value>jdbc:mysql://localhost:3306/hive</value>
        </property>
        <property>
            <name>Javax.jdo.option.ConnectionDriverName</name>
            <value>com.mysql.jdbc.Driver</value>
        </property>
        <property>
            <name>Javax.jdo.option.ConnectionUserName</name>
            <value>root</value>
        </property>
        <property>
            <name>Javax.jdo.option.ConnectionPassword</name>
            <value>admin</value>
        </property>
</configuration>
```

为了用于更低级别的日志记录，可能需要将日志配置添加到 conf/log4j.properties 中，内容如下：

```
log4j.logger.org.apache.spark.sql.hive.HiveUtils$=ALL
log4j.logger.org.apache.spark.sql.internal.SharedState=ALL
log4j.logger.org.apache.spark.sql.hive.client.HiveClientImpl=ALL
```

2. 复制 JDBC 驱动

因为 PySpark SQL 需要访问 Hive Metastore，所以需要将 JDBC 驱动配置在 driver 及所有 executors 的 classpath 中。最简单的方式是在提交应用程序或启动 PySpark Shell 时，使用 --jars 选项指定要提供的 JAR 文件。或者，将 MySQL 的 JDBC 驱动 JAR 包复制到 Spark 安装目录的 jars 子目录下。

3. 启动 Hive Metastore 服务

Hive Metastore 服务是 PySpark SQL 应用程序将要连接的服务，用于获取 Hive 表的元数据。启动 Hive Metastore 服务的命令如下：

```
$ cd $Hive_home/bin
$ hive --service metastore
```

5.7.2 PySpark SQL 读写 Hive 表

可以使用 SparkSession 或 DataFrameReader 的 table()方法从一个 Hive Metastore 的注册表中加载一个 DataFrame。例如，从 Hive 中读取表 test 中的数据，代码如下：

```
df = spark.read.table("test")
df.show()
```

或者，也可以在 SparkSession 上直接调用 table()方法，代码如下：

```
df = spark.table("test")
df.show()
```

将 DataFrame 保存到 Hive 表中，使用 DataFrameWriter 的 saveAsTable()方法，它会将 DataFrame 数据保存到 Hive 表中，并在 Hive Metastore 中注册。例如，将一个 DataFrame 保存到表 test_tb 中，代码如下：

```
spark.range(0,10).write.saveAsTable("test_tb")
spark.table("test_tb").show()
```

默认情况下，如果不指定表路径，PySpark 将把数据写到 warehouse 目录下的默认表路径。当删除表时，默认的表路径也将被删除。

当创建一个 Hive 表时，需要定义这个表如何从/向文件系统读取/写入数据，即定义"输入格式"和"输出格式"。还需要定义此表应该如何将数据反序列化为 Row，或将 Row 序列化为数据，即 serde。例如，创建一个 Hive 表，以 parquet 格式存储，代码如下：

```
CREATE TABLE src(id int) USING hive OPTIONS(fileFormat 'parquet')
```

在使用 create 语句创建 Hive 表时，OPTIONS 可以用来指定存储格式（如"serde"、"input format"、"output format"）。可以指定的 OPTIONS 选项见表 5-5。

表 5-5 Spark SQL 所支持的 OPTIONS 选项

属 性 名	含 义
fileFormat	文件格式是一种存储格式规范的包，包括"serde"、"输入格式"和"输出格式"。目前支持 6 种文件格式："sequencefile"、"rcfile"、"orc"、"parquet"、"textfile"和"avro"
inputFormat outputFormat	这两个选项用于指定相应的' InputFormat '和' OutputFormat '类的名称，例如"org.apache.hadoop.hive.ql.io.orc.OrcInputFormat"。 这两个选项必须成对出现，如果已经指定了 fileFormat 选项，则不能指定它们

续表

属 性 名	含 义
serde	此选项用于指定 serde 类的名称。 当指定"fileFormat"选项时,如果给定的"fileFormat"已经包含 serde 的信息,则不要指定此选项。 目前"sequencefile"、"textfile"和"rcfile"不包含 serde 信息,可以对这 3 种文件格式使用此选项
fieldDelim escapeDelim collectionDelim mapkeyDelim lineDelim	这些选项只能用于"textfile"文件格式。它们定义如何将带分隔符的文件读入行

默认情况下,PySpark 将以纯文本的形式读取表文件。需要注意,在创建表时还不支持 Hive 存储处理程序,可以在 Hive 端使用存储处理程序创建一个表,并使用 PySpark SQL 读取它。

下面通过一个示例程序,演示如何使用 Spark SQL 读取 Hive 元数据及读写 Hive 表数据。

【例 5-8】使用 Spark SQL 读取 Hive 元数据及读写 Hive 表数据。

建议按以下步骤操作。

(1) 确保已经启动了 Hive Metastore Server。

如果没有启动,则可在终端窗口中使用如下命令启动:

```
$ hive --service metastore
```

(2) 查看 spark.sql.catalogImplementation 的内部属性,该属性应该是 hive,代码如下:

```
spark.conf.get("spark.sql.catalogImplementation")
```

(3) 查看当前有哪些数据库和数据表。

列出目前所有已知的数据库,代码如下:

```
spark.catalog.listDatabases()
```

列出所有已知的表(包括 Hive 表和临时表),代码如下:

```
spark.catalog.listTables()
```

列出 default 数据库中的表,代码如下:

```
spark.catalog.listTables("default")
```

(4) 通过 PySpark 创建 Hive 表 src,代码如下:

```
spark.sql("CREATE TABLE IF NOT EXISTS src (key INT, value STRING) USING hive")
```

（5）将 PySpark 自带的数据文件 kv1.txt 加载到上面所创建的 Hive 表中，代码如下：

```
file = "/data/spark/resources/kv1.txt"
spark.sql(f"LOAD DATA LOCAL INPATH '{file}' INTO TABLE src")
```

（6）用 HiveQL 进行查询，代码如下：

```
spark.sql("SELECT * FROM src").show(10)
```

执行以上查询代码，输出内容如下：

```
+---+-------+
|key|  value|
+---+-------+
|238|val_238|
| 86| val_86|
|311|val_311|
| 27| val_27|
|165|val_165|
|409|val_409|
|255|val_255|
|278|val_278|
| 98| val_98|
|484|val_484|
+---+-------+
only showing Top 10 rows
```

还支持聚合查询，代码如下：

```
spark.sql("SELECT COUNT(*) as cnt FROM src").show()
```

执行以上查询代码，输出内容如下：

```
+----+
| cnt|
+----+
|1500|
+----+
```

SQL 查询的结果本身就是 DataFrame，并且支持所有正常的函数，代码如下：

```
sqlDF = sql("SELECT key, value FROM src WHERE key < 10 ORDER BY key")
sqlDF.show()
```

执行以上代码，输出内容如下：

```
+-------------------+
|              value|
+-------------------+
|Key: 0, Value: val_0|
```

```
|Key: 0, Value: val_0|
|Key: 0, Value: val_0|
|Key: 0, Value: val_0|
|Key: 0, Value: val_0|
|Key: 0, Value: val_0|
|Key: 0, Value: val_0|
|Key: 0, Value: val_0|
|Key: 2, Value: val_2|
|Key: 2, Value: val_2|
|Key: 2, Value: val_2|
|Key: 4, Value: val_4|
|Key: 4, Value: val_4|
|Key: 4, Value: val_4|
|Key: 5, Value: val_5|
|Key: 5, Value: val_5|
|Key: 5, Value: val_5|
|Key: 5, Value: val_5|
|Key: 5, Value: val_5|
+--------------------+
only showing top 20 rows
```

（7）另外创建一个 DataFrame，并将其与已有的 src 表执行 join 连接，代码如下：

```
#创建另一个 DataFrame
recordsDF = spark.sparkContext \
    .parallelize(range(1,101)) \
    .map(lambda i: (i, f"val_{i}")) \
    .toDF(["key","value"])

#使用 DataFrame 在 SparkSession 中创建临时视图
recordsDF.createOrReplaceTempView("records")

#将新的 DataFrame 与存储在 Hive 中的数据连接起来
spark.sql("SELECT * FROM records r JOIN src s ON r.key = s.key").show()
```

执行以上代码，输出结果如下：

```
+---+------+---+------+
|key| value|key| value|
+---+------+---+------+
| 86|val_86| 86|val_86|
| 27|val_27| 27|val_27|
| 98|val_98| 98|val_98|
| 66|val_66| 66|val_66|
| 37|val_37| 37|val_37|
| 15|val_15| 15|val_15|
```

```
| 82|val_82| 82|val_82|
| 17|val_17| 17|val_17|
| 57|val_57| 57|val_57|
| 20|val_20| 20|val_20|
| 92|val_92| 92|val_92|
| 47|val_47| 47|val_47|
| 72|val_72| 72|val_72|
|  4| val_4|  4| val_4|
| 35|val_35| 35|val_35|
| 54|val_54| 54|val_54|
| 51|val_51| 51|val_51|
| 65|val_65| 65|val_65|
| 83|val_83| 83|val_83|
| 12|val_12| 12|val_12|
+---+------+---+------+
only showing top 20 rows
```

（8）将 DataFrame 写入 Hive 表中，代码如下：

```
#创建一个 Hive 托管的 Parquet 表,使用 HQL 语法而不是 Spark SQL 原生语法"USING hive"
spark.sql("CREATE TABLE hive_records(key int, value string) STORED AS PARQUET")

#将 src 表数据加载到 DataFrame
df = spark.table("src")

#saveAsTable()方法：将 DataFrame 数据保存到 Hive 表中
df.write.mode("overwrite").saveAsTable("hive_records")

#插入之后，Hive 托管表现在有数据了
spark.sql("SELECT * FROM hive_records").show()
#spark.table("hive_records").show()   #也可
```

执行以上代码，输出结果如下：

```
+---+-------+
|key|  value|
+---+-------+
|238|val_238|
| 86| val_86|
|311|val_311|
| 27| val_27|
|165|val_165|
|409|val_409|
|255|val_255|
|278|val_278|
| 98| val_98|
```

```
|484|val_484|
|265|val_265|
|193|val_193|
|401|val_401|
|150|val_150|
|273|val_273|
|224|val_224|
|369|val_369|
| 66| val_66|
|128|val_128|
|213|val_213|
+---+-------+
only showing top 20 rows
```

(9)也可以通过 PySpark SQL 读写 Hive 外部表,代码如下:

```
#准备一个 HDFS 目录,DataFrame 数据将以 parquet 格式写入该目录
dataDir = "/data/parquet_data"              #用作外部表目录
spark.range(10).write.parquet(dataDir)

#可以使用 hdfs 命令查看外部表目录:hdfs dfs -ls /data/parquet_data

#创建一个 Hive external Parquet 表
spark.sql(f"CREATE EXTERNAL TABLE hive_bigints(id bigint) STORED AS PARQUET
LOCATION '{dataDir}'")

#Hive 外部表应该有数据了
spark.sql("SELECT * FROM hive_bigints").show()
#spark.table("hive_bigints").show()
```

(10)PySpark SQL 也支持 Hive 动态分区表。从 PySpark 2.1 开始,持久数据源表将每个分区的元数据存储在 Hive Metastore 中。例如,将 DataFrame 按 key 列动态分区存储到 Hive 分区表中,代码如下:

```
#打开 Hive 动态分区的标志
spark.sql("SET hive.exec.dynamic.partition = true")
spark.sql("SET hive.exec.dynamic.partition.mode = nonstrict")
#spark.sql("SET hive.exec.max.dynamic.partitions.pernode = 400")

#使用 DataFrame API 创建一个 Hive 分区表
df.write.partitionBy("key").format("hive").saveAsTable("hive_part_tbl")

#可以使用 hdfs 命令查看物理分区:hdfs dfs -ls /user/hive/warehouse/hive_part_tbl

#分区列'key'将被移动到 schema 的末尾
spark.sql("SELECT * FROM hive_part_tbl").show()
```

```
#按 key 查找
sql("SELECT * FROM hive_part_tbl where key=33").show()
```

也可以在创建 SparkSession 时启用 Hive 动态分区配置，代码如下：

```
spark = SparkSession.builder \
    .master("spark://localhost:7077") \
    .appName("pyspark sql demo") \
    .config("spark.hadoop.hive.exec.dynamic.partition", "true") \
    .config("spark.hadoop.hive.exec.dynamic.partition.mode","nonstrict") \
    .enableHiveSupport() \
    .getOrCreate()
```

也可以像下面这样启用 Hive 动态分区，代码如下：

```
from pyspark.sql import HiveContext

sqlContext = HiveContext(spark.sparkContext)
sqlContext.setConf("hive.exec.dynamic.partition", "true")
sqlContext.setConf("hive.exec.dynamic.partition.mode", "nonstrict")
```

注意：

（1）当将任何表作为分区表处理时，分区列将区分大小写。

（2）分区列应该在 DataFrame 中以相同的名称出现（区分大小写）。

【例 5-9】 使用 PySpark SQL 读取 Hive 元数据及读写 Hive 表数据。

本例中将使用的数据是 MovieLens 数据集。将使用其中的 movies、ratings 和 tags 一共 3 个数据集。建议按以下步骤操作。

（1）确保已经启动了 Hive Metastore Server。如果没有启动，则应该在终端窗口中使用如下命令启动：

```
$ hive --service metastore
```

（2）查看 spark.sql.catalogImplementation 的内部属性，该属性应该是 hive，代码如下：

```
spark.conf.get("spark.sql.catalogImplementation")
```

（3）查看当前有哪些数据库和数据表。

列出目前所有已知的数据库，代码如下：

```
spark.catalog.listDatabases()
```

列出所有已知的表（包括 Hive 表和临时表），代码如下：

```
spark.catalog.listTables()
```

列出 default 数据库中的表，代码如下：

```
spark.catalog.listTables("default")
```

查看数据库信息，代码如下：

```
spark.sql("show databases").show()
```

查看数据表信息，代码如下：

```
spark.sql("show tables in default").show()
#spark.sql("show tables").show()
```

（4）创建一个新的数据库 movies_db，代码如下：

```
spark.sql("create database if not exists movies_db")
```

（5）查看数据库信息，代码如下：

```
spark.sql("DESCRIBE DATABASE EXTENDED movies_db").show(truncate=False)
```

（6）切换到数据库 movies_db，代码如下：

```
spark.sql("use movies_db")
```

（7）在 movies_db 数据库中创建数据表 movies，并加载数据，代码如下：

```
sql = """
  create table if not exists movies (movieId int, title string, genres string)
  row format delimited fields terminated by ','
  stored as textfile
  tblproperties("skip.header.line.count"="1")
"""
spark.sql(sql)

#查看表的详细信息
spark.sql("describe formatted movies").show(truncate = False)

#将数据加载到 Hive 表中
file = "/data/spark/ml-latest-small/movies.csv"
spark.sql(f"load data local inpath '{file}' into table movies")

#用 HiveQL 进行查询
spark.sql("SELECT * FROM movies").show(truncate=False)
```

（8）将文件 ratings.csv 加载到 DataFrame 中，然后写入 ratings 表中，代码如下：

```
#以 ORC 格式创建另一张表
spark.sql("create table if not exists ratings (userId int,movieId int,rating float,timestamp string) stored as orc")

#查看表的详细信息
spark.sql("describe formatted ratings").show(truncate=False)

#查看现在有什么数据表
```

```
spark.sql("show tables in movies_db").show()

#列出所有Spark SQL已知的表
spark.catalog.listTables()

#指定Schema
from pyspark.sql.types import *

#指定Schema
ratingsFields = [
        StructField("userId", IntegerType(), nullable=True),
        StructField("movieId", IntegerType(), nullable=True),
        StructField("rating", DoubleType(), nullable=True),
        StructField("timestamp", StringType(), nullable=True)
]
ratingsSchema = StructType(ratingsFields)

#创建DataFrame
ratingsFile = "/data/spark/ml-latest-small/ratings.csv"
ratingsDF = spark.read \
   .option("header","true") \
   .schema(ratingsSchema) \
   .csv(ratingsFile)

ratingsDF.printSchema()
ratingsDF.show(5)

#saveAsTable()方法：将DataFrame数据保存到Hive表中
ratingsDF.write.mode("overwrite").saveAsTable("ratings")     #覆盖写入

#读取保存的表数据
spark.table("ratings").show()
```

（9）在Hive表中查询数据，并将查询结果保存到另一个Hive表，代码如下：

```
countByGenresSql = """
  select genres, count(*) as count
  from movies
  group by genres
  having count(*) > 500
  order by count desc
"""

genresByCountDF = spark.sql(countByGenresSql)

genresByCountDF.printSchema()
genresByCountDF.show()
```

```
#以 parquet 格式创建另一张表。稍后将按影片类型插入电影数量统计
spark.sql("create table if not exists genres_by_count( genres string,count
int) stored as parquet" )

#将统计结果保存到 Hive 表中
genresByCountDF.write.mode("overwrite").saveAsTable("genres_by_count")

#查看数据是否都已经正确地插入 Hive 表中了
spark.table("genres_by_count").show()
```

（10）从 tags 数据集创建一个临时表，然后将它与 Hive 中的 movies 表和 rating 表连接（join 连接）起来，代码如下：

```
#指定字段和 Schema
tagsFields = [
        StructField("userId", IntegerType(), nullable=True),
        StructField("movieId", IntegerType(), nullable=True),
        StructField("tag", StringType(), nullable=True),
        StructField("timestamp", StringType(), nullable=True),
]
tagsSchema = StructType(tagsFields)

#创建 DataFrame
tagsFile = "file://home/hduser/data/spark/ml-latest-small/tags.csv"
tagsDF = spark.read.option("header","true").schema(tagsSchema).csv(tagsFile)

tagsDF.printSchema()
tagsDF.show(5)

#接下来，将该 DataFrame 注册为临时表
tagsDF.createOrReplaceTempView("tags_tb")

#3 张表执行内连接
joinSql = """
    select m.title, m.genres, r.movieId, r.userId, r.rating,
        r.timestamp as ratingTimestamp, t.tag, t.timestamp as tagTimestamp
    from ratings as r
        inner join tags_tb t on r.movieId = t.movieId and r.userId = t.userId
        inner join movies as m on r.movieId = m.movieId
"""

joinedDF = spark.sql(joinSql)

joinedDF.printSchema()
joinedDF.show(10)
```

```
#查看前 5 条记录
joinedDF.select("title","genres","rating","tag").show(5, truncate = False)
```

5.7.3　分桶、分区和排序

在使用 PySpark SQL 读写 Hive 表数据时，对于基于文件的输出，还可以进行分桶、分区和排序。DataFrameWriter 提供了 bucketBy()和 sortBy()方法，用于控制输出文件的目录结构。如果要对输出结果分桶存储，则可以使用"write.bucketBy（<分桶数>,<分桶字段>）"方法。

修改上面的例子代码，让 movies 结果集以 parquet 格式按电影名称（title 列）分桶存储，代码如下：

```
from pyspark.sql import SparkSession
from pyspark.sql.functions import *

spark = SparkSession.builder \
    .master("spark://localhost:7077") \
    .appName("pyspark sql demo") \
    .config("spark.hadoop.hive.exec.dynamic.partition", "true") \
    .config("spark.hadoop.hive.exec.dynamic.partition.mode","nonstrict") \
    .enableHiveSupport() \
    .getOrCreate()

#将电影数据集文件加载到 DataFrame 中
file = "/data/spark/ml-latest-small/movies.csv"
movies = spark.read \
    .option("header","true") \
    .option("inferSchema","true") \
    .csv(file)

#输出 movies DataFrame，使用 parquet 格式并按 title 列存储在 5 个桶中
#不能直接将 bucket 数据保存到任何其他文件中，只能持久化到 Hive 的表中
movies.write \
    .format("parquet") \
    .mode("overwrite") \
    .bucketBy(5,"title") \
    .saveAsTable("movies_tb_3")

#可以查看 HDFS 文件系统
#hdfs dfs -ls /user/hive/warehouse/movies_tb_3
```

执行以上代码，将 movies 数据集以 parquet 格式存储到 movies_tb 表中，分为 5 个桶文件存储，然后打开浏览器，查看该表对应的桶文件，如图 5-10 所示。

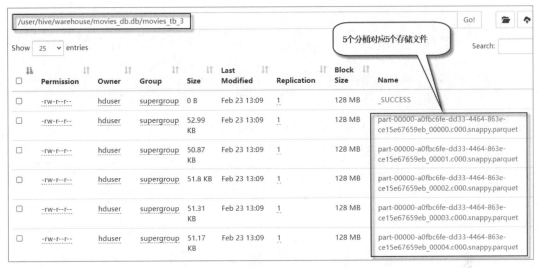

图 5-10　数据集在 Hive 中分桶存储

可以看到，将数据存储到 5 个桶中，对应 HDFS 中的 5 个存储文件。

有可能对单个表同时使用分区、分桶及排序。例如，存储 movies 电影数据集，按电影风格（genres 字段）分区存储，同一分区下按电影名称（title 字段）分 5 个桶存储，并且按 movieId 排序，修改之后的代码如下：

```
from pyspark.sql import SparkSession
from pyspark.sql.functions import *

spark = SparkSession.builder \
    .master("spark://localhost:7077") \
    .appName("pyspark sql demo") \
    .config("spark.hadoop.hive.exec.dynamic.partition", "true") \
    .config("spark.hadoop.hive.exec.dynamic.partition.mode","nonstrict") \
    .enableHiveSupport() \
    .getOrCreate()

#将电影数据集文件加载到 DataFrame 中
file = "/data/spark/movies2/movies.csv"
movies = spark.read \
    .option("header","true") \
    .option("inferSchema","true") \
    .csv(file)

#输出 movies DataFrame
movies \
    .where("year== 2012") \
    .write \
```

```
      .format("parquet") \
      .mode("overwrite") \
      .partitionBy("year") \
      .bucketBy(5,"title") \
      .sortBy("actor") \
      .saveAsTable("movies_tb_4")
```

执行以上代码,然后打开浏览器,查看对应的存储目录结构,结果如图5-11所示。

图5-11 数据集先分区后分桶进行存储

注意:
(1)目前bucketBy需要和saveAsTable()结合使用,而不能和save()一起使用。
(2)目前只支持sortBy()和bucketBy()结合使用,并且排序列不可以是分区列的一部分。

5.8 PySpark SQL 编程案例

本节通过对几个案例的学习,掌握使用PySpark SQL进行大数据分析的复杂用法。

5.8.1 电商订单数据分析

本例使用著名的电商数据集Northwind。

【例 5-10】Northwind 是一个最初由 Microsoft 创建的示例数据库,包含一个名为 Northwind Traders 的虚拟公司的销售数据,该公司从世界各地进口和出口特色食品。

现要求通过分析该电商数据集,回答以下问题:
(1)每个客户下了多少订单?
(2)每个国家的订单有多少?

(3) 每月/年有多少订单?
(4) 每个客户的年销售总额是多少?
(5) 客户每年的平均订单是多少?

要回答以上问题,需要用到其中的订单表和订单明细表。建议按以下步骤操作。

(1) 首先构造一个 SparkSession 实例,代码如下:

```
#第5章/sparksql_example1.ipynb

from pyspark.sql import SparkSession
from pyspark.sql.functions import *

#创建 SparkSession 的实例
spark = SparkSession.builder \
    .master("spark://localhost:7077") \
    .appName("pyspark sql demo") \
    .config("spark.hadoop.hive.exec.dynamic.partition", "true") \
    .config("spark.hadoop.hive.exec.dynamic.partition.mode","nonstrict") \
    .enableHiveSupport() \
    .getOrCreate()
```

(2) 加载订单数据集,代码如下:

```
#读取源数据文件
filePath = "/data/spark/nw/"

#加载订单数据集到 DataFrame 中
orders = spark.read \
    .option("header","true") \
    .option("inferSchema","true") \
    .csv(filePath + "NW-Orders-01.csv") \

print(f"订单有{orders.count()}行")

orders.printSchema()
orders.show(3)
```

执行以上代码,输出结果如下:

```
订单有830 行
root
 |-- OrderID: integer (nullable = true)
 |-- CustomerID: string (nullable = true)
 |-- EmployeeID: integer (nullable = true)
 |-- OrderDate: string (nullable = true)
 |-- ShipCountry: string (nullable = true)
```

```
+-------+----------+----------+---------+------------+
|OrderID|CustomerID|EmployeeID|OrderDate|ShipCountry|
+-------+----------+----------+---------+------------+
|  10248|     VINET|         5|   1996-7-2|     France|
|  10249|     TOMSP|         6|   1996-7-3|    Germany|
|  10250|     HANAR|         4|   1996-7-6|     Brazil|
+-------+----------+----------+---------+------------+
only showing top 3 rows
```

（3）加载订单明细数据集，代码如下：

```
#将订单明细数据集加载到 DataFrame 中
orderDetails = spark.read \
    .option("header","true") \
    .option("inferSchema","true") \
    .csv(filePath + "NW-Order-Details.csv") \

print(f"订单明细有{orderDetails.count()}行")
orderDetails.printSchema()
orderDetails.show(3)
```

执行以上代码，输出结果如下：

```
订单明细有 2155 行
root
 |-- OrderID: integer (nullable = true)
 |-- ProductId: integer (nullable = true)
 |-- UnitPrice: double (nullable = true)
 |-- Qty: integer (nullable = true)
 |-- Discount: double (nullable = true)

+-------+---------+---------+---+--------+
|OrderID|ProductId|UnitPrice|Qty|Discount|
+-------+---------+---------+---+--------+
|  10248|       11|     14.0| 12|     0.0|
|  10248|       42|      9.8| 10|     0.0|
|  10248|       72|     34.8|  5|     0.0|
+-------+---------+---------+---+--------+
only showing top 3 rows
```

（4）回答第 1 个问题：每个客户下了多少订单？代码如下：

```
from pyspark.sql.functions import col

orderByCustomer = orders.groupBy("CustomerID").count()
orderByCustomer.sort(col("count").desc()).show(3)
```

执行以上代码，输出结果如下：

```
+----------+-----+
|CustomerID|count|
+----------+-----+
|     SAVEA|   31|
|     ERNSH|   30|
|     QUICK|   28|
|     HUNGO|   19|
|     FOLKO|   19|
|     HILAA|   18|
|     RATTC|   18|
|     BERGS|   18|
|     BONAP|   17|
|     WARTH|   15|
+----------+-----+
only showing Top 10 rows
```

（5）回答第 2 个问题：每个国家的订单有多少？代码如下：

```
orderByCountry = orders.groupBy("ShipCountry").count()
orderByCountry.sort(col("count").desc()).show(3)
```

执行以上代码，输出结果如下：

```
+-----------+-----+
|ShipCountry|count|
+-----------+-----+
|    Germany|  122|
|        USA|  122|
|     Brazil|   82|
|     France|   77|
|         UK|   56|
|  Venezuela|   46|
|    Austria|   40|
|     Sweden|   37|
|     Canada|   30|
|     Mexico|   28|
+-----------+-----+
only showing Top 10 rows
```

对于后面的 3 个问题，需要对数据进行转换：①向 Orders DataFrame 增加一个 OrderTotal 列，为此，需要计算每个明细的实际金额，然后根据 Order ID 统计每张订单的总金额，并对 order details & orders 进行等值内连接，增加订单总金额。另外还要检查是否有任何 null 列；②增加一个 date 列；③增加 month 和 year 列，以便按月进行统计。

（6）向 Orders DataFrame 增加一个 OrderTotal 列。首先，向 order details 中增加每行的小计（每个订单明细的实际金额），代码如下：

```
orderDetails1 = orderDetails.select(
    "OrderID",
    (col("UnitPrice") * col("Qty") - col("UnitPrice") * col("Qty") *
col("Discount")).alias("OrderPrice")
)

orderDetails1.show(10)
```

执行以上代码，输出结果如下：

```
+-------+----------+
|OrderID|OrderPrice|
+-------+----------+
|  10248|     168.0|
|  10248|      98.0|
|  10248|     174.0|
|  10249|     167.4|
|  10249|    1696.0|
|  10250|      77.0|
|  10250|    1261.4|
|  10250|     214.2|
|  10251|     95.76|
|  10251|     222.3|
+-------+----------+
only showing Top 10 rows
```

然后根据 Order ID 统计每张订单的总金额，代码如下：

```
from pyspark.sql.functions import sum,bround

orderTot = orderDetails1 \
    .groupBy("OrderID") \
    .agg(sum("OrderPrice").alias("OrderTotal"))

orderTot \
    .select("OrderID",bround("OrderTotal",2).alias("OrderTotal")) \
    .sort("OrderID") \
    .show(5)
#orderTot.sort(col("OrderTotal").desc()).show(10)
```

执行以上代码，输出结果如下：

```
+-------+----------+
|OrderID|OrderTotal|
+-------+----------+
|  10248|     440.0|
|  10249|    1863.4|
|  10250|    1552.6|
```

```
|  10251|    654.06|
|  10252|    3597.9|
|  10253|    1444.8|
|  10254|    556.62|
|  10255|    2490.5|
|  10256|     517.8|
|  10257|    1119.9|
+-------+----------+
only showing Top 10 rows
```

接下来，对 orderTot & orders 进行等值内连接，为订单表增加一个订单总金额属性列（Total），代码如下：

```
orders1 = orders \
    .join(orderTot, "OrderID", "inner") \
    .select(
        orders.OrderID,
        orders.CustomerID,
        orders.OrderDate,
        orders.ShipCountry,
        orderTot.OrderTotal.alias("Total")
    )

orders1.sort("CustomerID").show()
```

执行以上代码，输出结果如下：

```
+-------+----------+----------+-----------+-----------------+
|OrderID|CustomerID| OrderDate|ShipCountry|            Total|
+-------+----------+----------+-----------+-----------------+
|  11011|     ALFKI|  1998-4-7|    Germany|            933.5|
|  10692|     ALFKI| 1997-10-1|    Germany|            878.0|
|  10702|     ALFKI|1997-10-11|    Germany|            330.0|
|  10835|     ALFKI| 1998-1-13|    Germany|            845.8|
|  10643|     ALFKI| 1997-8-23|    Germany|            814.5|
|  10952|     ALFKI| 1998-3-14|    Germany|            471.2|
|  10308|     ANATR| 1996-9-16|     Mexico|             88.8|
|  10926|     ANATR|  1998-3-2|     Mexico|            514.4|
|  10759|     ANATR|1997-11-26|     Mexico|            320.0|
|  10625|     ANATR|  1997-8-6|     Mexico|           479.75|
|  10507|     ANTON| 1997-4-13|     Mexico|          749.0625|
|  10365|     ANTON|1996-11-25|     Mexico|403.20000000000005|
|  10535|     ANTON| 1997-5-11|     Mexico|           1940.85|
|  10573|     ANTON| 1997-6-17|     Mexico|            2082.0|
|  10682|     ANTON| 1997-9-23|     Mexico|             375.5|
|  10677|     ANTON| 1997-9-20|     Mexico|           813.365|
|  10856|     ANTON| 1998-1-26|     Mexico|             660.0|
```

```
|  10383|     AROUT|1996-12-14|               UK|              899.0|
|  10355|     AROUT|1996-11-13|               UK|              480.0|
|  10453|     AROUT| 1997-2-19|               UK|              407.7|
+-------+----------+----------+-----------------+-------------------+
only showing top 20 rows
```

最后,检查 orders 是否有空值,代码如下:

```
orders1.filter(col("Total").isNull()).show()
```

执行以上代码,输出结果如下:

```
+-------+----------+---------+-----------+-----+
|OrderID|CustomerID|OrderDate|ShipCountry|Total|
+-------+----------+---------+-----------+-----+
+-------+----------+---------+-----------+-----+
```

可以看到没有空值,说明不需要处理空值。

(7) 对 OrderDate 列进行规范化,转换为一个新的 Date 属性列,代码如下:

```
from pyspark.sql.functions import to_date

orders2 = orders1.withColumn("Date",to_date("OrderDate"))
orders2.printSchema()
orders2.show(5)
```

执行以上代码,输出结果如下:

```
root
 |-- OrderID: integer (nullable = true)
 |-- CustomerID: string (nullable = true)
 |-- OrderDate: string (nullable = true)
 |-- ShipCountry: string (nullable = true)
 |-- Total: double (nullable = true)
 |-- Date: date (nullable = true)

+-------+----------+---------+-----------+------------------+----------+
|OrderID|CustomerID|OrderDate|ShipCountry|             Total|      Date|
+-------+----------+---------+-----------+------------------+----------+
|  10248|     VINET| 1996-7-2|     France|             440.0|1996-07-02|
|  10249|     TOMSP| 1996-7-3|    Germany|            1863.4|1996-07-03|
|  10250|     HANAR| 1996-7-6|     Brazil|1552.6000000000001|1996-07-06|
|  10251|     VICTE| 1996-7-6|     France|            654.06|1996-07-06|
|  10252|     SUPRD| 1996-7-7|    Belgium|            3597.9|1996-07-07|
+-------+----------+---------+-----------+------------------+----------+
only showing top 5 rows
```

(8) 再从 OrderDate 列中分别抽取月份和年份部分,增加两个新的属性列 month 和 year,代码如下:

```
from pyspark.sql.functions import month,year

orders3 = orders2 \
    .withColumn("Month",month("OrderDate")) \
    .withColumn("Year",year("OrderDate"))
orders3.show(10)
```

执行以上代码,输出结果如下:

```
+-------+----------+---------+-----------+-----------------+----------+-----+-----+
|OrderID|CustomerID|OrderDate|ShipCountry|            Total|      Date|Month| Year|
+-------+----------+---------+-----------+-----------------+----------+-----+-----+
|  10248|     VINET| 1996-7-2|     France|            440.0|1996-07-02|    7| 1996|
|  10249|     TOMSP| 1996-7-3|    Germany|           1863.4|1996-07-03|    7| 1996|
|  10250|     HANAR| 1996-7-6|     Brazil|1552.6000000000001|1996-07-06|    7| 1996|
|  10251|     VICTE| 1996-7-6|     France|           654.06|1996-07-06|    7| 1996|
|  10252|     SUPRD| 1996-7-7|    Belgium|           3597.9|1996-07-07|    7| 1996|
|  10253|     HANAR| 1996-7-8|     Brazil|1444.8000000000002|1996-07-08|    7| 1996|
|  10254|     CHOPS| 1996-7-9|Switzerland| 556.6199999999999|1996-07-09|    7| 1996|
|  10255|     RICSU|1996-7-10|Switzerland|           2490.5|1996-07-10|    7| 1996|
|  10256|     WELLI|1996-7-13|     Brazil|            517.8|1996-07-13|    7| 1996|
|  10257|     HILAA|1996-7-14|  Venezuela|           1119.9|1996-07-14|    7| 1996|
+-------+----------+---------+-----------+-----------------+----------+-----+-----+
only showing Top 10 rows
```

整理好相关的属性之后,现在可以回答本例开始所提出的剩余问题了。

(9) 现在可以回答第 3 个问题:每月/年有多少订单金额? 代码如下:

```
ordersByYM = orders3 \
    .groupBy("Year","Month") \
    .agg(sum("Total").alias("Total"))

ordersByYM.select(
    "Year",
    "Month",
    bround("Total",2).alias("Total")
).sort("Year","Month").show(10)
```

执行以上代码,输出结果如下:

```
+----+-----+-------+
|Year|Month|  Total|
+----+-----+-------+
|1996|    7|30741.9|
```

```
|1996|    8| 22726.88|
|1996|    9|  27691.4|
|1996|   10| 38380.12|
|1996|   11| 45694.44|
|1996|   12| 52494.33|
|1997|    1| 51612.97|
|1997|    2| 38483.64|
|1997|    3| 40918.82|
|1997|    4| 57116.71|
|1997|    5| 50270.33|
|1997|    6| 34392.08|
|1997|    7| 52744.68|
|1997|    8| 46991.78|
|1997|    9| 57723.23|
|1997|   10| 62253.63|
|1997|   11| 51294.81|
|1997|   12| 67920.23|
|1998|    1|107049.96|
|1998|    2| 85240.83|
+----+----+---------+
only showing top 20 rows
```

（10）回答第 4 个问题：每个客户的年销售总额是多少？代码如下：

```
ordersByCY = orders3 \
    .groupBy("CustomerID","Year") \
    .agg(sum("Total").alias("Total"))

ordersByCY.sort("CustomerID","Year").show(10)
```

执行以上代码，输出结果如下：

```
+----------+----+------------------+
|CustomerID|Year|             Total|
+----------+----+------------------+
|     ALFKI|1997|            2022.5|
|     ALFKI|1998|            2250.5|
|     ANATR|1996|              88.8|
|     ANATR|1997|            799.75|
|     ANATR|1998|             514.4|
|     ANTON|1996| 403.20000000000005|
|     ANTON|1997|          5960.7775|
|     ANTON|1998|             660.0|
|     AROUT|1996|            1379.0|
|     AROUT|1997| 6406.900000000001|
|     AROUT|1998|           5604.75|
|     BERGS|1996|            4324.4|
```

```
|    BERGS|1997|  13849.015000000001|
|    BERGS|1998|              6754.1625|
|    BLAUS|1997|                1079.8|
|    BLAUS|1998|                2160.0|
|    BLONP|1996|                9986.2|
|    BLONP|1997|               7817.88|
|    BLONP|1998|                 730.0|
|    BOLID|1996|                 982.0|
+---------+----+-------------------+
only showing top 20 rows
```

（11）回答第 5 个问题：客户每年的平均订单金额是多少?代码如下：

```
from pyspark.sql.functions import avg

ordersByCY = orders3 \
    .groupBy("CustomerID","Year") \
    .agg(avg("Total").alias("Avg"))

ordersByCY.select(
    "CustomerID",
    "Year",
    bround("Avg",2)
).sort("CustomerID","Year").show(10)
```

执行以上代码，输出结果如下：

```
+----------+----+--------------+
|CustomerID|Year|bround(Avg, 2)|
+----------+----+--------------+
|     ALFKI|1997|        674.17|
|     ALFKI|1998|        750.17|
|     ANATR|1996|          88.8|
|     ANATR|1997|        399.88|
|     ANATR|1998|         514.4|
|     ANTON|1996|         403.2|
|     ANTON|1997|       1192.16|
|     ANTON|1998|         660.0|
|     AROUT|1996|         689.5|
|     AROUT|1997|        915.27|
|     AROUT|1998|       1401.19|
|     BERGS|1996|       1441.47|
|     BERGS|1997|        1384.9|
|     BERGS|1998|       1350.83|
|     BLAUS|1997|        269.95|
|     BLAUS|1998|         720.0|
|     BLONP|1996|       3328.73|
|     BLONP|1997|       1116.84|
```

```
|    BLONP|1998|              730.0|
|    BOLID|1996|              982.0|
+---------+----+-------------------+
only showing top 20 rows
```

5.8.2 电影评分数据集分析

6min

本节使用 PySpark SQL 实现对电影数据集进行分析。在这里使用推荐领域的一个著名的开放测试数据集 MovieLens。MovieLens 数据集包括电影元数据信息和用户属性信息。本例将使用其中的 users.dat 和 ratings.dat 两个数据集。

【例 5-11】使用 PySpark DataFrame API 统计看过电影"Lord of the Rings,The(1978)"的用户的年龄和性别分布（提示该影片的 ID 是 2116）。

建议按以下步骤执行。

（1）首先构造一个 SparkSession 实例，代码如下：

```
#第5章/sparksql_example2.ipynb

from pyspark.sql import SparkSession
from pyspark.sql.functions import *

#创建 SparkSession 的实例
spark = SparkSession.builder \
    .master("spark://localhost:7077") \
    .appName("pyspark sql demo") \
    .config("spark.hadoop.hive.exec.dynamic.partition", "true") \
    .config("spark.hadoop.hive.exec.dynamic.partition.mode","nonstrict") \
    .enableHiveSupport() \
    .getOrCreate()
```

（2）读取用户数据集 users.dat，加载到 DataFrame 中，代码如下：

```
from pyspark.sql.types import *

#定义文件路径
usersFile = "/data/spark/ml-1m/users.dat"

#指定一个 schema(模式)
fields = [
    StructField("userID", LongType(), True),
    StructField("gender", StringType(), True),
    StructField("age", IntegerType(), True),
    StructField("occupation", StringType(), True),
    StructField("zipcode", StringType(), True)
]
```

```
schema = StructType(fields)

#将数据文件加载到DataFrame,并指定schema
usersDF = spark.read \
   .options(sep="::",header="false") \
   .schema(schema) \
   .csv(usersFile)

#查看用户数据
usersDF.printSchema()
usersDF.show(5)
```

执行以上代码,输出结果如下:

```
root
 |-- userID: long (nullable = false)
 |-- gender: string (nullable = true)
 |-- age: integer (nullable = true)
 |-- occupation: string (nullable = true)
 |-- zipcode: string (nullable = true)

+------+------+---+----------+-------+
|userID|gender|age|occupation|zipcode|
+------+------+---+----------+-------+
|     1|     F|  1|        10|  48067|
|     2|     M| 56|        16|  70072|
|     3|     M| 25|        15|  55117|
|     4|     M| 45|         7|  02460|
|     5|     M| 25|        20|  55455|
+------+------+---+----------+-------+
only showing top 5 rows
```

(3)读取评分数据集 ratings.dat,加载到 DataFrame 中,代码如下:

```
#定义文件路径
ratingsFile = "/data/spark/ml-1m/ratings.dat"

#指定一个schema(模式)
fields = [
   StructField("userID", LongType(), True),
   StructField("movieID", LongType(), True),
   StructField("rating", IntegerType(), True),
   StructField("timestamp", LongType(), True)
]
schema = StructType(fields)

#将数据文件加载到DataFrame,并指定schema
ratingsDF = spark.read \
```

```
        .options(sep="::",header="false") \
        .schema(schema) \
        .csv(ratingsFile)

#查看用户数据
ratingsDF.printSchema()
ratingsDF.show(5)
```

执行以上代码,输出结果如下:

```
root
 |-- userID: long (nullable = false)
 |-- movieID: long (nullable = false)
 |-- rating: integer (nullable = true)
 |-- timestamp: long (nullable = false)

+------+-------+------+---------+
|userID|movieID|rating|timestamp|
+------+-------+------+---------+
|     1|   1193|     5|978300760|
|     1|    661|     3|978302109|
|     1|    914|     3|978301968|
|     1|   3408|     4|978300275|
|     1|   2355|     5|978824291|
+------+-------+------+---------+
only showing top 5 rows
```

(4) 将两个 DataFrame 注册为临时表,对应的表名分别为 users 和 ratings,代码如下:

```
usersDF.createOrReplaceTempView("users")
ratingsDF.createOrReplaceTempView("ratings")
```

(5) 通过 SQL 处理临时表 users 和 ratings 中的数据,并输出最终结果。为了简单起见,避免 3 张表连接操作,这里直接使用了 movieID,代码如下:

```
MOVIE_ID = "2116"
sqlStr = f"""select age,gender,count(*) as total_people
            from users as u
                join ratings as r on u.userid=r.userid
            where movieid={MOVIE_ID} group by gender,age"""
resultDF = spark.sql(sqlStr)

#显示resultDF的内容
resultDF.show()
```

执行以上代码,输出结果如下:

```
+---+------+------------+
|age|gender|total_people|
```

```
+---+------+--------------+
| 18|     M|            72|
| 18|     F|             9|
| 56|     M|             8|
| 45|     M|            26|
| 45|     F|             3|
| 25|     M|           169|
| 56|     F|             2|
|  1|     M|            13|
|  1|     F|             4|
| 50|     F|             3|
| 50|     M|            22|
| 25|     F|            28|
| 35|     F|            13|
| 35|     M|            66|
+---+------+--------------+
```

（6）以交叉表的形式统计不同年龄不同性别用户数，代码如下：

```
from pyspark.sql.functions import *
resultDF \
   .groupBy("age") \
   .pivot("gender") \
   .agg(sum("total_people").alias("cnt")) \
   .show()
```

执行以上代码，输出结果如下：

```
+---+---+---+
|age|  F|  M|
+---+---+---+
|  1|  4| 13|
| 35| 13| 66|
| 50|  3| 22|
| 45|  3| 26|
| 25| 28|169|
| 56|  2|  8|
| 18|  9| 72|
+---+---+---+
```

（7）在 Zeppelin 中，支持查询结果的可视化显示。在 Zeppelin Notebook 的单元格中，执行以下 SQL 语句，可视化显示数据（注意，第 1 行必须输入%sql，用来指定使用 SQL 解释器），代码如下：

```
%sql
select age,
       gender,
```

```
       count(*) as total_people
from users as u join ratings as r
on u.userid=r.userid
where movieid=${MOVIE_ID=2116}
group by gender,age
```

执行以上代码，输出结果如图 5-12 所示。

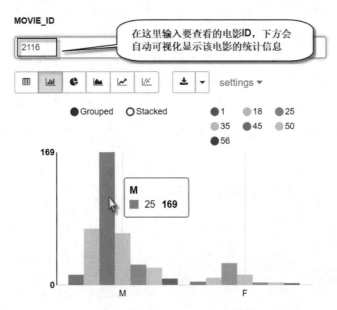

图 5-12　在 Zeppelin Notebook 中可视化查询结果

用户可以在输入框输入要查看的影片 ID，下方会自动可视化显示该影片的统计信息。

第 6 章 PySpark 结构化流(初级)

在很多领域,如股市走向分析、气象数据测控、网站用户行为分析等,由于数据产生快,实时性强,数据量大,所以很难统一采集并入库存储后再进行处理,这便导致传统的数据处理架构不能满足需要。流计算的出现,就是为了更好地解决这类数据在处理过程中遇到的问题。与传统架构不同,流计算模型在数据流动的过程中实时地进行捕捉和处理,并根据业务需求对数据进行计算分析,最终把结果保存或者分发给需要的组件。

另外,随着物联网的爆发,互联设备通过传感器、摄像头、加速度计、激光雷达和深度传感器收集越来越多的信息。从制造业到汽车、医疗技术、能源、公用事业和可穿戴技术等各个行业都存在联网产品。在人工智能和 5G 融合的帮助下,被收集的数据量只会不断扩大。据统计,一辆完全自动驾驶的汽车将包含超过 60 个微处理器和传感器,每年产生超过 300TB 的数据。或者,反过来讲,在一小时的长途旅行中,联网车辆之间将发送多达 25GB 的信息(相当于 100 小时的视频)。

有了这些海量数据,捕获、聚合和分析数据就成了一个挑战,而 Apache Spark 的统一数据处理平台能够同时执行流数据处理和批数据处理。

6.1 PySpark DStream 流简介

14min

第 1 代 Spark 流式处理引擎 PySpark Streaming 是在 2012 年引入的,这个引擎的主要编程抽象称为离散流,即 DStream,它表示连续的数据流。DStream 的工作方式如图 6-1 所示。

图 6-1 DStream 的工作方式

PySpark Streaming 是对核心 PySpark API 的扩展,支持可扩展、高吞吐量、容错的实时数据流处理。一个 DStream 可以从 Kafka、AWS Kinesis 或 TCP 套接字等许多输入源获取并创建，也可以通过对其他 DStream 应用高级操作创建，并且可以使用复杂的算法来处理，这些算法由高级函数（如 map、reduce、join 和 window）表示，如图 6-2 所示。

图 6-2　DStream 支持多种数据源

对于实时数据处理，PySpark Streaming 使用"微批处理"模型。它的工作方式是使用微批处理模型将传入的数据流分成批处理，这意味着 PySpark 流会将特定时间段内的数据块打包成 RDD，然后由 PySpark 批处理引擎处理。在内部，DStream 被表示为 RDD 的序列，如图 6-3 所示。

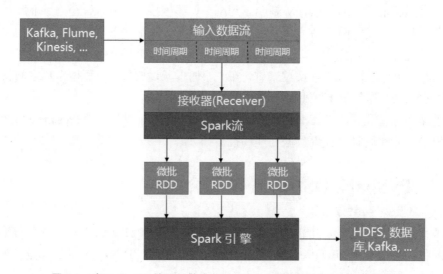

图 6-3　在 PySpark 处理引擎内部，DStream 被表示为 RDD 的序列

在创建 DStream 时，需要的关键信息之一是批间隔，这个批间隔可以是几秒或几毫秒。有了 DStream，就可以在数据输入流上应用一个高级的数据处理函数，如 map、filter、reduce 或 reduceByKey。此外，还可以执行窗口操作，例如通过提供窗口长度和滑动间隔来减少和计算一个固定/滚动或滑动窗口。一个重要的注意事项是，窗口长度和滑动间隔必须是批处理间隔的倍数。例如，如果批间隔是 3s，并且使用了固定/滚动间隔，则窗口长度和滑动间隔可以是 6s。在 DStream 中支持跨批次数据在执行计算时保持任意状态，但这是一个手动过程，而且有点麻烦。一个 DStream 还可以与另一个 DStream 或代表静态数据的 RDD

连接（join）起来。在所有处理逻辑完成之后，可以使用 DStream 将数据写到外部系统，如数据库、文件系统或 HDFS。

下面是一个较小的单词记数 PySpark DStream 应用程序。通过这个程序可以了解一个典型的 PySpark DStream 应用程序。

【例 6-1】实时统计文本流的单词数量（单词计数的实时处理版本）。

首先，导入流执行环境对象 StreamingContext，它是所有流功能的主要入口点。创建一个 StreamingContext 对象实例，并将批处理间隔时间设置为 3s，代码如下：

```python
#第6章/streaming_demo01.py

from pyspark.sql import SparkSession
from pyspark.streaming import StreamingContext

spark = SparkSession.builder.appName("NetworkWordCount").getOrCreate()
sc = spark.sparkContext

#设置日志级别
sc.setLogLevel('WARN')

#创建StreamingContext，将批处理间隔设置为3s
ssc = StreamingContext(sc, 3)
```

使用这个上下文对象,可以创建一个 DStream 来表示来自 TCP 源的流数据,代码如下：

```python
#创建一个连接到hostname:port 的DStream
lines = ssc.socketTextStream("localhost", 9999)
```

这个 lines 表示将从数据服务器接收的 DStream 数据流，DStream 中的每条记录都是一行文本。接下来，要将行按空格分割成单词，代码如下：

```python
words = lines.flatMap(lambda line: line.split(" "))
```

这里的 flatMap() 函数是一个一对多 DStream 操作，它通过从源 DStream 中的每条记录生成多个新记录来创建一个新的 DStream。在本例中，每行将被分割为多个单词，单词流表示为单词 DStream。接下来，计算这些单词的数量，代码如下：

```python
#计算每一批单词的数量
pairs = words.map(lambda word: (word, 1))
wordCounts = pairs.reduceByKey(lambda x, y: x + y)

#将这个DStream中生成的每个RDD的前10个元素打印到控制台
wordCounts.pprint()
```

上面名为 words 的 DStream 流被进一步通过 map() 函数一对一地转换到一个元素为 Tuple(word, 1)元组对的 DStream，然后通过调用 reduceByKey() 方法进行规约，以获得每批数据中单词的数量。最后，wordCounts.pprint()将打印每秒生成的一些计数。

注意，代码执行到这里时，只是设置了它在启动时将执行的计算，而实际的处理还没有启动。为了在设置完所有转换之后开始处理，最后需要调用start()方法，代码如下：

```
ssc.start()                      #开始计算
ssc.awaitTermination()           #等待计算结束
```

以上例子的完整代码如下：

```python
#第6章/streaming_demo01.py

"""
    要运行这个程序，首先需要运行一个Netcat服务器：`$ nc -lk 9999`
    然后提交并运行这个程序：
      '$ bin/spark-submit --master spark://xueai8:7077 streaming_demo01.py localhost 9999'
"""
import sys

from pyspark.sql import SparkSession
from pyspark.streaming import StreamingContext

if __name__ == "__main__":
    if len(sys.argv) != 3:
        print("用法: network_wordcount.ipynb <hostname> <port>", file=sys.stderr)
        sys.exit(-1)
    spark = SparkSession.builder.appName("NetworkWordCount").getOrCreate()
    sc = spark.sparkContext

    #设置日志级别
    sc.setLogLevel('WARN')

    #创建StreamingContext，将批处理间隔设置为3s
    ssc = StreamingContext(sc, 3)

lines = ssc.socketTextStream(sys.argv[1], int(sys.argv[2]))
words = lines.flatMap(lambda line: line.split(" "))
pairs = words.map(lambda word: (word, 1))
wordCounts = pairs.reduceByKey(lambda x, y: x + y)

wordCounts.pprint()

ssc.start()
  ssc.awaitTermination()
```

在编写一个 PySpark DStream 应用程序时,有以下几个重要的步骤:

(1) DStream 应用程序的入口点是 StreamingContext,其中一个必需的输入是批间隔,它定义了持续时间,PySpark 用来将输入的数据输入 RDD 进行处理。它也代表了一个触发点,在该触发点,PySpark 应该执行流应用程序计算逻辑。例如,如果批处理间隔是 3s,就会触发所有到达 3s 间隔内的数据;在该间隔之后,它将把这批数据转换为 RDD,并根据提供的处理逻辑对其进行处理。

(2) 一旦创建了 StreamingContext,下一步将通过定义输入源来创建实例 DStream。前面的例子将输入源定义为读取文本行的 Socket 套接字。

(3) 接下来,将为新创建的 DStream 提供处理逻辑。前面例子中的处理逻辑并不复杂。一旦一个系列的 RDD 在一秒后可用,那么 PySpark 就会执行将每行代码分割成单词的逻辑,将每个单词转换成单词的元组和 1 的计数,最后总结同一个单词的计数。

(4) 最后,计数将在控制台上打印出来。

(5) 流应用程序是一个长期运行的应用程序,因此,它需要一个信号来启动接收和处理传入的数据流的任务。这个信号是通过调用 StreamingContext 的 start()函数实现的,这通常是在文件的末尾完成的。awaitTermination()函数用于等待流应用程序的停止,以及在流应用程序运行时防止驱动程序退出。在一个典型的程序中,一旦执行最后一行代码,它就会退出,然而,一个长时间运行的流应用程序需要在启动时继续运行,并且只有当显式地停止它时才会结束。

要运行该程序,建议按以下步骤执行。

(1) 首先需要运行一个 Netcat 服务器。打开一个终端窗口,执行的命令如下:

```
$ nc -lk 9999
```

(2) 打开第 2 个终端窗口,将 streaming_demo01.py 程序提交到 PySpark 集群上运行(需要确保已经启动了 Spark 集群,并假设 streaming_demo01.py 源文件位于~/pyspark_demos/chapter06/目录下),命令如下:

```
$ cd ~/bigdata/spark-3.1.2
$ ./bin/spark-submit --master spark://localhost:7077
~/pyspark_demos/chapter06/streaming_demo01.py localhost 9999
```

执行过程如图 6-4 所示。

```
[hduser@xueai8 ~]$ cd bigdata/spark-3.1.2/
[hduser@xueai8 spark-3.1.2]$ ./bin/spark-submit --master spark://xueai8:7077 ~/pyspark_demos/chapter06/network_wordcount.py localhost 9999
```

图 6-4 将 streaming_demo01.py 程序提交到 PySpark 集群上运行

(3) 回到第 1 个终端窗口,任意输入一些内容,单词之间以空格分割。例如,输入以下内容:

```
good good study
day day up
```

执行过程如图 6-5 所示。

图 6-5　在 Netcat 服务器端输入几行文本

(4) 在第 2 个终端窗口，查看流程序输出的结果，如图 6-6 所示。

图 6-6　Streaming 流程序输出的结果

6.2　PySpark 结构化流简介

PySpark 结构化流提供了快速、可扩展、容错、端到端的精确一次性流处理，而用户无须对流进行推理。结构化流操作直接工作在 DataFrame 上。不再有"流"的概念，只有流式 DataFrame 和普通 DataFrame。流式 DataFrame 是作为 append-only 表实现的。在流数据上的查询会返回新的 DataFrame，使用它们就像在批处理程序中一样使用。

使用 PySpark 结构化流的模型如图 6-7 所示。

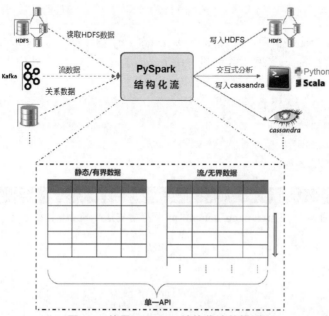

图 6-7　使用 PySpark 结构化流的模型

在结构化流处理模型中,将实时数据流视为一个不断增长的表。触发器会指定检查输入新数据到达的时间间隔。查询表示输入上的查询或操作,如映射、筛选和合并,结果表示在每个触发器间隔中根据指定的操作更新的最终表。输出定义了在每个时间间隔内将结果写入数据接收器的部分。

当一组新的数据到达时,将这些新到达的数据作为一组新的行添加到输入表。考虑传入数据流的方式,只不过是一个不断追加的表。这样就能够利用现有的用于 DataFrame 的结构化 API 执行流计算,并且随着新的流数据的到来,由结构化流引擎负责增量地、持续地运行它们。

以单词计数为例,当查询启动时,PySpark 将持续检查来自套接字连接的新数据。如果有新数据,则 PySpark 将运行一个增量查询,将以前的运行计数与新数据结合起来,计算更新的计数,如图 6-8 所示。

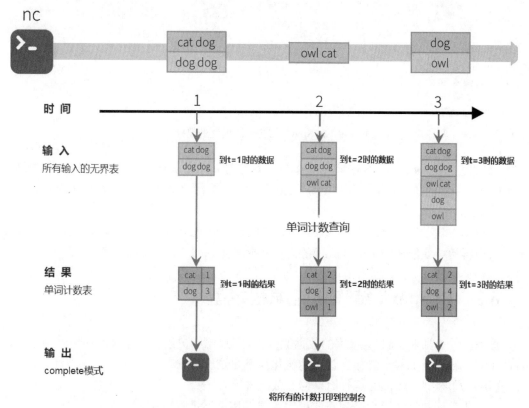

图 6-8　将实时数据流视为一个不断增长的输入表

因此,无论是静态数据还是流式数据,只需像在静态数据表上那样启动类似于批处理的查询,PySpark 就会在无界输入表上作为增量查询运行它。因此,开发人员在输入表上定义一个查询,对于静态有界表和动态无界表都是一样的。

在结构化流处理模型中,用户使用批处理 API 在输入表上进行查询,Spark SQL Planner(规划器)在流数据上增量执行,整个过程如图 6-9 所示。

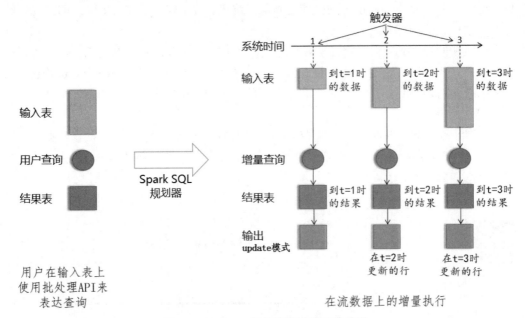

图 6-9 PySpark 结构化流增量处理模型

可以使用统一入口点 SparkSession 从流源创建流 DataFrame,并对它们应用与静态 DataFrame 相同的操作。有一些内置的源,例如文件,支持的文件格式是文本、CSV、JSON、ORC、parquet。

从根本上说,结构化流由 PySpark SQL 的 Catalyst 优化器负责优化,因此,它使开发人员不用再担心底层的管道,在处理静态或实时数据流时,使查询更高效。

6.3 PySpark 结构化流编程模型

假设有一个监听 TCP 套接字的数据服务器,用户想维护该服务器接收的文本数据的运行时单词计数。下面看一看如何使用结构化流满足这一需求。

【例 6-2】实现 PySpark 运行时单词计数流程序。

下面先分步执行,最后是完整的程序,建议按以下步骤操作。

(1)先导入必要的类并创建本地 SparkSession,这是所有与 PySpark 相关的功能的起点,代码如下:

```
from pyspark.sql import SparkSession
from pyspark.sql.functions import explode
```

```python
from pyspark.sql.functions import split

#构造一个SparkSession实例
spark = SparkSession \
    .builder \
    .appName("StructuredNetworkWordCount") \
    .getOrCreate()
```

（2）接下来，创建一个流 DataFrame（它表示从所监听的 localhost:9999 服务器接收的文本数据），并对该 DataFrame 进行转换以进行单词计数，代码如下：

```python
#创建表示从所监听的localhost:9999服务器接收的文本数据的DataFrame
lines = spark.readStream \
    .format("socket") \
    .option("host", "localhost") \
    .option("port", 9999) \
    .load()

#将行拆分为单词
words = lines.select(
   explode(
       split(lines.value, " ")
   ).alias("word")
)

#生成运行时单词计数
wordCounts = words.groupBy("word").count()
```

上面名为 lines 的 DataFrame 表示一个包含流文本数据的无界表，这个表包含一列名为 value 的字符串，流文本数据中的每行都成为表中的一行。接下来，使用了两个内置的 SQL 函数，即 split() 和 explode()，将每行内容分割为多行，分割后每行包含一个单词。最后，定义了名为 wordCounts 的 DataFrame，方法是根据数据集中唯一的值进行分组并对它们进行计数的。注意，这是一个流 DataFrame，它表示流的运行时单词数。

（3）现在已经在流数据上设置了查询，剩下的就是开始接收数据并计数。为此，设置它在每次更新计数时将完整的计数集（由 outputMode("complete")指定）打印到控制台，然后使用 start() 启动流计算，代码如下：

```python
#启动流计算，将运行时单词计数打印到控制台
query = wordCounts.writeStream \
    .outputMode("complete") \
    .format("console") \
    .start()

query.awaitTermination()
```

此代码执行后，流计算将在后台启动。注意，必须调用 start()来执行查询。这将返回一个 StreamingQuery 对象，它是持续运行执行的一个句柄。用户可以使用这个对象来管理查询。

StreamingQuery 对象是活动的流查询的句柄，在执行作业时需要使用 awaitTermination() 等待查询的终止，以防止在查询处于活动状态时进程退出。对于生产和长期运行的流应用程序，有必要调用 StreamingQuery.awaitTermination()函数，这是一个阻塞调用，它会防止 Driver 驱动程序退出，并允许流查询持续运行，并且当新数据到达数据源时处理新数据。

完整的程序代码如下：

```python
#第6章/structed_wordcount.py

import sys

from pyspark.sql import SparkSession
from pyspark.sql.functions import explode
from pyspark.sql.functions import split

if __name__ == "__main__":
    if len(sys.argv) != 3:
        print("用法: structed_wordcount.ipynb <hostname> <port>", file=sys.stderr)
        sys.exit(-1)

    spark = SparkSession \
        .builder \
        .appName("StructuredNetworkWordCount") \
        .getOrCreate()

    #创建表示从所监听的localhost:9999服务器接收的文本数据的DataFrame
    lines = spark.readStream \
        .format("socket") \
        .option("host", sys.argv[1]) \
        .option("port", sys.argv[2]) \
        .load()

    #将行拆分为单词
    words = lines.select(
        explode(
            split(lines.value, " ")
        ).alias("word")
    )

    #生成运行时单词计数
```

```
wordCounts = words.groupBy("word").count()

#启动流计算,将运行时单词计数打印到控制台
query = wordCounts.writeStream \
    .outputMode("complete") \
    .format("console") \
    .start()

query.awaitTermination()
```

要运行这个结构化流程序,建议按以下步骤操作。

(1)首先需要使用 Netcat 作为数据服务器。在 Linux 的第 1 个终端窗口中,启动 Netcat 服务器,使其保持运行,命令如下:

```
$ nc -lk 9999
```

如果没有 Netcat 服务器,则可以使用如下命令先安装:

```
$ sudo yum install -y nc
```

(2)打开第 2 个终端窗口,将 structed_wordcount.py 程序提交到 PySpark 集群上运行(确保启动了 Spark 集群,并假设 structed_wordcount.py 位于~/pyspark_demos/chapter06/目录下),命令如下:

```
$ cd ~/bigdata/spark-3.1.2/
$ ./bin/spark-submit --master spark://xueai8:7077
~/pyspark_demos/chapter06/structed_wordcount.py localhost 9999
```

执行过程如图 6-10 所示。

```
[hduser@xueai8 ~]$ cd bigdata/spark-3.1.2/
[hduser@xueai8 spark-3.1.2]$ ./bin/spark-submit --master spark://xueai8:7077 ~/pyspark_demos/chapter06/structed_wordcount.py localhost 9999
```

图 6-10 PySpark 结构化流程序提交运行过程

(3)切换到第 1 个终端窗口,任意输入一些内容,单词之间以空格分隔。例如,输入以下内容:

```
good good study
day day up
```

执行过程如图 6-11 所示。

图 6-11 在 Netcat 服务器端输入几行文本内容

(4)切换回流程序执行窗口,查看输出结果,可以看到输出结果如图 6-12 所示。

```
Batch: 1
-------------------------------------------
22/02/25 12:10:17 INFO CodeGenerator: Code generated in 34.165973 ms
22/02/25 12:10:17 INFO CodeGenerator: Code generated in 10.95491 ms
+-----+-----+
| word|count|
+-----+-----+
|study|    1|
| good|    2|
```

图 6-12 单词计数流程序输出内容

有时用户希望停止流查询来改变输出模式、触发器或其他配置。可以使用 StreamingQuery.stop() 函数来阻止数据源接收新数据,并在流查询中停止逻辑的连续执行,代码如下:

```
#这是阻塞调用
mobileSQ.awaitTermination()

#停止流查询
mobileSQ.stop()
```

6.4 PySpark 结构化流核心概念

9min

PySpark 结构化流应用程序包括以下几个主要部分:
(1) 指定一个或多个流数据源。
(2) 提供了以 DataFrame 转换的形式操作传入数据流的逻辑。
(3) 定义输出模式和触发器(都有默认值,所以是可选的)。
(4) 最后指定一个数据接收器(Data Sink)。
以上步骤如图 6-13 所示。

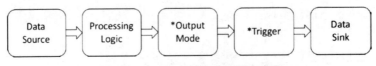

图 6-13 PySpark 结构化流程序步骤

其中,用星号标记的步骤表示非必须执行的步骤。下面将详细描述这些概念。

6.4.1 数据源

对于批处理,数据源是驻留在某些存储系统上的静态数据集,如本地文件系统、HDFS 或 S3,而结构化流的数据源是完全不同的。它们生产的数据是连续的,可能永远不会结束,而且生产速率也会随着时间的变化而变化。

PySpark 结构化流提供了以下开箱即用的数据源：

（1）Kafka 源：要求 Apache Kafka 的版本是 0.10 或更高版本。这是生产环境中最流行的数据源。连接和读取来自 Kafka 主题的数据需要提供一组特定的设置。

（2）文件源：文件位于本地文件系统、HDFS 或 S3 上。当新的文件被放入一个目录中时，这个数据源将会把它们挑选出进行处理。支持常用的文件格式，如文本、CSV、JSON、ORC 和 parquet。在处理这个数据源时，一个好的实践是先完全地写出输入文件，然后将它们移动到这个数据源的路径中，例如，流程序监控的是 HDFS 上的 A 目录，那么先将输入文件写到 HDFS 的 B 目录中，再从 B 目录将它们移动到 A 目录。

（3）Socket 源：这仅用于测试目的。它从一个监听特定的主机和端口的 Socket 上读取 UTF-8 数据。

（4）Rate 源：这仅用于测试和基准测试。这个源可以被配置为每秒产生的事件，其中每个事件由时间戳和一个单调递增的值组成。这是学习结构化流时使用的最简单的源。

数据源需要提供的一个重要的属性是一种跟踪流中的读位置的方法，用于结构化的流来传递端到端、精确一次性保证。例如，Kafka 的数据源提供了一个 Kafka 的偏移量来跟踪一个主题分区的读位置。这个属性决定了一个特定的数据源是否具有容错能力。

开箱即用 Spark 结构化流数据源，见表 6-1。

表 6-1　Spark 结构化流数据源

数据源	是否容错	配　　置
File	是	path：输入目录的路径 maxFilesPerTrigger：每个触发器读取新行的最大数量 latestFirst：是否处理最新的文件（根据 modification time）
Socket	否	要求有以下参数 host：要连接到的主机 port：要连接到的端口号
Rate	是	rowsPerSecond：每秒生成的行的数量 rampUpTime：在到达 rowsPerSecond 之前的时间，以秒为单位 numPartitions：分区的数量
Kafka	是	kafka.Bootstrap.servers：Kafka brokers 列表，以逗号分隔，即 host:port subscribe：主题列表，以逗号分隔

6.4.2　输出模式

输出模式是一种方法，可以告诉结构流如何将输出数据写入 Sink 中。这个概念对于 PySpark 中的流处理来讲是独一无二的。输出模式有 3 个选项，分别解释如下。

（1）append 模式：如果没有指定输出模式，这是默认模式。在这种模式下，只有追加到结果表的新行才会被发送到指定的输出接收器。只有自上次触发后在结果表中附加的新行将被写入外部存储器。这仅适用于结果表中的现有行不会更改的查询。

（2）complete 模式：此模式将数据完全从内存写入接收器，即整个结果表将被写到输出接收器。当对流数据执行聚合查询时，就需要这种模式。

（3）update 模式：只有自上次触发后在结果表中更新的行才会被写到输出接收器中。对于那些没有改变的行，它们将不会被写出来。注意，这与 complete 模式不同，因为此模式不输出未更改的行。

6.4.3 触发器类型

触发器是另一个需要理解的重要概念。结构化流引擎使用触发器信息来确定何时在流应用程序中运行提供的流计算逻辑。不同的触发器类型见表 6-2。

表 6-2 Spark 结构化流触发器类型

类型	描述
未指定（默认）	对于默认类型，PySpark 将使用微批模型，并且当前一批数据完成处理后，立即处理下一批数据
固定周期	对于这种类型，PySpark 将使用微批模型，并基于用户提供的周期处理这批数据。如果因为任何原因导致上一批数据的处理超过了该周期，则前一批数据完成处理后，立即处理下一批数据。换句话说，PySpark 将不会等到下一个周期区间的边界
一次性	这个触发器类型意味着用于一次性处理可用的批数据，并且一旦该处理完成，PySpark 将立即停止流程序。当数据量特别低时，这个触发器很有用，因此，构建一个集群并每天处理几次数据更划算
持续	这个触发器类型会调用新的持续处理模型，该模型被设计用于有非常低延迟需求的特定流应用程序。这是 PySpark 2.3 中新的实验性处理模式。这时将支持"最少一次性"保证

6.4.4 数据接收器

数据接收器（Data Sink）用来存储流应用程序的输出。不同的 Sink 可以支持不同的输出模式，并且具有不同的容错能力，了解这一点很重要。PySpark 结构化流支持以下几种数据接收器。

（1）Kafka Sink：要求 Apache Kafka 的版本是 0.10 或更高版本。有一组特定的设置可以连接到 Kafka 集群。

（2）File Sink：这是文件系统、HDFS 或 S3 的目的地。支持常用的文件格式，如文本、CSV、JSON、ORC、parquet。

（3）Foreach Sink：这是为了在输出中的行上运行任意计算。

（4）Console Sink：这仅用于测试和调试目的，以及在处理低容量数据时使用。每个触发器的输出会被打印到控制台。

（5）Memory Sink：这是在处理低容量数据时进行测试和调试目的的。它使用驱动程序的内存来存储输出。

每个 Data Sink 配置选项说明，见表 6-3。

表 6-3 PySpark 结构化流 Data Sink 配置选项

名 称	支持的输出模式	是否容错	配 置
File	append	是	path：这是输入目录的路径。支持所有流行的文件格式。详细信息可查看 DataFrameWriter API
Foreach	append update complete	依情况而定	这是一个非常灵活的接收器，它是特定于实现的
Console	append update complete	否	numRows：这是每个触发器输出的行的数量。默认为 20 行 truncate：如果每行太长，是否截断。默认为 true
Memory	append complete	否	N/A
Kafka	append update complete	是	kafka.Bootstrap.servers：Kafka brokers 列表，以逗号分隔，即 host:port topic：这是写入数据的 Kafka 主题

数据接收器必须支持的一个重要的属性（用于结构化的流交付端到端、精确一次性保证）是处理重做的幂等性。换句话说，它必须能够处理使用相同数据的多个写（在不同的时间发生），结果就像只有一个写一样。多重写是在故障场景中重新处理数据的结果。

6.4.5 水印

水印是流处理引擎中常用的一种技术，用于处理迟到的数据。流应用程序开发人员可以指定一个阈值，让结构化的流引擎知道数据在事件时间（Event Time）内的预期延迟时间。有了这个信息，超过这个预期延迟时间到达的迟到数据会被丢弃。

更重要的是，结构化流使用指定的阈值来确定何时可以丢弃旧状态。没有这些信息，结构化流将需要无限期地维护所有状态，这将导致流应用程序的内存溢出问题。任何执行某种聚合或连接的生产环境下的结构化流应用程序都需要指定水印。

注意：水印是一个重要的概念，关于这个主题的更多细节将在后面的部分中讨论和说明。

6.5 使用各种流数据源

6.4 节描述了结构化流提供的各个数据源。本节将更详细地介绍这些数据源，并将提供使用它们的示例代码。

6.5.1 使用 Socket 数据源

Socket（套接字）数据源很容易使用，只需提供主机和端口号。不过该数据源仅限于学习和测试使用，不能在生产环境中使用。下面这个例子应用 Socket 数据源。

【例 6-3】编写 PySpark 结构化流应用程序，使用 Socket 数据源。

建议按以下步骤操作。

（1）在启动 Socket 数据源的流式查询之前，首先使用一个网络命令行实用工具，如 Mac 上的 nc 或 Windows 上的 Netcat，启动一个套接字服务器。打开一个终端窗口，启动带有端口号 9999 的套接字服务器，命令如下：

```
$ nc -lk 9999
```

（2）另外打开第 2 个终端，启动 PySpark Shell，命令如下：

```
$ pyspark --master spark://localhost:7077
```

（3）在 PySpark Shell 中，执行结构化流处理程序，代码如下：

```python
#第6章/source_socket.py

from pyspark.sql.functions import explode
from pyspark.sql.functions import split

#将shuffle后的分区数设置为10（测试环境下）
spark.conf.set("spark.sql.shuffle.partitions",10)

#从Socket数据源读取流数据
lines = spark.readStream \
   .format("socket") \
   .option("host", "localhost") \
   .option("port", "9999") \
   .load()

#将行拆分为单词
words = lines.select(
   explode(split(lines.value, " ")).alias("word")
)

#生成运行时单词计数
wordCounts = words.groupBy("word").count()

query = wordCounts.writeStream \
   .format("console") \
   .outputMode("complete") \
   .start()
```

```
#等待流程序执行结束（作为作业文件提交时启用）
#query.awaitTermination()
```

（4）回到第 1 个终端窗口，任意输入一些单词，以空格分隔，并按 Enter 键。多输入几行单词，然后在第 2 个终端窗口观察流计算输出，如图 6-14 所示。

图 6-14　在 Netcat 运行窗口输入文本内容

在第 2 个终端窗口观察到的输出结果如图 6-15 所示。

图 6-15　流计算结果

（5）当完成测试 Socket 数据源时，可以通过调用 stop()函数来停止流查询。在停止流查询之后，在第 1 个终端中输入任何单词都不会导致在 PySpark Shell 中显示这些单词。停止流查询的代码如下：

```
query.stop()
```

6.5.2　使用 Rate 数据源

与 Socket 数据源类似，Rate 数据源也是为了测试和学习目的而设计的。Rate 数据源支持以下选项。

（1）rowsPerSecond：每秒应该生成多少行，例如，指定为 100。默认为 1。如果这个数值很高，就可以提供下一个可选配置 rampUpTime。

（2）rampUpTime：在生成速度变为 rowsPerSecond 之前需要多长时间用来提升，例如，5s。默认为 0s。使用比秒更细的粒度将被截断为整数秒。

(3) numPartitions: 生成行的分区数。默认为 Spark 默认并行度。

Rate 数据源将尽力达到 rowsPerSecond，但是查询可能受到资源限制，可以调整 numPartitions 以帮助达到所需的速度。

Rate 数据源产生的每段数据只包含两列：时间戳和自动增加的值。下面的例子包含打印 Rate 数据源数据的代码。

【例 6-4】编写 PySpark 结构化流应用程序，使用 Rate 数据源。

启动 PySpark Shell，执行代码如下：

```
#第6章/source_rate.py

from pyspark.sql.functions import explode
from pyspark.sql.functions import split

#将 shuffle 后的分区数设置为10（测试环境下）
spark.conf.set("spark.sql.shuffle.partitions",10)

#从 Socket 数据源读取流数据
lines = spark.readStream \
    .format("socket") \
    .option("host", "localhost") \
    .option("port", "9999") \
    .load()

#从 Rate 数据源读取流数据，配置它每秒产生10行
rateSourceDF = spark.readStream \
    .format("rate") \
    .option("rowsPerSecond","10") \
    .load()

#以 update 模式将结果写到控制台，并启动流计算
query = rateSourceDF.writeStream \
    .outputMode("update") \
    .format("console") \
    .option("truncate", "false") \
    .start()

#等待流程序执行结束（作为作业文件提交时启用）
#query.awaitTermination()
```

从控制台可观察到每秒输出10条数据，其中部分批次数据如下：

```
-------------------------------------------
Batch: 1
-------------------------------------------
```

```
+-----------------------+-----+
|timestamp              |value|
+-----------------------+-----+
|2021-02-02 17:32:01.264|0    |
|2021-02-02 17:32:01.664|4    |
|2021-02-02 17:32:02.064|8    |
|2021-02-02 17:32:01.364|1    |
|2021-02-02 17:32:01.764|5    |
|2021-02-02 17:32:02.164|9    |
|2021-02-02 17:32:01.464|2    |
|2021-02-02 17:32:01.864|6    |
|2021-02-02 17:32:01.564|3    |
|2021-02-02 17:32:01.964|7    |
+-----------------------+-----+
```

值得注意的一件事是，value 列中的数保证在所有分区中都是连续的。例如，要查看 3 个分区的输出结果，代码如下：

```
#第6章/source_rate2.py

from pyspark.sql.functions import *

#从 Rate 数据源读取流数据，配置它每秒产生 10 行，分 3 个分区
rateSourceDF = spark.readStream \
    .format("rate") \
    .option("rowsPerSecond","10") \
    .option("numPartitions",3) \
    .load()

#添加分区 ID 列，以便检查
rateWithPartitionDF = rateSourceDF \
   .withColumn("partition_id", spark_partition_id())

#以 update 模式将结果写到控制台，并启动流计算
query = rateWithPartitionDF.writeStream \
    .outputMode("update") \
    .format("console") \
    .option("truncate", "false") \
    .start()

#等待流程序执行结束（作为作业文件提交时启用）
#query.awaitTermination()
```

输出结果如下：

```
-------------------------------------------
Batch: 1
-------------------------------------------
```

```
+-----------------------+-----+------------+
|timestamp              |value|partition_id|
+-----------------------+-----+------------+
|2021-02-02 17:35:43.461|0    |0           |
|2021-02-02 17:35:43.761|3    |0           |
|2021-02-02 17:35:44.061|6    |0           |
|2021-02-02 17:35:44.361|9    |0           |
|2021-02-02 17:35:43.561|1    |1           |
|2021-02-02 17:35:43.861|4    |1           |
|2021-02-02 17:35:44.161|7    |1           |
|2021-02-02 17:35:43.661|2    |2           |
|2021-02-02 17:35:43.961|5    |2           |
|2021-02-02 17:35:44.261|8    |2           |
+-----------------------+-----+------------+
```

前面的输出显示了这 10 行分布在 3 个分区上，并且这些值是连续的，就好像它们是为单个分区生成的一样。

6.5.3 使用 File 数据源

File 数据源是最容易理解和使用的。PySpark 结构化流开箱即用地支持所有常用的文件格式，包括文本、CSV、JSON、ORC 和 parquet。

File 数据源支持以下选项配置。

（1）path：输入目录的路径，对所有文件格式都通用。

（2）maxFilesPerTrigger：每个触发器中考虑处理的最大新文件数（默认为 no max）。

（3）latestFirst：是否先处理最新的文件，当有大量文件积压时很有用（默认为 False）。

（4）fileNameOnly：是否仅根据文件名而不是根据完整路径检查新文件（默认为 False）。将此值设置为 True 后，具有一样的文件名但是不同前缀协议的文件将被认为是相同的文件。例如，下面的这些文件被认为是相同的文件，均被认为是 dataset.txt：

```
"file://dataset.txt"
"s3://a/dataset.txt"
"s3n://a/b/dataset.txt"
"s3a://a/b/c/dataset.txt"
```

使用 File 数据源的流程序模板代码如下：

```
#使用 File 数据源，读取 JSON 文件
mobileSSDF = spark.readStream \
.schema(mobileDataSchema) \
.json("<directoryname>")

#如果将 maxFilesPerTrigger 指定为 1，则表示一个文件一个文件地处理
mobileSSDF = spark.readStream \
```

```
.schema(mobileDataSchema) \
.option("maxFilesPerTrigger",1) \
.json("<directory name>")

#如果将 latestFirst 指定为 True,则表示首先处理新产生的文件
mobileSSDF = spark.readStream \
.schema(mobileDataSchema) \
.option("latestFirst", "true") \
.json("<directory name>")
```

下面通过一个示例程序来演示如何使用结构化流读取文件数据源。

【例 6-5】假设移动电话的开关机等事件会保存在 JSON 格式的文件中。现在编写 PySpark 结构化流处理程序来读取这些事件并处理。

建议按以下步骤操作。

（1）准备数据。

在本例中，使用 File 数据源，该数据源以 JSON 文件的格式记录了一小组移动电话动作事件。每个事件由 3 个字段组成。

① id：表示手机的唯一 ID。在样例数据集中，手机 ID 将类似于 phone1、phone2、phone3 这样的字符串。

② action：表示用户所采取的操作。该操作的可能值是"open"或"close"。

③ ts：表示用户 action 发生时的时间戳。这是事件时间（Event Time）。

下面准备了 3 个存储手机事件数据的 JSON 文件，这 3 个文件的内容如下。

file1.json 文件的内容如下：

```
{"id":"phone1","action":"open","ts":"2018-03-02T10:02:33"}
{"id":"phone2","action":"open","ts":"2018-03-02T10:03:35"}
{"id":"phone3","action":"open","ts":"2018-03-02T10:03:50"}
{"id":"phone1","action":"close","ts":"2018-03-02T10:04:35"}
```

file2.json 文件的内容如下：

```
{"id":"phone3","action":"close","ts":"2018-03-02T10:07:35"}
{"id":"phone4","action":"open","ts":"2018-03-02T10:07:50"}
```

file3.json 文件的内容如下：

```
{"id":"phone2","action":"close","ts":"2018-03-02T10:04:50"}
{"id":"phone5","action":"open","ts":"2018-03-02T10:10:50"}
```

为了模拟数据流的行为，将把这 3 个 JSON 文件复制到指定的目录下。

（2）先导入相关的依赖包，代码如下：

```
from pyspark.sql.types import *
from pyspark.sql.functions import *
```

(3) 为手机事件数据创建模式(Schema)。

默认情况下,结构化流在从基于文件的数据源读取数据时需要一个模式(因为最初目录可能是空的,因此结构化的流无法推断模式),但是,可以将配置参数 spark.sql.streaming.schemaInference 的值设置为 True 来启用模式推断。在这个例子中,将显式地创建一个模式,代码如下:

```
#为手机事件数据创建一个Schema
fields = [
    StructField("id", StringType(), nullable = False),
    StructField("action", StringType(), nullable = False),
    StructField("ts", TimestampType(), nullable = False)
]
mobileDataSchema = StructType(fields)
```

(4) 创建流 DataFrame。

读取流 File 数据源,创建流 DataFrame,并将 action 列值转换为大写,代码如下:

```
#指定监听的文件目录
dataPath = "/data/spark/stream/mobile"

#读取指定目录下的源数据文件,一次一个
mobileDF = spark.readStream \
    .option("maxFilesPerTrigger", 1) \
    .option("mode","failFast") \
    .schema(mobileDataSchema) \
    .json(dataPath)

#mobileDF.printSchema()

#将所有"action"列值转换为大写
upperDF = mobileDF.select("id",upper("action").alias("action"),"ts")
```

(5) 输出结果。

将结果 DataFrame 输出到控制台显示,代码如下:

```
#将结果输出到控制台
query = upperDF.writeStream \
    .format("console") \
    .option("truncate","false") \
    .outputMode("append") \
    .start()

#等待流程序执行结束(作为作业文件提交时启用)
#query.awaitTermination()
```

（6）执行流处理程序，输出结果如下：

```
-------------------------------------------
Batch: 0
-------------------------------------------
+------+------------+-------------------+
|id    |action      |ts                 |
+------+------------+-------------------+
|phone1|OPEN        |2018-03-02 10:02:33|
|phone2|OPEN        |2018-03-02 10:03:35|
|phone3|OPEN        |2018-03-02 10:03:50|
|phone1|CLOSE       |2018-03-02 10:04:35|
+------+------------+-------------------+

-------------------------------------------
Batch: 1
-------------------------------------------
+------+------------+-------------------+
|id    |action      |ts                 |
+------+------------+-------------------+
|phone3|CLOSE       |2018-03-02 10:07:35|
|phone4|OPEN        |2018-03-02 10:07:50|
+------+------------+-------------------+

-------------------------------------------
Batch: 2
-------------------------------------------
+------+------------+-------------------+
|id    |action      |ts                 |
+------+------------+-------------------+
|phone2|CLOSE       |2018-03-02 10:04:50|
|phone5|OPEN        |2018-03-02 10:10:50|
+------+------------+-------------------+
```

完整的实现代码如下：

```
#第6章/source_file.py

from pyspark.sql import SparkSession
from pyspark.sql.types import *
from pyspark.sql.functions import *

#创建 SparkSession 实例
spark = SparkSession \
        .builder \
        .appName("file source") \
        .getOrCreate()

#为手机事件数据创建一个 Schema
```

```
fields = [
    StructField("id", StringType(), nullable = False),
    StructField("action", StringType(), nullable = False),
    StructField("ts", TimestampType(), nullable = False)
]
mobileDataSchema = StructType(fields)

#指定监听的文件目录
dataPath = "/data/spark/stream/mobile"

#读取指定目录下的源数据文件，一次一个
mobileDF = spark.readStream \
    .option("maxFilesPerTrigger", 1) \
    .option("mode","failFast") \
    .schema(mobileDataSchema) \
    .json(dataPath)

#mobileDF.printSchema()

#将所有"action"列值转换为大写
upperDF = mobileDF.select("id",upper("action").alias("action"),"ts")

#将结果输到控制台
query = upperDF.writeStream \
    .format("console") \
    .option("truncate","false") \
    .outputMode("append") \
    .start()

#等待流程序执行结束
query.awaitTermination()
```

6.5.4 使用 Kafka 数据源

Kafka 通常用于构建实时流数据管道，以可靠地在系统之间移动数据，还用于转换和响应数据流。Kafka 作为集群运行在一个或多个服务器上。Kafka 的一些关键概念描述如下。

（1）Topic：主题。消息发布到的类别或流名称的高级抽象。主题可以有 0、1 或多个消费者，这些消费者订阅发布到该主题的消息。用户为每个新的消息类别定义一个新主题。

（2）Producers：生产者。向主题发布消息的客户端。

（3）Consumers：消费者。使用来自主题的消息的客户端。

（4）Brokers：服务器。复制和持久化消息数据的一个或多个服务器。

此外，生产者和消费者可以同时对多个主题进行读写。每个 Kafka 主题都是分区的，写入每个分区的消息都是顺序排序的。分区中的消息具有一个偏移量，用来唯一标识分区内的每条消息。

Kafka 中主题（Topic）的分区是分布式的，每个 Broker 处理对分区共享的请求。每个分区在 Brokers（数量可配置）之间复制。Kafka 集群在一段时间内（可配置）保留所有已发布的消息。Kafka 使用 ZooKeeper 作为其分布式进程的协调服务。

注意：Kafka 的数据源可能是在生产型流应用程序中最常用的数据源。为了有效地处理这个数据源，读者需要一定的 Kafka 基本知识。

在使用 Kafka 数据源时，PySpark 流程序实际上充当了 Kafka 的消费者，因此，流程序所需要的信息与 Kafka 的消费者所需要的信息相似。Kafka 数据源的一些配置选项见表 6-4。

表 6-4　Kafka 数据源的配置选项

Option	值	描述
kafka.Bootstrap.servers	host1:port1, host2:port2	Kafka 服务器列表，以逗号分隔
subscribe	topic1, topic2	这个数据源要读取的主题名列表，以逗号分隔
subscribePattern	topic.*	使用正则模式表示要读取数据的主题，比 subscribe 要灵活
assign	{topic1:[1,2], topic2:[3,4]}	指定要读取数据的主题的分区。这个信息必须是 JSON 格式

其中必需的信息是要连接的 Kafka 服务器的列表，以及一个或多个从其读取数据的主题。为了支持选择从哪个主题和主题分区来读取数据的各种方法，它支持 3 种不同的方式来指定这些信息。用户只需选择最适合自身用例的那个。

还有一些可选配置选项，它们都有自己的默认值，见表 6-5。

表 6-5　Kafka 数据源的可选配置选项

Option	默认值	值	描述

startingOffsets	latest	earliest, latest
		每个主题的开始偏移位置，JSON 格式字符串，例如
		{
		"topic1":{"0":45, "1":-1},
		"topic2":{"0":-2}
		}
			earliest：意味着主题的开始处
			latest：意味着主题中的任何最新数据
			当使用 JSON 字符串格式时，-2 代表在一个特定分区中的 earliest offset，-1 代表在一个特定分区中的 latest offset

endingOffsets	latest	Latest
		JSON 格式字符串，例如
		{
		"topic1":{"0":45, "1":-1},
		"topic2":{"0":-1}
		}
			latest：意味着主题中的最新数据
			当使用 JSON 字符串格式时，-1 代表在一个特定分区中的 latest offset。当然-2 不适用于此选项

续表

Option	默认值	值	描 述
maxOffsetsPerTrigger	none	Long，例如，500	此选项是一种速率限制机制，用于控制每个触发器间隔要处理的记录数量。如果指定了一个值，则它表示所有分区的记录总数，而不是每个分区的记录总数

要设置 Kafka，首先需要下载它，下载网址为 http://kafka.apache.org/downloads.html。选择与 PySpark 版本相兼容的版本，然后，按以下步骤设置 Kafka。

（1）将下载的 Kafka 压缩包解压缩到 ~/bigdata/ 目录下，命令如下：

```
$ cd ~/bigdata
$ tar -xvfz kafka_2.12-2.4.1.tgz
```

（2）Kafka 依赖于 Apache ZooKeeper，所以在启动 Kafka 之前，需要先启动它，命令如下：

```
$ cd ~/bigdata/kafka_2.12-2.4.1
$ ./bin/zookeeper-server-start.sh config/zookeeper.properties
```

这将在 2181 端口启动 ZooKeeper 进程，并让 ZooKeeper 在后台工作。

（3）接下来，启动 Kafka 服务器，命令如下：

```
$ ./bin/kafka-server-start.sh config/server.properties
```

（4）创建主题，命令如下：

```
$ ./bin/kafka-topics.sh --create --zookeeper localhost:2181 --replication-factor 1 --partitions 1 --topic test
```

（5）查看已有的主题，命令如下：

```
$ ./bin/kafka-topics.sh --list --zookeeper localhost:2181
```

（6）删除一个主题，命令如下：

```
$ ./bin/kafka-topics.sh --delete --Bootstrap-server localhost:9092 --topic test
```

需要修改启动的配置文件 server.properties，设置 delete.topic.enable=true（默认设置为 false）。

默认情况下，Kafka 的数据源并不是 PySpark 的内置数据源，因此如果要开发读取 Kafka 数据的 PySpark 结构化流处理程序，就必须将 Kafka 的依赖包添加到 classpath 中。

如果要从 PySpark Shell 使用 Kafka 数据源，则需要在启动 PySpark Shell 时将依赖的 JAR 包添加到 classpath 中。有以下两种方式可以做到：

（1）手动将依赖包添加到 classpath。

首先将 kafka_2.12-2.4.1.jar、spark-streaming-kafka-0-10-assembly_2.12-3.1.2.jar 和 spark-sql-kafka-0-10_2.12-3.1.2.jar 包复制到~/bigdata/spark-3.1.2/jars/目录下，并重启 Spark 集群（如果已经启动），然后启动 PySpark Shell，命令如下：

```
$ cd ~/bigdata/spark-3.1.2
$ ./spark-shell --master spark://localhost:7077
```

（2）使用 package 参数让 PySpark 自动下载这些文件，命令如下：

```
$ cd ~/bigdata/spark-3.1.2
$ ./bin/pyspark --master spark://localhost:7077 --packages org.apache.spark:spark-sql-kafka-0-10_2.12:3.1.2,org.apache.kafka:kafka_2.12:2.4.1,org.apache.spark:spark-streaming-kafka-0-10-assembly_2.12:3.1.2
```

下面通过一个例 6-6 演示如何编写 Spark 结构化流处理程序来读取 Kafka 中的数据。

【例 6-6】编写 PySpark 结构化流程序作为 Kafka 的消费者程序，将 Kafka 作为结构化流程序的数据源。

在这个示例中，使用 Kafka 自带的生产者脚本向 Kafka 的 test 主题发送内容，而 PySpark 结构化流程序会订阅该主题。一旦它收到了订阅的消息，马上输到控制台中。程序处理流程如图 6-16 所示。

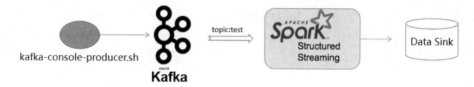

图 6-16　程序处理流程

编写 PySpark 结构化流程序代码，代码如下：

```python
#第 6 章/source_kafka.py

from pyspark.sql import SparkSession
from pyspark.sql.types import *
from pyspark.sql.functions import *

#创建 SparkSession 实例
spark = SparkSession.builder.appName("Kafka Source").getOrCreate()

#创建一个流来监听 test 主题的消息
dataDF = spark.readStream \
    .format("kafka") \
    .option("kafka.Bootstrap.servers", "localhost:9092") \
```

```
        .option("subscribe", "test") \
        .option("startingOffsets", "earliest") \
        .load()

#查看这个 DataFrame 的 Schema
dataDF.printSchema()

#将该流转换为 String 数据类型（key 和 value 都是字节数组形式）
#kvstream = dataDF.selectExpr("CAST(key as string)", "CAST(value as string)")
kvstream = dataDF.selectExpr("CAST(value as string)", "topic", "partition", "offset")

#将该流写到控制台
query = kvstream.writeStream \
    .outputMode("append") \
    .format("console") \
    .start()

#等待流程序执行结束（作为作业文件提交时启用）
#query.awaitTermination()
```

要运行这个流程序，建议按以下步骤进行操作。

(1) 启动 ZooKeeper 服务。Kafka 依赖于 Apache ZooKeeper，所以在启动 Kafka 之前，要先启动它。打开一个终端窗口，执行的命令如下：

```
$ cd ~/bigdata/kafka_2.12-2.4.1
$ ./bin/zookeeper-server-start.sh config/zookeeper.properties
```

等待 30s 左右（ZooKeeper 启动）。

(2) 接下来，启动 Kafka 服务器。另打开一个终端窗口，执行的命令如下：

```
$ cd ~/bigdata/kafka_2.12-2.4.1
$ ./bin/kafka-server-start.sh config/server.properties
```

等待 30s 左右（Kafka 启动）。

(3) 查看和创建 Kafka 主题（如果已经有了 test 主题，则此步略过）。另外打开第 3 个终端窗口，创建 test 主题，执行的命令如下：

```
$ cd ~/bigdata/kafka_2.12-2.4.1
$ ./bin/kafka-topics.sh --create --zookeeper localhost:2181 --replication-factor 1 --partitions 1 --topic test
```

查看已有的主题，执行的命令如下：

```
$ ./bin/kafka-topics.sh --list --zookeeper localhost:2181
```

(4) 启动流程序，开始接收从 Kafka test 主题订阅的消息。

（5）向 Kafka 的 test 主题发送消息。另外打开第 4 个终端窗口，产生消息并发布给 test 主题，执行的命令如下：

```
$ cd ~/bigdata/kafka_2.12-2.4.1
$ ./bin/kafka-console-producer.sh --broker-list localhost:9092 --topic test
```

然后，随意键入一些消息。例如，输入以下内容：

```
> good good study
> day day up
```

（6）回到流程序执行窗口，如果一切正常，则应该可以在控制台看到输出了收到的订阅消息，内容如下：

```
root
 |-- key: binary (nullable = true)
 |-- value: binary (nullable = true)
 |-- topic: string (nullable = true)
 |-- partition: integer (nullable = true)
 |-- offset: long (nullable = true)
 |-- timestamp: timestamp (nullable = true)
 |-- timestampType: integer (nullable = true)

-------------------------------
Batch: 0
-------------------------------
+-----+-----+---------+------+
|value|topic|partition|offset|
+-----+-----+---------+------+
+-----+-----+---------+------+


-------------------------------------
Batch: 1
-------------------------------------
+---------------+-----+---------+------+
|          value|topic|partition|offset|
+---------------+-----+---------+------+
|good good study| test|        0|     0|
+---------------+-----+---------+------+


-----------------------------------
Batch: 2
-----------------------------------
+----------+-----+---------+------+
|     value|topic|partition|offset|
+----------+-----+---------+------+
|day day up| test|        0|     1|
+----------+-----+---------+------+
```

从上面的输出内容可以看出，从 Kafka 中读取的数据每列都有固定的格式，见表 6-6。

表 6-6　Kafka 消息格式

列	类型
key	binary
value	binary
topic	string
partition	int
offset	long
timestamp	long
timestampType	int

在从 Kafka 读取消息时，有多种不同的方式。指定 Kafka 的主题、分区和从 Kafka 读取消息的偏移量的不同变化方式的模板代码如下：

```
#指定 Kafka 的主题、分区和偏移量的各种变化

#订阅 1 个主题
kafkaDF = spark \
  .readStream \
  .format("kafka") \
  .option("kafka.Bootstrap.servers", "host1:port1,host2:port2") \
  .option("subscribe", "topic1") \
  .load()
kafkaDF.selectExpr("CAST(key AS STRING)", "CAST(value AS STRING)")

#订阅 1 个主题，带有标题
kafkaDF = spark \
  .readStream \
  .format("kafka") \
  .option("kafka.Bootstrap.servers", "host1:port1,host2:port2") \
  .option("subscribe", "topic1") \
  .option("includeHeaders", "true") \
  .load()
kafkaDF.selectExpr("CAST(key AS STRING)", "CAST(value AS STRING)", "headers")

#订阅多个主题，使用默认的 startingOffsets 和 endingOffsets
kafkaDF = spark \
  .readStream \
  .format("kafka") \
  .option("kafka.Bootstrap.servers", "host1:port1,host2:port2") \
  .option("subscribe", "topic1,topic2") \
  .load()
kafkaDF.selectExpr("CAST(key AS STRING)", "CAST(value AS STRING)")
```

```
#订阅模式,使用subscribePattern
kafkaDF = spark \
  .readStream \
  .format("kafka") \
  .option("kafka.Bootstrap.servers", "host1:port1,host2:port2") \
  .option("subscribePattern", "topic.*") \
  .load()
kafkaDF.selectExpr("CAST(key AS STRING)", "CAST(value AS STRING)")

#使用JSON格式从一个特定的偏移量读取
#三引号格式用于转义JSON字符串中的双引号
kafkaDF = spark \
  .readStream \
  .format("kafka") \
  .option("kafka.Bootstrap.servers", "host1:port1,host2:port2") \
  .option("subscribe", "topic1,topic2") \
  .option("startingOffsets", """{"topic1": {"0":51} } """) \
  .load()
```

6.6 流 DataFrame 操作

13min

前面的例子表明,一旦配置和定义了数据源,DataStreamReader 将返回一个 DataFrame 的实例。这意味着可以使用大多数熟悉的 DataFrame 关系操作和 PySpark SQL 函数来表达应用程序流计算逻辑,但是要注意,并不是所有的 DataFrame 操作都受流式 DataFrame 支持,例如 limit、distinct 和 sort 就不能在流式 DataFrame 上使用,这是因为它们在流数据处理的上下文中不适用。

6.6.1 选择、投影和聚合操作

PySpark 结构化流的一个优点是具有一组统一的 API,用于 PySpark 的批处理和流处理。使用流数据格式的 DataFrame,可以应用任何 select()和 filter()转换,以及任何作用在列上的 PySpark SQL 函数。此外,基本聚合和高级分析函数也可用于流 DataFrame。下面通过一个移动电话事件数据流分析例子程序来演示这些用法。

【例 6-7】移动电话事件数据流分析。移动电话的开关机等事件会保存在 JSON 格式的文件中。现在编写 PySpark 结构化流处理程序来读取这些事件并处理。

建议按以下步骤操作。

(1)准备数据。

在本示例中,使用文件数据源,该数据源以 JSON 文件的格式记录了一小组移动电话动作事件。每个事件由 3 个字段组成。

① id: 表示手机的唯一 ID。在样例数据集中，电话 ID 将类似于 phone1、phone2、phone3 等字符串。

② action: 表示用户所采取的操作。该操作的可能值是 open 或 close。

③ ts: 表示用户 action 发生时的时间戳。这是事件时间（Event Time）。

已经准备了 3 个存储移动电话事件数据的 JSON 文件，这 3 个文件的内容如下。

file1.json 文件的内容如下：

```
{"id":"phone1","action":"open","ts":"2018-03-02T10:02:33"}
{"id":"phone2","action":"open","ts":"2018-03-02T10:03:35"}
{"id":"phone3","action":"open","ts":"2018-03-02T10:03:50"}
{"id":"phone1","action":"close","ts":"2018-03-02T10:04:35"}
```

file2.json 文件的内容如下：

```
{"id":"phone3","action":"close","ts":"2018-03-02T10:07:35"}
{"id":"phone4","action":"open","ts":"2018-03-02T10:07:50"}
```

file3.json 文件的内容如下：

```
{"id":"phone2","action":"close","ts":"2018-03-02T10:04:50"}
{"id":"phone5","action":"open","ts":"2018-03-02T10:10:50"}
```

为了模拟数据流的行为，应把这 3 个 JSON 文件复制到 HDFS 的指定目录下。

(2) 初始化代码。

先导入相关的依赖包，并构造一个 SparkSession 实例，代码如下：

```
from pyspark.sql import SparkSession
from pyspark.sql.types import *
from pyspark.sql.functions import *

#创建 SparkSession 实例
spark = SparkSession \
        .builder \
        .appName("streaming demo") \
        .getOrCreate()

#将 shuffle 后的分区数设置为 10（测试环境下）
spark.conf.set("spark.sql.shuffle.partitions",10)
```

(3) 为手机事件数据创建模式（Schema）。

默认情况下，结构化流在从基于文件的数据源读取数据时需要一个模式（因为最初目录可能是空的，因此结构化的流无法推断模式），但是，可以将配置参数 spark.sql.streaming.schemaInference 的值设置为 True 来启用模式推断。在这个例子中，将显式地创建一个模式，代码如下：

```
#为手机事件数据创建一个schema
fields = [
    StructField("id", StringType(), nullable = False),
    StructField("action", StringType(), nullable = False),
    StructField("ts", TimestampType(), nullable = False)
]
mobileDataSchema = StructType(fields)
```

(4) 创建 DataFrame。

读取流文件数据源, 创建 DataFrame, 并将 action 列值转换为大写, 代码如下:

```
#指定监听的文件目录
dataPath = "/data/spark/stream/mobile"

#读取指定目录下的源数据文件, 一次一个
mobileDF = spark.readStream \
    .option("maxFilesPerTrigger", 1) \
    .option("mode","failFast") \
    .schema(mobileDataSchema) \
    .json(dataPath)

#mobileSSDF.printSchema()
```

(5) 转换操作。

将 action 列值转换为大写, 执行过滤、投影、聚合等转换操作, 代码如下:

```
mobileDF2 = mobileDF \
    .where("action='open' or action='close'") \
    .withColumn("action",upper(col("action"))) \
    .select("id","action","ts") \
    .groupBy("action") \
    .count()
```

(6) 输出结果。

将结果 DataFrame 输到控制台显示, 代码如下:

```
#将结果输到控制台
query = mobileDF2.writeStream \
    .format("console") \
    .option("truncate","false") \
    .outputMode("complete") \
    .start()

#等待流程序执行结束(作为作业文件提交时启用)
#query.awaitTermination()
```

（7）执行流处理程序，输出结果如下：

```
-------------------------------------------
Batch: 0
-------------------------------------------
+------+-----+
|action|count|
+------+-----+
|CLOSE |1    |
|OPEN  |3    |
+------+-----+


-------------------------------------------
Batch: 1
-------------------------------------------
+------+-----+
|action|count|
+------+-----+
|CLOSE |2    |
|OPEN  |4    |
+------+-----+


-------------------------------------------
Batch: 2
-------------------------------------------
+------+-----+
|action|count|
+------+-----+
|CLOSE |3    |
|OPEN  |5    |
+------+-----+
```

完整的实现代码如下：

```python
#第6章/streaming_operator.py

from pyspark.sql import SparkSession
from pyspark.sql.types import *
from pyspark.sql.functions import *

#创建SparkSession实例
spark = SparkSession \
    .builder \
    .appName("streaming demo") \
    .getOrCreate()

#将shuffle后的分区数设置为10（测试环境下）
spark.conf.set("spark.sql.shuffle.partitions",10)
```

```python
#为手机事件数据创建一个schema
fields = [
    StructField("id", StringType(), nullable = False),
    StructField("action", StringType(), nullable = False),
    StructField("ts", TimestampType(), nullable = False)
]
mobileDataSchema = StructType(fields)

#指定监听的文件目录
dataPath = "/data/spark/stream/mobile"

#读取指定目录下的源数据文件，一次一个
mobileDF = spark.readStream \
    .option("maxFilesPerTrigger", 1) \
    .option("mode","failFast") \
    .schema(mobileDataSchema) \
    .json(dataPath)

mobileDF2 = mobileDF \
    .where("action='open' or action='close'") \
    .withColumn("action",upper(col("action"))) \
    .select("id","action","ts") \
    .groupBy("action") \
    .count()

#将结果输到控制台
query = mobileDF2.writeStream \
    .format("console") \
    .option("truncate","false") \
    .outputMode("complete") \
    .start()

#等待流程序执行结束（作为作业文件提交时启用）
query.awaitTermination()
```

在这个示例中，采用的输出模式是 complete。在没有聚合操作的情况下，不能使用 complete 输出模式；在有聚合操作的情况下，不能使用 append 模式。

在 PySpark 结构化流程序中，也可以创建一个视图来应用 SQL 查询。修改上面的示例，代码如下：

```python
#第6章/streaming_operator2.py

from pyspark.sql import SparkSession
from pyspark.sql.types import *
from pyspark.sql.functions import *
```

```python
#创建 SparkSession 实例
spark = SparkSession \
    .builder \
    .appName("streaming demo") \
    .getOrCreate()

#将 shuffle 后的分区数设置为 10（测试环境下）
spark.conf.set("spark.sql.shuffle.partitions",10)

#为手机事件数据创建一个 schema
fields = [
    StructField("id", StringType(), nullable = False),
    StructField("action", StringType(), nullable = False),
    StructField("ts", TimestampType(), nullable = False)
]
mobileDataSchema = StructType(fields)

#指定监听的文件目录
dataPath = "file://home/hduser/data/spark/stream/mobile"

#读取指定目录下的源数据文件，一次一个
mobileDF = spark.readStream \
    .option("maxFilesPerTrigger", 1) \
    .option("mode","failFast") \
    .schema(mobileDataSchema) \
    .json(dataPath)

#也可以创建一个视图来应用 SQL 查询
mobileDF.createOrReplaceTempView("mobile_tb")

#选择、投影、聚合等操作
sqlDF = spark.sql("""
select upper('action') as action,count(*) as cnt
from mobile_tb
where action='open' or action='close'
group by action
""")

#将结果输出到控制台（注意，输出模式设置）
query = sqlDF.writeStream \
    .format("console") \
    .option("truncate","false") \
    .outputMode("complete") \
    .start()

#等待流程序执行结束（作为作业文件提交时启用）
query.awaitTermination()
```

执行以上代码，会看到相同的输出结果。

需要注意，在流 DataFrame 中，不支持以下 DataFrame 转换（因为它们太过复杂，无法维护状态，或者由于流数据的无界性）：

（1）在流 DataFrame 上的多个聚合或聚合链。

（2）limit 和 take N 行。

（3）distinct 转换。

（4）在没有任何聚合的情况下对流 DataFrame 进行排序。

使用任何不受支持的操作的尝试都会导致一个 AnalysisException 异常及类似"XYZ 操作不受流 streaming DataFrame 支持"的消息。

6.6.2 执行 join 连接操作

从 PySpark 2.3 开始，结构化流支持对两个流 DataFrame 执行 join 连接操作。考虑到流 DataFrame 的无界性，结构化的流必须维护两个流 DataFrame 的历史数据，以匹配任何未来的尚未收到的数据。

可以用一个流式 DataFrame 来 join 连接另一个静态的 DataFrame 或者流式 DataFrame，然而，join 连接是一个复杂的操作，其中最棘手的问题在于并非所有的数据在连接时都是可用的流 DataFrame，因此，join 连接的结果是在每个触发器点上增量地生成。

下面通过一个 IoT（物联网）示例来学习两个流 DataFrame 的连接操作。

【例 6-8】假设在某个数据中心，通过不同的传感器采集不同类型的实时数据，其中第 1 个传感器采集不同机架的实时温度读数；第 2 个传感器采集不同机架的实时负载信息。

这些数据都存储在 JSON 格式的数据文件中，包含数据中心中不同位置机架的温度读数，数据如下所示。

file1_temp.json 文件的内容如下：

```
{"temp_location_id":"rack1","temperature":99.5,"temp_taken_time":"2017-06-02T08:01:01"}
{"temp_location_id":"rack2","temperature":100.5,"temp_taken_time":"2017-06-02T08:06:02"}
{"temp_location_id":"rack3","temperature":101.0,"temp_taken_time":"2017-06-02T08:11:03"}
{"temp_location_id":"rack4","temperature":102.0,"temp_taken_time":"2017-06-02T08:16:04"}
```

包含同一数据中心中每台计算机的负载信息，数据如下所示。

file2_load.json 文件的内容如下：

```
{"load_location_id":"rack1","load":199.5,"load_taken_time":"2017-06-02T08:01:02"}
```

```
{"load_location_id":"rack2","load":1105.5,"load_taken_time":"2017-06-
02T08:06:04"}
{"load_location_id":"rack3","load":2104.0,"load_taken_time":"2017-06-
02T08:11:06"}
{"load_location_id":"rack4","load":1108.0,"load_taken_time":"2017-06-
02T08:16:08"}
{"load_location_id":"rack4","load":1108.0,"load_taken_time":"2017-06-
02T08:21:10"}
```

现在需要编写一个 PySpark 流处理程序，连接这两个流数据集，统计每个机架实时的温度和负载。实现代码如下（注意其中的连接条件和时间约束条件的设置）：

```python
#第6章/streaming_join.py

#导入依赖包
from pyspark.sql import SparkSession
from pyspark.sql.types import *
from pyspark.sql.functions import *

#创建 SparkSession 实例
spark = SparkSession \
    .builder \
    .appName("streaming demo") \
    .getOrCreate()

#将 shuffle 后的分区数设置为10（测试环境下）
spark.conf.set("spark.sql.shuffle.partitions",10)

#为 IoT 温度数据创建一个 schema
tempDataSchema = StructType() \
    .add("temp_location_id", StringType(), nullable = False) \
    .add("temperature", DoubleType(), nullable = False) \
    .add("temp_taken_time", TimestampType(), nullable = False)

#为 IoT 负载数据创建一个 schema
loadDataSchema = StructType() \
    .add("load_location_id", StringType(), nullable = False) \
    .add("load", DoubleType(), nullable = False) \
    .add("load_taken_time", TimestampType(), nullable = False)

#读取流数据源
dataPath1 = "file://home/hduser/data/spark/stream/iot/temp"
tempDataDF = spark.readStream \
    .option("maxFilesPerTrigger", 1) \
    .option("timestampFormat","yyyy-MM-dd hh:mm:ss") \
    .option("mode","failFast") \
```

```python
    .schema(tempDataSchema) \
    .json(dataPath1)

dataPath2 = "file://home/hduser/data/spark/stream/iot/load"
loadDataDF = spark.readStream \
    .option("maxFilesPerTrigger", 1) \
    .option("timestampFormat","yyyy-MM-dd hh:mm:ss") \
    .option("mode","failFast") \
    .schema(loadDataSchema) \
    .json(dataPath2)

#基于 location id 连接，以及事件时间约束
sqlExpr = \
    """temp_location_id = load_location_id AND
       load_taken_time >= temp_taken_time AND
       load_taken_time <= temp_taken_time + interval 10 minutes
    """
tempWithLoadDataDF = tempDataDF.join(loadDataDF, expr(sqlExpr))

#将结果输到控制台（注意，因为有聚合操作，所以输出模式必须是"complete"）
query = tempWithLoadDataDF.writeStream \
    .format("console") \
    .option("truncate","false") \
    .outputMode("append") \
    .start()

#等待流程序执行结束（作为作业文件提交时启用）
query.awaitTermination()
```

注意：在上面的代码中 timestampFormat 选项不是必需的。如果是非标准时间戳格式，则可使用这个 option 来指定解析格式。

执行以上代码，得到的输出结果如图 6-17 所示。

```
Batch: 0
-------------------------------------------
+----------------+-----------+-------------------+----------------+------+-------------------+
|temp_location_id|temperature|temp_taken_time    |load_location_id|load  |load_taken_time    |
+----------------+-----------+-------------------+----------------+------+-------------------+
|rack2           |100.5      |2017-06-02 08:06:02|rack2           |1105.5|2017-06-02 08:06:04|
|rack1           |99.5       |2017-06-02 08:01:01|rack1           |199.5 |2017-06-02 08:01:02|
|rack3           |101.0      |2017-06-02 08:11:03|rack3           |2104.0|2017-06-02 08:11:06|
|rack4           |102.0      |2017-06-02 08:16:04|rack4           |1108.0|2017-06-02 08:16:08|
|rack4           |102.0      |2017-06-02 08:16:04|rack4           |1108.0|2017-06-02 08:21:10|
```

图 6-17　IoT 流连接结果

在这个结果表上，可以进一步执行 PySpark SQL 查询操作。

当连接一个静态 DataFrame 和一个流 DataFrame 时,以及当连接两个流 DataFrame 时,外连接会受到更多的限制,相关的一些细节见表 6-7。

表 6-7 两个流 DataFrame 在执行 join 连接时的限制说明

左侧 + 右侧	连接类型	说 明
静态数据 + 流数据	内连接	支持
静态数据 + 流数据	左外连接	不支持
静态数据 + 流数据	右外连接	支持
静态数据 + 流数据	全外连接	不支持
流数据 + 流数据	内连接	支持
流数据 + 流数据	左外连接	有条件地支持。必须在右侧指定水印及时间约束
流数据 + 流数据	右外连接	有条件地支持。必须在左侧指定水印及时间约束
流数据 + 流数据	全连接	不支持

6.7 使用数据接收器

27min

流应用程序的最后一步通常是将计算结果写入外部系统或存储系统。PySpark 结构化流提供了 5 个内置数据接收器(Data Sink),其中 3 个是用于生产的,两个是用于测试目的,见表 6-8。

表 6-8 PySpark 内置数据接收器

Data Sink	支持的模式	选 项 参 数	是否容错	备 注
File Data Sink	append	path:输出目录路径,必须指定。 retention:输出文件的生命周期(TTL)。批量提交的比 TTL 时间早的输出文件,最终将被排除在元数据日志中。这意味着读取接收器输出目录的读取器查询可能无法处理它们。可以提供时间的字符串格式的值(如"12h""7d"等)。默认情况下是禁用的	是 (exactly-once)	支持写入分区表。按时间进行分区可能是有用的
Kafka Data Sink	append update complete	参见 6.7.2 节的示例	是 (at-least-once)	
Foreach Data Sink	append update complete	无	是 (at-least-once)	

续表

Data Sink	支持的模式	选项参数	是否容错	备注
Console Data Sink	append update complete	numRows：每个触发器打印的行数，默认为 20 truncate：是否截断过长的输出，默认为 true	否	
Memory Data Sink	append complete	无	否 在 complete 模式下，重启查询将重新创建全表	

下面的部分将详细介绍每个 Data Sink。

6.7.1 使用 File Data Sink

File Data Sink 是一个非常简单的数据接收器，需要提供的唯一必须选项是输出目录。由于 File Data Sink 是容错的，结构化的流将需要一个检查点位置来写进度信息和其他元数据，以帮助在出现故障时进行恢复。

【例 6-9】配置 Rate 数据源，每秒产生 10 行数据，将生成的数据行发送到两个分区，并将数据以 JSON 格式写到指定的目录。

实现代码如下：

```python
#第6章/sink_file.py

#导入依赖包
from pyspark.sql import SparkSession

#创建 SparkSession 实例
spark = SparkSession \
    .builder \
    .appName("streaming demo") \
    .getOrCreate()

#将 shuffle 后的分区数设置为 10（测试环境下）
spark.conf.set("spark.sql.shuffle.partitions",10)

#设置日志级别
#spark.sparkContext.setLogLevel("WARN")

#将数据从 Rate 数据源写到 File Sink
rateSourceDF = spark.readStream \
    .format("rate") \
```

```
        .option("rowsPerSecond","10") \
        .option("numPartitions","2") \
        .load()

#将流 DataFrame 写到指定的目录,并指定 checkpoint
#"path"选项:设置输出目录;"checkpointLocation"选项:设置检查点位置
query = rateSourceDF.writeStream \
        .outputMode("append") \
        .format("json") \
        .option("path", "/tmp/output") \
        .option("checkpointLocation", "/tmp/ck") \
        .start()

#等待流程序执行结束(作为作业文件提交时启用)
#query.awaitTermination()
```

由于分区的数量被配置为两个分区,所以每当结构化流在每个触发点上写出数据时,就会将两个文件写到输出目录中,因此,如果检查输出目录,将会看到带有名称的文件,这些名称以 part-00000 或 part-00001 开头。Rate 数据源被配置为每秒 10 行,并且有两个分区;因此,每个输出包含 5 行。

part-00000-*.json 文件的内容如下:

```
{"timestamp":"2021-02-03T17:56:46.283+08:00","value":0}
{"timestamp":"2021-02-03T17:56:46.483+08:00","value":2}
{"timestamp":"2021-02-03T17:56:46.683+08:00","value":4}
{"timestamp":"2021-02-03T17:56:46.883+08:00","value":6}
{"timestamp":"2021-02-03T17:56:47.083+08:00","value":8}
{"timestamp":"2021-02-03T17:56:47.283+08:00","value":10}
{"timestamp":"2021-02-03T17:56:47.483+08:00","value":12}
{"timestamp":"2021-02-03T17:56:47.683+08:00","value":14}
{"timestamp":"2021-02-03T17:56:47.883+08:00","value":16}
{"timestamp":"2021-02-03T17:56:48.083+08:00","value":18}
```

part-00001-*.json 文件的内容如下:

```
{"timestamp":"2021-02-03T17:56:46.383+08:00","value":1}
{"timestamp":"2021-02-03T17:56:46.583+08:00","value":3}
{"timestamp":"2021-02-03T17:56:46.783+08:00","value":5}
{"timestamp":"2021-02-03T17:56:46.983+08:00","value":7}
{"timestamp":"2021-02-03T17:56:47.183+08:00","value":9}
{"timestamp":"2021-02-03T17:56:47.383+08:00","value":11}
{"timestamp":"2021-02-03T17:56:47.583+08:00","value":13}
{"timestamp":"2021-02-03T17:56:47.783+08:00","value":15}
{"timestamp":"2021-02-03T17:56:47.983+08:00","value":17}
{"timestamp":"2021-02-03T17:56:48.183+08:00","value":19}
```

6.7.2 使用 Kafka Data Sink

在结构化的流中，将流 DataFrame 的数据写入 Kafka 的 Data Sink，要比从 Kafka 的数据源中读取数据简单得多。

19min

Kafka 的 Data Sink 配置选项，见表 6-9。

表 6-9 Kafka Data Sink 配置选项

Option	值	描述
kafka.Bootstrap.servers	host1:port1 host2:port2	Kafka 服务器列表，用逗号分隔
topic	字符串，如"topic1"	这是单个的主题(topic)名称
key	一个字符串，或二进制	这个 key 用来决定一个 Kafka 消息应该被发送到哪个分区。所有具有相同 key 的 Kafka 消息将被发送到同一分区。这是一个可选项
value	一个字符串，或二进制	这是消息的内容。对于 Kafka，它只是一字节数组，对 Kafka 没有任何意义

其中有 3 个选项是必需的。重点要理解的是 key 和 value，它们与 Kafka 消息的结构有关。正如前面提到的，Kafka 的数据单元是一条消息，本质上是一个键-值对。这个 value 的作用就是保存消息的实际内容，而它对 Kafka 没有任何意义。就 Kafka 而言，value 只是一堆字节，然而，key 被 Kafka 认为是一个元数据，它和 value 一起被保存在 Kafka 的信息中。当一条消息被发送到 Kafka 并且一个 key 被提供时，Kafka 将其作为一种路由机制来确定一个特定的 Kafka 消息应该被发送到哪一个分区，按照对该 key 哈希并对 topic 的分区数求余。这意味着所有具有相同 key 的消息都将被路由到同一个分区。如果消息中没有提供 key，则 Kafka 就不能保证消息被发送到哪个分区，而 Kafka 使用了一个循环算法平衡分区之间的消息。

提供 topic 主题名称有两种方法。第 1 种方法是在设置 Kafka Data Sink 时在配置中提供主题名称，第 2 种方法是在流 DataFrame 中定义一个名为 topic 的列，该列的值将用作 topic 主题的名称。

如果名为 key 的列存在于流 DataFrame 中，则该列的值将用作消息的 key。因为该 key 是一个可选的元数据，所以在流 DataFrame 中不必有这一列。

另一方面，必须提供 value 值，而 Kafka 的 Data Sink 则期望在流 DataFrame 中有一个名为 value 的列。

注意：如果要开发以 Kafka 为 Data Sink 的 PySpark 结构化流处理程序，则必须将 Kafka 的依赖包添加到 classpath 中。可参考 6.5.4 节中的说明。

下面编写一个 PySpark 结构化流应用程序，用于读取 Rate 数据源，并将数据写到 Kafka 的指定主题中。

【例6-10】编写 PySpark 结构化流应用程序作为 Kafka 的生产者，将从 Rate 数据源读取的消息写入 Kafka 的 rates 主题中。

在这个示例中，PySpark 结构化流程序会向 Kafka 的 rates 主题发送消息（本例为读取自 Rate 数据源的数据），用 Kafka 自带的消费者脚本程序订阅该主题。一旦它收到了订阅的消息，就马上输出。程序处理流程如图6-18所示。

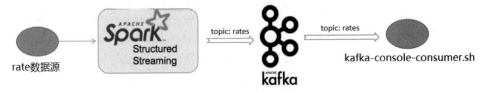

图6-18 **PySpark 结构化流程序将计算结果写入 Kafka**

首先，编写 PySpark 结构化流程序代码，代码如下：

```
#第6章/sink_kafka.py

#导入依赖包
from pyspark.sql import SparkSession
from pyspark.sql.functions import *

#创建 SparkSession 实例
spark = SparkSession \
    .builder \
    .appName("kafka data sink demo") \
    .getOrCreate()

#将 shuffle 后的分区数设置为10（测试环境下）
spark.conf.set("spark.sql.shuffle.partitions",10)

#设置日志级别
#spark.sparkContext.setLogLevel("WARN")

#以每秒10行的速度设置 Rate 数据源，并使用两个分区
ratesSinkDF = spark.readStream \
    .format("rate") \
    .option("rowsPerSecond","10") \
    .option("numPartitions","2") \
    .load()

#转换 ratesSinkDF 以创建一个"key"列和"value"列
#value 列包含一个 JSON 字符串，该字符串包含两个字段：timestamp 和 value
ratesSinkForKafkaDF = ratesSinkDF.select(
    col("value").cast("string").alias("key"),
```

```
            to_json(struct("timestamp","value")).alias("value")
)

#设置一个流查询,使用topic "rates",将数据写到Kafka
query = ratesSinkForKafkaDF.writeStream \
    .outputMode("append") \
    .format("kafka") \
    .option("kafka.Bootstrap.servers", "xueai8:9092") \
    .option("topic","rates") \
    .option("checkpointLocation", "/tmp/rates") \
    .start()

#等待流程序执行结束(作为作业文件提交时启用)
#query.awaitTermination()          #在PySpark Shell交互环境下,不需要这句
```

要运行这个流程序,建议按以下步骤进行操作。

(1)启动 ZooKeeper 服务。

Kafka 依赖于 Apache ZooKeeper,所以在启动 Kafka 之前,要先启动它。打开一个终端窗口,执行的命令如下:

```
$ cd ~/bigdata/kafka_2.12-2.4.1
$ ./bin/zookeeper-server-start.sh config/zookeeper.properties
```

等待 30s 左右后 ZooKeeper 启动。

(2)接下来,启动 Kafka 服务器。

另打开一个终端窗口,执行的命令如下:

```
$ cd ~/bigdata/kafka_2.12-2.4.1
$ ./bin/kafka-server-start.sh config/server.properties
```

等待 30s 左右(Kafka 启动)。

(3)查看和创建 Kafka 主题(如果已经有了 rates 主题,则此步略过)。

创建 rates 主题。另外打开第 3 个终端窗口,执行的命令如下:

```
$ cd ~/bigdata/kafka_2.12-2.4.1
$ ./bin/kafka-topics.sh --create --zookeeper localhost:2181
--replication-factor 1 --partitions 1 --topic rates
```

查看已有的主题,执行的命令如下:

```
$ ./bin/kafka-topics.sh --list --zookeeper localhost:2181
```

(4)在第 3 个终端窗口,运行 Kafka 自带的消费者脚本,订阅 rates 主题消息,执行的命令如下:

```
$ ./bin/kafka-console-consumer.sh --Bootstrap-server 192.168.190.145:9092
--topic rates
```

(5)启动上面的流处理程序,执行过程如图6-19所示。

```
>>> from pyspark.sql.functions import *
>>> spark.conf.set("spark.sql.shuffle.partitions",10)
>>> ratesSinkDF = spark.readStream \
...        .format("rate") \
...        .option("rowsPerSecond","10") \
...        .option("numPartitions","2") \
...        .load()
>>> ratesSinkForKafkaDF = ratesSinkDF.select(
...          col("value").cast("string").alias("key"),
...          to_json(struct("timestamp","value")).alias("value")
...     )
>>> query = ratesSinkForKafkaDF.writeStream \
...        .outputMode("append") \
...        .format("kafka") \
...        .option("kafka.bootstrap.servers", "xueai8:9092") \
...        .option("topic","rates") \
...        .option("checkpointLocation", "/tmp/rates") \
...        .start()
query.stop()
```

图 6-19 执行 PySpark 结构化流程序

(6)回到第 3 个终端窗口(运行消费者脚本的窗口),如果一切正常,则应该可以看到在终端输出了收到的订阅消息,内容如下(部分):

```
{"timestamp":"2021-02-03T19:14:59.352+08:00","value":1960}
{"timestamp":"2021-02-03T19:14:59.552+08:00","value":1962}
{"timestamp":"2021-02-03T19:14:59.752+08:00","value":1964}
{"timestamp":"2021-02-03T19:14:59.952+08:00","value":1966}
{"timestamp":"2021-02-03T19:15:00.152+08:00","value":1968}
{"timestamp":"2021-02-03T19:14:59.452+08:00","value":1961}
{"timestamp":"2021-02-03T19:14:59.652+08:00","value":1963}
{"timestamp":"2021-02-03T19:14:59.852+08:00","value":1965}
...
```

6.7.3 使用 Foreach Data Sink

与结构化流提供的其他内置 Data Sinks 相比,foreach 和 foreachBatch Data Sink 是很有意思的数据接收器。它根据数据应该如何被写出、何时写出数据及将数据写入何处,具有很高的灵活性,但这种灵活性和可扩展性是有要求的,即由用户自己来负责在使用这个数据接收器时写出数据的逻辑。

foreach 和 foreachBatch 操作允许用户对流查询的输出应用任意操作和编写逻辑,但它们的应用场景略有不同,即 foreach 允许在每行上自定义写逻辑,而 foreachBatch 允许对每个微批处理的输出进行任意操作和自定义逻辑。

1. foreachBatch

使用 foreachBatch(),可指定一个函数,该函数在流查询的每个微批的输出数据上执行。

它接受两个参数：微批输出数据的 DataFrame 和微批的唯一 ID。其基本模式如下：

```
def foreach_batch_function(df, epoch_id):
    #转换并写df
    pass

streamingDF.writeStream.foreachBatch(foreach_batch_function).start()
```

使用 foreachBatch，用户可以执行以下操作。

（1）重用现有的批处理数据源：对于许多存储系统来讲，可能还没有可用的流接收器，但可能已经存在用于批处理查询的数据写入器。使用 foreachBatch()方法，可以在每个微批处理的输出上使用批处理数据写入器。

（2）写入多个位置：如果想将流查询的输出写入多个位置，则可以简单地输出 DataFrame 多次，但是，每次尝试写出时都可能导致重新计算输出数据（包括可能重新读取输入数据）。为了避免重新计算，应该缓存输出 DataFrame，将其写入多个位置，然后取消缓存。模板代码如下：

```
def foreach_batch_function(df, epoch_id):
df.persist()
df.write.format(...).save(...)        //location 1
df.write.format(...).save(...)        //location 2
df.unpersist()
    pass

streamingDF.writeStream.foreachBatch(foreach_batch_function).start()
```

（3）应用额外的 DataFrame 操作：许多 DataFrame 操作在流式 DataFrame 中是不支持的，因为在这些情况下 PySpark 不支持生成增量计划。使用 foreachBatch()方法，可以对每个微批处理输出应用其中的一些操作，但是，用户必须自己推理执行该操作的端到端语义。

默认情况下，foreachBatch()方法只提供 at-least-once（至少一次）写保证，但是，可以使用提供给该函数的 batchId 消除输出的重复，并获得 exactly-once（精确一次）的保证。另外，foreachBatch()方法不能用于连续（continuous）处理模式，因为它从根本上依赖于流查询的微批处理执行。如果以连续模式写入数据，则可使用 foreach()方法代替（从 Spark 2.4 开始，foreachBatch()方法可以在 Python 中使用）。

2. foreach

在不能选择使用 foreachBatch()方法时，例如，对应的批处理数据写入器不存在，或者连续处理模式不存在，则可以使用 foreach()方法来表达自定义写入器逻辑。具体来讲，可以通过将数据写入逻辑分为 3 种方法来表示：open()、process()和 close()（从 Spark 2.4 开始，foreach()方法可以在 Python 中使用）。

在 Python 中，可以用两种方式调用 foreach()方法：在一个函数中或在一个对象中。使用函数的方式提供了一种简单的方法来表达处理逻辑,但不允许在由于失败导致重新处理某些输入数据时对生成的数据进行去重。对于这种情况,必须在对象中指定处理逻辑。

使用函数方式时,该函数接收一行作为输入,其基本模式如下:

```
def process_row(row):
    #将 row 写到存储中
    pass

query = streamingDF.writeStream.foreach(process_row).start()
```

在使用对象方式时,该对象具有一个 process()方法,以及可选的 open()和 close()方法,其基本模式如下:

```
class ForeachWriter:

    def open(self, partition_id, epoch_id):
        #打开连接,这种方法是可选的
        pass

    def process(self, row):
        #将 row 写入连接,这种方法是必需的
        pass

    def close(self, error):
        #关闭连接,这种方法是可选的
        pass

query = streamingDF.writeStream.foreach(ForeachWriter()).start()
```

ForeachWriter 包含 3 种方法:open()、process()和 close()。只要有一个触发器的输出生成一系列的行,这些方法就会被调用。下面是使用这个 Data Sink 的一些细节。

(1) ForeachWriter 类的一个实例将在驱动程序端被创建,它将被发送到 Spark 集群中的 executors 执行。这有两个条件。首先,ForeachWriter 的实现必须是可序列化的,否则它的实例不能通过网络发送到 executors。其次,如果在创建过程中有任何初始化都将发生在驱动程序端。

(2) 在流 DataFrame 中分区的数量决定了有多少个 ForeachWriter 实现的实例被创建。

(3) 在 ForeachWriter 抽象类中定义的 3 种方法将在 executors 上被调用。

(4) 执行初始化(例如打开数据库连接或套接字连接)的最佳位置是在 open()方法中,然而,每当有数据被写出来时,就会调用 open()方法,因此,这种逻辑必须是智能和高效的。

(5) open()方法签名有两个输入参数:分区 ID 和版本。这两个参数的组合唯一地表示一组需要被写出来的行。这个版本的值是一个单调递增的 ID,随着每个触发器的增加而增

加。根据分区 ID 的值和版本参数,open()方法需要决定它是否需要写出行序列,并将适当的布尔值返回结构流引擎。

(6)如果 open()方法的返回值为 true,则对于触发器输出的每行就会调用 process()方法。

(7)无论何时调用 open()方法,不管它返回什么值,close()方法也都会被调用。如果在调用 process()方法时出现错误,则该错误将被传递到 close()方法中。调用 close()方法的目的是给用户一个机会来清理在 open()或 process()方法调用期间创建的任何必要状态。只有当 executor 的 JVM 崩溃或者 open()方法抛出一个 Throwable 异常时,才不会调用 close()方法。

简而言之,这个数据接收器为用户提供了一个在写出流 DataFrame 的数据时足够的灵活性。下面通过一个示例来演示如何使用这种类型的 Data Sink。

【例 6-11】编写 PySpark 结构化流应用程序,通过将 Rate 数据源中的数据写入控制台,包含了一个 ForeachWriter 类的简单实现。

首先定义一个 ForeachWriter 类的实现,并实现它的 3 种方法:open()、process()和 close(),代码如下:

```
#第6章/sink_foreach.py

#自定义一个ForeachWriter
class ForeachConsoleWriter:
    def __init__(self):
        self.p_id = 0
        self.e_id = 0

    def open(self, partition_id, epoch_id):
        self.p_id = partition_id
        self.e_id = epoch_id
        print(f"open => ({partition_id},{epoch_id})")
        return True

    def process(self, row):
        print(f"writing => {row}")

    def close(self, error):
        print(f"close => ({self.p_id}, {self.e_id})")
```

然后,在流数据写出时,指定 foreach 方法使用上面自定义的 ForeachConsoleWriter,代码如下:

```
#第6章/sink_foreach.py

#导入依赖包
```

```python
from pyspark.sql import SparkSession
from pyspark.sql.functions import *

#创建 SparkSession 实例
spark = SparkSession \
    .builder \
    .appName("foreach data sink demo") \
    .getOrCreate()

#将 shuffle 后的分区数设置为 2（测试环境下）
spark.conf.set("spark.sql.shuffle.partitions",2)

#设置日志级别
#spark.sparkContext.setLogLevel("WARN")

#以每秒 10 行的速度设置 Rate 数据源，并使用两个分区
ratesSourceDF = spark.readStream \
    .format("rate") \
    .option("rowsPerSecond","10") \
    .option("numPartitions","2") \
    .load()

#设置 Foreach Data Sink
query = ratesSourceDF \
    .writeStream \
    .foreach(ForeachConsoleWriter()) \
    .outputMode("append") \
    .start()

#等待流程序执行结束（作为作业文件提交时启用）
#query.awaitTermination()
```

注意：以上代码，应在 Local 模式下执行，因为 print()打印函数在集群模式下的 driver 端是看不到内容的。

当开始执行时，可以看到控制台的输出，类似的内容如下：

```
open => (0, 1)
writing => [2021-02-03 19:52:53.194,0]
writing => [2021-02-03 19:52:53.394,2]
writing => [2021-02-03 19:52:53.594,4]
writing => [2021-02-03 19:52:53.794,6]
writing => [2021-02-03 19:52:53.994,8]
close => (0, 1)
open => (1, 1)
writing => [2021-02-03 19:52:53.294,1]
writing => [2021-02-03 19:52:53.494,3]
```

```
writing => [2021-02-03 19:52:53.694,5]
writing => [2021-02-03 19:52:53.894,7]
writing => [2021-02-03 19:52:54.094,9]
close => (1, 1)
open => (0, 2)
writing => [2021-02-03 19:52:54.194,10]
writing => [2021-02-03 19:52:54.394,12]
writing => [2021-02-03 19:52:54.594,14]
writing => [2021-02-03 19:52:54.794,16]
writing => [2021-02-03 19:52:54.994,18]
close => (0, 2)
open => (1, 2)
writing => [2021-02-03 19:52:54.294,11]
writing => [2021-02-03 19:52:54.494,13]
writing => [2021-02-03 19:52:54.694,15]
writing => [2021-02-03 19:52:54.894,17]
writing => [2021-02-03 19:52:55.094,19]
close => (1, 2)
...
```

6.7.4 使用 Console Data Sink

这个数据接收器非常简单，但它不是一个容错的 Data Sink。它主要用于学习和测试，不能在生产环境下使用。它只有两种选项配置：要显示的行数，以及输出太长时是否截断。这些选项都有一个默认值，见表 6-10。

表 6-10 Console Data Sink 配置选项

Option	默 认 值	描 述
numRows	20	在控制台输出的行的数量
truncate	true	当每行的内容超过 20 个字符时，是否截断显示

这个数据接收器的底层实现使用了与 DataFrame 的 show()方法相同的逻辑来显示流 DataFrame 中的数据。下面通过一个示例来了解这些 option 参数的用法。

【例 6-12】编写 PySpark 结构化流程序，读取 rate 数据源流数据，并将流数据的处理结果写到控制台，每次输出不超过 3 行。

实现代码如下：

```
#第6章/sink_console.py

#导入依赖包
from pyspark.sql import SparkSession
from pyspark.sql.functions import *

#创建 SparkSession 实例
```

```python
spark = SparkSession \
    .builder \
    .appName("console data sink demo") \
    .getOrCreate()

#将 shuffle 后的分区数设置为 2（测试环境下）
spark.conf.set("spark.sql.shuffle.partitions",2)

#设置日志级别
#spark.sparkContext.setLogLevel("WARN")

#以每秒 10 行的速度设置 Rate 数据源，并使用两个分区
ratesSourceDF = spark.readStream \
    .format("rate") \
    .option("rowsPerSecond","10") \
    .option("numPartitions","2") \
    .load()

#将结果 DataFrame 写到控制台
query = ratesSourceDF.writeStream \
    .outputMode("append") \
    .format("console") \
    .option("truncate",value = False) \
    .option("numRows",3) \
    .start()

#等待流程序执行结束（作为作业文件提交时启用）
#query.awaitTermination()        #在 Spark-Shell 交互环境下不需要
```

执行上面的程序，输出结果如图 6-20 所示。

```
Batch: 1
+-----------------------+-----+
|timestamp              |value|
+-----------------------+-----+
|2022-02-26 10:36:36.648|0    |
|2022-02-26 10:36:36.848|2    |
|2022-02-26 10:36:37.048|4    |
+-----------------------+-----+
only showing top 3 rows

Batch: 2
+-----------------------+-----+
|timestamp              |value|
+-----------------------+-----+
|2022-02-26 10:36:37.648|10   |
|2022-02-26 10:36:37.848|12   |
|2022-02-26 10:36:38.048|14   |
+-----------------------+-----+
only showing top 3 rows
```

图 6-20　将流数据处理结果写到控制台，每次输出不超过 3 行

6.7.5　使用 Memory Data Sink

与 Console Data Sink 类似，这个数据接收器也很容易理解和使用。事实上，它非常简单，不需要任何配置。它也不是一个容错的 Data Sink，主要用于学习和测试，不要在生产环境中使用。它收集的数据被发送给 Driver，并作为内存中的表存储在 Driver 中。换句话说，可以发送到 Memory Data Sink 的数据量是由 Driver 中 JVM 拥有的内存大小决定的。在设置这个 Data Sink 时，可以指定一个查询名称作为 DataStreamWriter.queryName 函数参数，然后就可以对内存中的表进行 SQL 查询。与 Console Data Sink 不同的是，一旦数据被发送到内存中的表，就可以使用绝大多数在 PySpark SQL 组件中可用的特性进一步分析或处理数据（如果数据量很大，并且不适合内存，则最好的选择就是使用 File Data Sink 以 Parquet 格式写出数据）。

下面通过一个示例来了解 Memory 数据接收器的用法。

【例 6-13】编写 PySpark 结构化流程序，读取 rate 数据源流数据，并将流数据的处理结果写到内存表中，然后将该内存表发出。

实现代码如下：

```
#第6章/sink_memory.py

#导入依赖包
from pyspark.sql import SparkSession
from pyspark.sql.functions import *

#创建 SparkSession 实例
spark = SparkSession \
    .builder \
    .appName("console data sink demo") \
    .getOrCreate()

#将 shuffle 后的分区数设置为2（测试环境下）
spark.conf.set("spark.sql.shuffle.partitions",2)

#设置日志级别
#spark.sparkContext.setLogLevel("WARN")

#以每秒10行的速度设置 Rate 数据源，并使用两个分区
ratesSourceDF = spark.readStream \
    .format("rate") \
    .option("rowsPerSecond","10") \
    .option("numPartitions","2") \
    .load()
```

```
#将数据写到 Memory Data Sink, 内存表名为"rates"
query = ratesSourceDF.writeStream \
    .outputMode("append") \
    .format("memory") \
    .queryName("rates_tb") \
    .start()

#针对"rates"内存表发出 SQL 查询
spark.sql("select * from rates_tb order by value desc").show(truncate=False)

#统计"rates"内存表中的行数
spark.sql("select count(*) from rates_tb").show(truncate=False)

#等待流程序执行结束（作为作业文件提交时启用）
#query.awaitTermination() #在 PySpark Shell 交互环境下，不需要这一句
```

需要注意的一点是，即使在流查询 ratesSQ 停止之后，内存中的 rates 仍然会存在，然而，一旦一个新的流查询以相同的名称开始，那么来自内存中的数据就会被截断。

6.7.6 Data Sink 与输出模式

在了解了有哪些 Data Sink 之后，还有很重要的一点就是要了解每种类型的 Data Sink 支持哪些输出。关于 Data Sink 和所支持的输出模式，见表 6-11。

表 6-11 Data Sink 与输出模式

sink	支持的输出模式	备 注
File	append	只支持写出新行，没有更新
Kafka	append、update、complete	
Foreach	append、update、complete	依赖于 ForeachWriter 实现
Console	append、update、complete	
Memory	append、complete	不支持 in-place 更新

关于输出模式的详细信息将在 6.8 节中介绍。

13min

6.8 深入研究输出模式

在 PySpark 结构化流中，输出模式可以是 complete、update 或 append，其中 complete 输出模式意味着每次都要写入全部的结果表，update 输出模式写入从上批处理中已更改的行，而 append 输出模式仅写入新行。

一般来讲，有两种类型的流查询：

（1）第 1 种类型称为"无状态类型"，它只对流入的流数据进行基本的转换，然后将数据写到一个 Data Sink 上。

（2）第 2 种类型称为"有状态类型"，它需要保持一定数量的状态，不管它是隐式地还是显式地完成。

有状态类型通常执行某种聚合或使用像 mapGroupsWithState 或 flatMapGroupsWithState 这样的结构化流 API，可以维护特定用例所需的任意状态，例如，维护用户会话数据。

6.8.1 无状态流查询

无状态流查询的典型应用场景是实时流 ETL，它可以连续读取实时流数据，例如在线服务连续生成的 PV 事件，以捕获哪些页面正在被哪些用户浏览。在这种用例中，它通常执行以下操作：

（1）过滤、转换和清洗。实际获取的数据是混乱的脏数据，而且这种结构可能不太适合对数据进行重复分析。

（2）转换为更有效的存储格式。像 CVS 和 JSON 这样的文本文件格式易读性虽然好，但对于重复分析来讲是低效的，特别是如果数据量很大，例如几百太字节。更有效的二进制格式，如 PRC、Parquet 或 Avro，通常用于减少文件大小和提高分析速度。

（3）按某些列划分数据。在将数据写到 Data Sink 时，可以根据常用列的值对数据进行分区，以加快组织中不同团队的重复分析。

以前的任务在将数据写到 Data Sink 之前不需要流查询来维护任何类型的状态，PySpark 也不需要记住以前微批处理的任何数据来处理当前记录。随着新数据的出现，它被清理、转换，并可能进行重组，然后立即被写出，因此，这种无状态流类型的唯一适用的输出模式是 append。complete 输出模式是不适用的，因为这需要结构化的流来维护所有以前的数据，这些数据可能太大而无法维护。update 输出模式也不适用，因为只有新数据被写出来，然而，当这种 update 输出模式被用于无状态流查询时，结构化流就会识别这个并将其与 append 输出模式相同对待。当将不适当的输出模式用于流查询时，结构化的流引擎会抛出异常。

下面的示例展示了在使用不适当的输出模式时（在使用无状态流查询时使用了 complete 输出模式）会发生什么情况，代码如下：

```
#使用 rate 数据源
ratesDF = spark.readStream \
.format("rate") \
.option("rowsPerSecond","10") \
.option("numPartitions","2") \
.load()
```

```
#简单转换
from pyspark.sql.functions import col

ratesOddEvenDF = ratesDF.withColumn("even_odd", col("value") % 2 == 0)

#写到 Console Data Sink，使用 complete 输出模式
ratesSQ = ratesOddEvenDF.writeStream \
    .outputMode("complete") \
    .format("console") \
    .option("truncate",False) \
    .start()
```

当提交以上代码执行时，会抛出异常信息，内容如下：

```
...
pyspark.sql.utils.AnalysisException: Complete output mode not supported when there are no streaming aggregations on streaming DataFrames/Datasets;
Project [timestamp#6303, value#6304L, ((value#6304L % cast(2 as bigint)) = cast(0 as bigint)) AS even_odd#6307]
+- StreamingRelationV2 org.apache.spark.sql.execution.streaming.sources.RateStreamProvider@5a189f75, rate, org.apache.spark.sql.execution.streaming.sources.RateStreamTable@421856e0, [rowsPerSecond=10, numPartitions=2], [timestamp#6303, value#6304L]
```

6.8.2 有状态流查询

在执行诸如计数、平均、求和等聚合运算时，PySpark 将跨先前的多个微批组合信息。为此，流程序需要维护每个 Executor 上部分计数的一些元数据信息，这就是所谓的"状态"。例如，假设正在计算已经解析或过滤的记录的数量，这里的计数就是状态，每个选中的记录增加计数，代码如下：

```
df.where(col("data.type") == lit("type1")).count()
```

有状态流处理面临着两个挑战：
（1）确保容错。
（2）精确一次性交付语义。

结构化流在底层默认作为微批处理进行处理。流查询运行时，状态在微批之间进行版本控制，因此，在生成一系列增量执行计划时，每个执行计划都知道它需要从哪个版本的状态中读取数据。每个微批处理读取状态数据的前一个版本，即前一个运行计数，然后更新它并创建一个新版本。每个版本的检查点都位于在查询中提供的相同检查点位置。这样，Spark 引擎就可以从故障中恢复，并确保精确地一次交付，因为它确切地知道需要从哪里重新启动。

这些分布式状态文件存储在每个 Executor 的内存中，对这些状态文件的所有更改都通过将 WAL（Write Ahead Log，预写日志）文件写到检查点位置（如 HDFS 或 S3）来支持。

在有状态流中最简单和最常用的是流聚合。流聚合又包含多种聚合类型。

（1）按 key 进行聚合，模板代码如下：

```
eventsDF.groupBy(col("data.type")).count()
```

（2）按事件时间窗口聚合，模板代码如下：

```
eventsDF
    .groupBy(window("timestamp", "5 mins"))
    .agg(avg(col("data.clicks")))
```

（3）同时按 key 和事件时间窗口聚合，模板代码如下：

```
eventsDF
    .groupBy(col("data.type"), window("timestamp", "5 mins"))
    .agg(avg(col("data.clicks")))
```

（4）用户定义的聚合函数(UDAF)。

当一个有状态的流查询通过一个 group by 转换执行一个聚合时，这个聚合的状态是由结构化的流引擎隐式地维护的。随着更多的数据到达，新数据聚合的结果被更新到结果表中。

在每个触发点上，根据输出模式，更新后的数据或结果表中的所有数据都被写到一个 Data Sink 上。这意味着使用 append 输出模式是不合适的，因为这违反了输出模式的语义，该模式指定只有附加到结果表的新行将被发送到指定的输出接收器。换句话说，只有 complete 和 update 输出模式适合于有状态查询类型，并且聚合状态隐式地由结构化流引擎负责维护。使用 complete 输出模式的流查询的输出总是等于或超过使用 update 输出模式的相同流查询的输出。

下面的示例用来说明 update 和 complete 模式之间的输出差异。

【例 6-14】移动电话的开关机等事件会保存在 JSON 格式的文件中。现在编写 PySpark 结构化流处理程序来读取这些事件并统计不同的 action 发生的数量。

建议按以下步骤操作。

（1）准备数据。

在本示例中使用文件数据源，该数据源以 JSON 文件的格式记录了一小组移动电话动作事件。每个事件由 3 个字段组成。

① id：表示手机的唯一 ID。在样例数据集中，电话 ID 将类似于 phone1、phone2、phone3 等这样的字符串。

② action：表示用户所采取的操作。该操作的可能值是 open 或 close。

③ ts：表示用户 action 发生时的时间戳。这是事件时间。

本例准备了两个存储移动电话事件数据的 JSON 文件：action.json 和 newaction.json。action.json 文件的内容如下：

```
{"id":"phone1","action":"open","ts":"2018-03-02T10:02:33"}
{"id":"phone2","action":"open","ts":"2018-03-02T10:03:35"}
{"id":"phone3","action":"open","ts":"2018-03-02T10:03:50"}
{"id":"phone1","action":"close","ts":"2018-03-02T10:04:35"}
{"id":"phone3","action":"close","ts":"2018-03-02T10:07:35"}
```

newaction.json 文件的内容如下：

```
{"id":"phone4","action":"open","ts":"2018-03-02T10:07:50"}
{"id":"phone2","action":"crash","ts":"2018-03-02T11:09:13"}
{"id":"phone5","action":"swipe","ts":"2018-03-02T11:17:29"}
```

注意在这两个数据文件中，action 字段值的区别。在 action.json 文件中，包含两类 action 值，分别为 open 和 close，而在 newaction.json 文件中，包含三类 action 值，分别为 open、crash 和 swipe。

为了模拟数据流的行为，应把这两个 JSON 文件复制到 HDFS 指定目录下。

（2）编写 PySpark 结构化流程序，读取文件流数据源并执行聚合操作，然后输出结果。这里使用的输入模式是 complete，代码如下：

```python
#第6章/without_state.py

from pyspark.sql import SparkSession
from pyspark.sql.types import *
from pyspark.sql.functions import *

#创建 SparkSession 实例
spark = SparkSession \
      .builder \
      .appName("stream state demo") \
      .getOrCreate()

#将 shuffle 后的分区数设置为 2（测试环境下）
spark.conf.set("spark.sql.shuffle.partitions",2)

#设置日志级别
#spark.sparkContext.setLogLevel("WARN")

#为手机事件数据创建一个 schema
fields = [
    StructField("id", StringType(), nullable = False),
    StructField("action", StringType(), nullable = False),
    StructField("ts", TimestampType(), nullable = False)
]
```

```
mobileDataSchema = StructType(fields)

#指定监听的文件目录
dataPath = "file://home/hduser/data/spark/stream/mobile2"

#读取指定目录下的源数据文件,一次一个
mobileDF = spark.readStream \
    .option("maxFilesPerTrigger", 1) \
    .option("mode","failFast") \
    .schema(mobileDataSchema) \
    .json(dataPath)

#mobileDF.printSchema()

#执行聚合等操作
actionCountDF = mobileDF.groupBy("action").count()

#输出模式是 complete
query = actionCountDF.writeStream \
    .format("console") \
    .option("truncate", "false") \
    .outputMode("complete")    \
    .start()

#等待流程序执行结束(作为作业文件提交时启用)
#query.awaitTermination() #在 PySpark Shell 命令行下不需要执行这一句
```

执行上面的流查询代码,输出结果如下:

```
-------------------------------------------
Batch: 0
-------------------------------------------
+------+-----+
|action|count|
+------+-----+
|close |2    |
|open  |3    |
+------+-----+


-------------------------------------------
Batch: 1
-------------------------------------------
+------+-----+
|action|count|
+------+-----+
|close |2    |
|swipe |1    |
```

```
|crash |1    |
|open  |4    |
+------+-----+
```

观察上面的输出结果，在 Batch 1 中输出的聚合结果包含了 Batch 0 中的状态。这说明在上面的代码中，带有 complete 输出模式的流查询的输出包含了结果表中的所有 action 类型。

（3）接下来，将在上面的代码中的输出改为使用 update 输出模式，其余代码保持不变，代码如下：

```
...
#输出模式设为update
query = actionCountDF.writeStream \
    .format("console") \
    .option("truncate", "false") \
    .outputMode("update") \
    .start()
...
```

执行上面的代码，流查询输出如下：

```
-------------------------------------------
Batch: 0
-------------------------------------------
+------+-----+
|action|count|
+------+-----+
|close |2    |
|open  |3    |
+------+-----+


-------------------------------------------
Batch: 1
-------------------------------------------
+------+-----+
|action|count|
+------+-----+
|swipe |1    |
|crash |1    |
|open  |4    |
+------+-----+
```

观察上面的输出结果，在 Batch 1 中输出的聚合结果并不包含 Batch 0 中的状态。这说明在上面的代码中，带有 update 输出模式的流查询的输出只包含 newaction.json 文件中的 actions，这些 actions 结果（包含更新的 Open Action）以前从未出现过。

同样地，如果有状态查询类型使用了不适当的输出模式，则结构化的流引擎会抛出异常信息。例如，继续修改上面的代码，将输出模式设为 append，代码如下：

```
...
#对有状态流查询使用不适当的输出模式
query = actionCountDF.writeStream \
    .format("console") \
    .option("truncate", "false") \
    .outputMode("append") \
    .start()
...
```

因为执行了 groupBy()这样的聚合操作，所以会产生异常，异常信息的内容如下：

```
pyspark.sql.utils.AnalysisException: Append output mode not supported when
there are streaming aggregations on streaming DataFrames/DataSets without
watermark;
Aggregate [action#1], [action#1, count(1) AS count#10L]
```

也有一种例外情况：如果向带有聚合的有状态的流查询提供一个水印，则所有的输出模式都是适用的。这是因为结构化的流引擎将删除旧的聚合状态数据，这些数据比指定的水印要"古老"，这意味着一旦水印被逾越，新的行就可以被添加到结果表中。这时 append 输出的语义是有意义的。

6.9 深入研究触发器

触发器设置决定了结构化的流引擎何时运行在流查询中表达的流计算逻辑，其中包括所有的转换，以及将数据写入 Data Sink。换句话说，触发器设置控制什么时候数据被写到 Data Sink 上，以及使用哪种处理模式。从 PySpark 2.3 开始，引入了一种称为"连续的（continuous）"新处理模式。

PySpark 结构化流所支持的不同类型的触发器，见表 6-12。

表 6-12 Data Sink 与输出模式

触发器类型	描述
未指定（默认）	如果没有显式地指定触发器设置，则默认情况下，查询将以微批处理模式执行，在这种模式下，上一个微批处理完成后将立即生成新的微批
固定间隔的微批	查询将以微批模式执行，其中微批将在用户指定的时间间隔启动。 如果前一个微批处理在间隔内完成，则引擎将等待直到间隔结束，然后启动下一个微批处理。 如果前一个微批需要的时间比间隔时间长(一个间隔边界丢失)，则下一个微批将在前一个微批完成后立即开始(不会等待下一个间隔边界)。 如果没有新数据可用，则不会启动微批处理

续表

触发器类型	描 述
一次性微批	查询将只执行一个微批处理来处理所有可用数据，然后自行停止。这在希望定期启动集群、处理自上一阶段以来可用的所有内容，然后关闭集群的场景中非常有用。在某些情况下，这可能会导致显著的成本节约
连续，具有固定的检查点间隔	查询将在新的低延迟、连续处理模式下执行

到目前为止，所有的流查询示例都没有指定触发器类型，而是使用了默认触发器类型，因为这个默认的触发器类型选择微批模式作为处理模式，而流查询中的逻辑执行不是基于时间的，而是在前一批数据完成处理后立即执行的。这意味着，在数据被写入的频率方面，可预测性会降低。

如果需要更强的可预测性，则可以指定固定的间隔触发器，示例代码如下：

```
#默认触发器(尽可能快地运行微批处理)
df.writeStream \
  .format("console") \
  .start()

#2s 微批处理间隔的 ProcessingTime 触发器（按处理时间）
df.writeStream \
  .format("console") \
  .trigger(processingTime='2 seconds') \
  .start()

#一次性触发器
df.writeStream \
  .format("console") \
  .trigger(once=True) \
  .start()

#具有1s检查点间隔的 continuous 触发器
df.writeStream \
  .format("console") \
  .trigger(continuous='1 second') \
  .start()
```

6.9.1 固定间隔触发器

固定间隔触发器可以根据用户提供的时间间隔（例如，每3s)，在特定的时间间隔内执行流查询中的逻辑。在处理模式方面，这个触发器类型使用微批处理模式。这个间隔可以用字符串格式指定。使用固定间隔触发器的示例，代码如下：

```python
#第 6 章/trigger_fixed.py

from pyspark.sql import SparkSession

#创建 SparkSession 实例
spark = SparkSession \
    .builder \
    .appName("trigger demo") \
    .getOrCreate()

#将 shuffle 后的分区数设置为 2（测试环境下）
spark.conf.set("spark.sql.shuffle.partitions",2)

#设置日志级别
#spark.sparkContext.setLogLevel("WARN")

#/ 使用 rate 数据源，设置每秒 2 行
ratesDF = spark.readStream \
    .format("rate") \
    .option("rowsPerSecond","2") \
    .option("numPartitions","2") \
    .load()

#每 3s 触发一次流查询执行，并将其写入控制台
query = ratesDF.writeStream \
    .format("console") \
    .outputMode("append") \
    .option("numRows",20) \
    .option("truncate",value = False) \
    .trigger(processingTime='3 seconds') \
    .start()

#等待流程序执行结束（作为作业文件提交时启用）
#query.awaitTermination()        #在 PySpark Shell 交互环境下不需要
```

提交执行上面的代码，输出结果如图 6-21 所示。

因为指定 rate 数据源每秒产生 2 行，而触发器指示每 3s 计算一批，所以可以看到每 3s 有 6 行输出。

固定间隔触发器并不总能保证流查询的执行会精确地在每个用户指定的时间间隔内发生。这有以下两个原因：

（1）第 1 个原因很明显，如果没有数据到达，就没有什么可处理的，因此没有任何东西被写入 Data Sink 中。

（2）第 2 个原因是，当前一批的处理时间超过指定间隔时间时，流查询的下一个执行将在处理完成后立即启动。换句话说，它不会等待下一个时间间隔边界。

```
Batch: 1
+-----------------------+-----+
|timestamp              |value|
+-----------------------+-----+
|2022-02-26 11:38:53.996|0    |
|2022-02-26 11:38:54.996|2    |
|2022-02-26 11:38:55.996|4    |
|2022-02-26 11:38:54.496|1    |
|2022-02-26 11:38:55.496|3    |
|2022 02 26 11:38:56.496|5    |
+-----------------------+-----+

Batch: 2
+-----------------------+-----+
|timestamp              |value|
+-----------------------+-----+
|2022-02-26 11:38:56.996|6    |
|2022-02-26 11:38:57.996|8    |
|2022-02-26 11:38:58.996|10   |
|2022-02-26 11:38:57.496|7    |
|2022-02-26 11:38:58.496|9    |
|2022-02-26 11:38:59.496|11   |
+-----------------------+-----+

Batch: 3
```

图 6-21　3s 固定间隔触发器，每次输出 6 行

6.9.2　一次性的触发器

顾名思义，一次性触发器以微批处理模式在流查询中执行逻辑，并将数据写到 Data Sink 一次，然后处理停止。这种触发类型在开发和生产环境中都很有用。在开发阶段，通常流计算逻辑是以迭代的方式开发的，在每个迭代中都希望测试逻辑。这个触发器类型简化了开发—测试—迭代流程。对于生产环境，这种触发器类型适合于流入流数据量较低的情况，这时只需每天运行几次数据处理逻辑。

指定这个一次性触发器类型非常简单。使用这种触发器类型，代码如下：

```
#第 6 章/trigger_once.py

from pyspark.sql import SparkSession

#创建 SparkSession 实例
spark = SparkSession \
        .builder \
        .appName("trigger demo") \
        .getOrCreate()

#将 shuffle 后的分区数设置为 2（测试环境下）
spark.conf.set("spark.sql.shuffle.partitions",2)
```

```
#设置日志级别
#spark.sparkContext.setLogLevel("WARN")

#设置每秒2行
ratesDF = spark.readStream \
    .format("rate") \
    .option("rowsPerSecond","2") \
    .option("numPartitions","2") \
    .load()

#触发一次流查询执行,并将其写入控制台
query = ratesDF.writeStream \
    .format("console") \
    .outputMode("append") \
    .option("numRows",20) \
    .option("truncate",value = False) \
    .trigger(once=True) \
    .start()

#等待流程序执行结束（作为作业文件提交时启用）
#query.awaitTermination()   #PySpark Shell 交互环境下不需要执行这一句
```

6.9.3 连续性的触发器

最后一个触发类型称为连续触发类型。这是在 PySpark 2.3 中引入的实验性的处理模式，以解决需要端到端的毫秒级延迟的情况。连续处理是 Apache Spark 的新执行引擎，每次处理事件的延迟非常低（以毫秒为单位）。在这个新的处理模式中，结构化流启动长时间运行的任务，以持续读取、处理和将数据写到一个 Data Sink。这意味着，一旦传入的数据到达数据源，它就会立即被处理并写入 Data Sink，所以端到端延迟是几毫秒。此外，还引入了一个异步检查点机制，用于记录流查询的进度，以避免中断长时间运行的任务，从而提供一致的毫秒级延迟。

利用这个连续触发器类型的一个比较好的案例是信用卡欺诈性交易检测。在较高的层次上，结构化流引擎根据触发器类型确定要使用哪种处理模式，如图 6-22 所示。

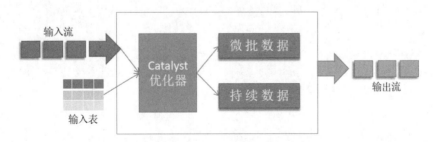

图 6-22　PySpark 连续触发器类型

PySpark 一直通过微批处理提供流处理能力，这种方法的主要缺点是每个任务/微批处理必须定期收集和调度，通过这种方式 PySpark 可以提供的最佳（最小）延迟大约是 1s。不存在单一事件/消息处理的概念。PySpark 试图通过连续处理来克服这些限制，以提供低延迟（毫秒级响应）的流处理。

为了启用这些特性，Spark/PySpark 对其底层代码进行两项主要改进：

（1）创建可以连续读取消息（而不是微批处理）的新数据源和数据接收器，称为 DataSourceV2。

（2）创建一个新的执行引擎，即 ContinuousProcessing，它使用 ContinuousTrigger 并使用 DataSourceV2 启动长运行任务。ContinuousTrigger 内部使用 ProcessingTimeExecutor（与 ProcessingTime 触发器相同）。

在持续处理模式中只有投影和选择操作是允许的，如 select()、where()、map()、flatmap() 和 filter()。在这种处理模式下，除了聚合函数之外，所有 PySpark SQL 函数都是受支持的。另外需要特别注意的是，在连续处理中不支持水印，因为这涉及收集数据。

例如，在使用连续流读取 Kakfa 时，其执行过程如图 6-23 所示。

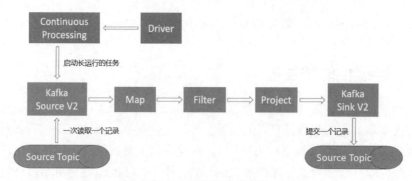

图 6-23 Spark 连续流消费 Kafka

要使用流查询的连续处理模式，需要指定一个连续触发器（Continuous Trigger），其中包含一个期望的检查点间隔，代码如下：

```
#第6章/trigger_continuous.py

from pyspark.sql import SparkSession

#创建 SparkSession 实例
spark = SparkSession \
    .builder \
    .appName("trigger demo") \
    .getOrCreate()

#将 shuffle 后的分区数设置为 2（测试环境下）
spark.conf.set("spark.sql.shuffle.partitions",2)
```

```
#设置日志级别
#spark.sparkContext.setLogLevel("WARN")

#使用 rate 数据源，设置每秒 2 行，分 2 个分区
#将数据写出到 console，并使用 Continuous Trigger
query = spark.readStream \
    .format("rate") \
    .option("rowsPerSecond","2") \
    .option("numPartitions","2") \
    .load() \
    .writeStream \
    .format("console") \
    .outputMode("update") \
    .option("truncate","false") \
    .trigger(continuous='2 second') \
    .start()

#等待流程序执行结束（作为作业文件提交时启用）
#query.awaitTermination()
```

提交执行以上代码，可以在控制台上看到的输出内容如图 6-24 所示。

图 6-24 连续处理模式的流程序输出

在上面的代码中的 ratesDF 流 DataFrame 被设置为两个分区，因此，结构化流在连续处理模式下启动了两个正在运行的任务，所以输出显示所有的偶数出现在一起，所有奇数出现在一起。

第 7 章 PySpark 结构化流（高级）

CHAPTER 7

第 6 章介绍了流处理的核心概念，PySpark 结构化流处理引擎提供的特性，以及将流应用程序组合在一起的基本步骤。本章将涵盖结构化流的事件时间处理和有状态处理特性，并解释结构化流提供的支持，以帮助流应用程序对故障进行容错，并监控流应用程序的状态和进展。

7.1 事件时间和窗口聚合

基于数据创建时间处理所传入的实时数据的能力是一个优秀的流处理引擎的必备功能。这一点很重要，因为要真正理解并准确地从流数据中提取见解或模式，需要能够根据数据或事件发生的时间来处理它们。

7.1.1 固定窗口聚合

一个固定的窗口（也就是一个滚动的窗口）操作本质上是根据一个固定的窗口长度将一个流入的数据流离散到非重叠的桶中。对于每一片输入的数据，根据它的事件时间将它放置到其中的一个桶中。执行聚合仅仅是遍历每个桶并在每个桶上应用聚合逻辑（例如计数或求和）。固定窗口聚合逻辑如图 7-1 所示。

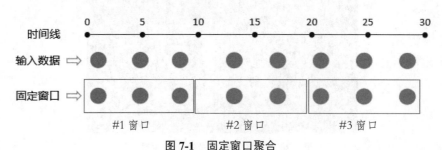

图 7-1 固定窗口聚合

下面通过一个示例程序来演示如何使用结构化流读取文件数据源。

【例 7-1】移动电话的开关机等事件会保存在 JSON 格式的文件中。现要求编写 PySpark 结构化流处理程序来读取并分析这些移动电话数据，统计每 10min 内不同电话操作（如 open

或 close) 发生的数量。

这实际上是在一个 10min 长的固定窗口上对移动电话操作事件的数量进行 count 聚合。建议按以下步骤操作。

（1）准备数据。

在本示例中使用文件数据源，该数据源以 JSON 文件的格式记录了一小组移动电话动作事件。每个事件由 3 个字段组成，分别说明如下。

① id：表示手机的唯一 ID。在样例数据集中，电话 ID 将类似于 phone1、phone2、phone3 等字符串。

② action：表示用户所采取的操作。该操作的可能值是 open 或 close。

③ ts：表示用户 action 发生时的时间戳，这是事件时间。

本示例准备了 4 个存储移动电话事件数据的 JSON 文件。

file1.json 文件的内容如下：

```
{"id":"phone1","action":"open","ts":"2018-03-02T10:02:33"}
{"id":"phone2","action":"open","ts":"2018-03-02T10:03:35"}
{"id":"phone3","action":"open","ts":"2018-03-02T10:03:50"}
{"id":"phone1","action":"close","ts":"2018-03-02T10:04:35"}
```

file2.json 文件的内容如下：

```
{"id":"phone3","action":"close","ts":"2018-03-02T10:07:35"}
{"id":"phone4","action":"open","ts":"2018-03-02T10:07:50"}
```

file3.json 文件的内容如下：

```
{"id":"phone2","action":"close","ts":"2018-03-02T10:04:50"}
{"id":"phone5","action":"open","ts":"2018-03-02T10:10:50"}
```

newaction.json 文件的内容如下：

```
{"id":"phone2","action":"crash","ts":"2018-03-02T11:09:13"}
{"id":"phone5","action":"swipe","ts":"2018-03-02T11:17:29"}
```

为了模拟数据流的行为，应把这 4 个 JSON 文件复制到 HDFS 的指定目录下。

（2）编辑源程序。

编辑流程序源代码，内容如下：

```python
#第7章/streaming_mobile.py

from pyspark.sql import SparkSession
from pyspark.sql.types import *
from pyspark.sql.functions import *

#创建 SparkSession 实例
spark = SparkSession \
```

```python
        .builder \
        .appName("fixed window demo") \
        .getOrCreate()

#将shuffle后的分区数设置为2（测试环境下）
spark.conf.set("spark.sql.shuffle.partitions",2)

#设置日志级别
#spark.sparkContext.setLogLevel("WARN")

#为手机事件数据创建一个schema
fields = [
    StructField("id", StringType(), nullable = False),
    StructField("action", StringType(), nullable = False),
    StructField("ts", TimestampType(), nullable = False)
]
mobileDataSchema = StructType(fields)

#指定监听的文件目录
dataPath = "/data/spark/stream/mobile3"

#读取指定目录下的源数据文件，一次一个
mobileDF = spark.readStream \
    .option("maxFilesPerTrigger", 1) \
    .option("mode","failFast") \
    .schema(mobileDataSchema) \
    .json(dataPath)

#在一个10min的窗口上执行聚合操作
windowCountDF = mobileDF.groupBy(window("ts", "10 minutes")).count()

windowCountDF.printSchema()

#执行流查询，将结果输到控制台
query = windowCountDF \
    .select("window.start","window.end","count") \
    .orderBy("start") \
    .writeStream \
    .format("console") \
    .option("truncate", "false") \
    .outputMode("complete") \
    .start()

#等待流程序执行结束（作为作业文件提交时启用）
#query.awaitTermination()           #在Spark-Shell交互环境下不需要执行这句
```

执行流处理程序，输出结果如下：

```
root
 |-- window: struct (nullable = false)
 |    |-- start: timestamp (nullable = true)
 |    |-- end: timestamp (nullable = true)
 |-- count: long (nullable = false)

+-------------------+-------------------+-----+
|start              |end                |count|
+-------------------+-------------------+-----+
|2018-03-02 10:00:00|2018-03-02 10:10:00|7    |
|2018-03-02 10:10:00|2018-03-02 10:20:00|1    |
|2018-03-02 11:00:00|2018-03-02 11:10:00|1    |
|2018-03-02 11:10:00|2018-03-02 11:20:00|1    |
+-------------------+-------------------+-----+
```

可以看出，当用窗口执行聚合时，输出窗口实际上是一个 struct 类型，它包含开始和结束时间。在上面的代码中，分别取 window 的 start 列和 end 列，并按 start 列（窗口开始时间）排序。可以看到，每 10min 作为一个窗口进行统计。

（3）除了可以在 groupBy() 转换中指定一个窗口外，还可以从事件本身指定额外的列。对上面的例子稍做修改，使用一个窗口并在 action 列上执行聚合，实现对每个窗口和该窗口中 action 类型的 count 计数，代码如下：

```python
#第7章/streaming_mobile2.py

from pyspark.sql import SparkSession
from pyspark.sql.types import *
from pyspark.sql.functions import *

#创建 SparkSession 实例
spark = SparkSession \
      .builder \
      .appName("fixed window demo") \
      .getOrCreate()

#将 shuffle 后的分区数设置为2（测试环境下）
spark.conf.set("spark.sql.shuffle.partitions",2)

#设置日志级别
#spark.sparkContext.setLogLevel("WARN")

#为手机事件数据创建一个 schema
fields = [
      StructField("id", StringType(), nullable = False),
```

```
        StructField("action", StringType(), nullable = False),
        StructField("ts", TimestampType(), nullable = False)
]
mobileDataSchema = StructType(fields)

#指定监听的文件目录
dataPath = "/data/spark/stream/mobile3"

#读取指定目录下的源数据文件,一次一个
mobileDF = spark.readStream \
     .option("maxFilesPerTrigger", 1) \
     .option("mode","failFast") \
     .schema(mobileDataSchema) \
     .json(dataPath)

#在一个10min 的窗口上执行聚合操作
windowCountDF = mobileDF \
    .groupBy(window("ts", "10 minutes"), "action")     \
    .count()

windowCountDF.printSchema()

#执行流查询,将结果输出到控制台
windowCountDF \
    .select("window.start", "window.end", "action", "count") \
    .orderBy("start") \
    .writeStream \
    .format("console") \
    .option("truncate", "false") \
    .outputMode("complete") \
    .start()

#等待流程序执行结束(作为作业文件提交时启用)
#query.awaitTermination()    #在 PySpark Shell 交互环境下不需要执行这句
```

执行以上代码,输出结果如下:

```
root
 |-- window: struct (nullable = false)
 |    |-- start: timestamp (nullable = true)
 |    |-- end: timestamp (nullable = true)
 |-- action: string (nullable = false)
 |-- count: long (nullable = false)


-------------------------------------------
Batch: 0
-------------------------------------------
```

```
+-------------------+-------------------+------+-----+
|start              |end                |action|count|
+-------------------+-------------------+------+-----+
|2018-03-02 10:00:00|2018-03-02 10:10:00|close |3    |
|2018-03-02 10:00:00|2018-03-02 10:10:00|open  |4    |
|2018-03-02 10:10:00|2018-03-02 10:20:00|open  |1    |
|2018-03-02 11:00:00|2018-03-02 11:10:00|crash |1    |
|2018-03-02 11:10:00|2018-03-02 11:20:00|swipe |1    |
+-------------------+-------------------+------+-----+
```

7.1.2 滑动窗口聚合

除了固定窗口类型外,还有另一种称为滑动窗口(Sliding Window)的窗口类型。定义一个滑动窗口需要两个信息,即窗口长度和一个滑动间隔,滑动间隔通常比窗口的长度小。由于聚合计算在传入的数据流上滑动,因此结果通常比固定窗口类型的结果更平滑,因此,这种窗口类型通常用于计算移动平均。关于滑动窗口,需要注意的一点是,由于重叠的原因,同一数据块可能会落入多个窗口,如图 7-2 所示。

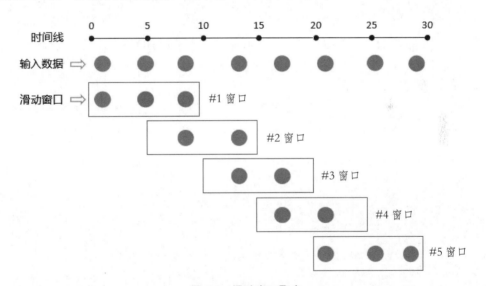

图 7-2 滑动窗口聚合

下面通过一个示例程序使用和理解滑动窗口。

【例 7-2】应用滑动窗口聚合解决一个 IOT 流数据分析问题:在一个数据中心中,按一定的时间间隔周期性地检测每个服务器机架的温度,并生成一个报告,显示每个机架在窗口长度 10min、滑动间隔 5min 的平均温度。

建议按以下步骤操作。

(1) 准备数据。

在本示例中使用文件数据源，该数据源以 JSON 文件的格式记录了某数据中心两个机架的温度数据。每个事件由 3 个字段组成。

① rack：表示机器的唯一 ID，字符串类型。

② temperature：表示采集到的温度值，double 类型。

③ ts：表示该事件发生时的时间戳，这是事件时间。

本示例准备了两个 IOT 事件数据的 JSON 文件。

file1.json 文件的内容如下：

```
{"rack":"rack1","temperature":99.5,"ts":"2017-06-02T08:01:01"}
{"rack":"rack1","temperature":100.5,"ts":"2017-06-02T08:06:02"}
{"rack":"rack1","temperature":101.0,"ts":"2017-06-02T08:11:03"}
{"rack":"rack1","temperature":102.0,"ts":"2017-06-02T08:16:04"}
```

file2.json 文件的内容如下：

```
{"rack":"rack2","temperature":99.5,"ts":"2017-06-02T08:01:02"}
{"rack":"rack2","temperature":105.5,"ts":"2017-06-02T08:06:04"}
{"rack":"rack2","temperature":104.0,"ts":"2017-06-02T08:11:06"}
{"rack":"rack2","temperature":108.0,"ts":"2017-06-02T08:16:08"}
```

为了模拟数据流的行为，应把这两个 JSON 文件复制到 HDFS 的指定目录下。

（2）代码编写。

实现的流查询代码如下：

```
#第7章/streaming_iot.py

from pyspark.sql import SparkSession
from pyspark.sql.types import *
from pyspark.sql.functions import *

#创建 SparkSession 实例
spark = SparkSession \
     .builder \
     .appName("sliding window demo") \
     .getOrCreate()

#将 shuffle 后的分区数设置为 2 （测试环境下）
spark.conf.set("spark.sql.shuffle.partitions",2)

#设置日志级别
#spark.sparkContext.setLogLevel("WARN")

#为手机事件数据创建一个 schema
fields = [
     StructField("rack", StringType(), nullable = False),
```

```
            StructField("temperature", DoubleType(), nullable = False),
            StructField("ts", TimestampType(), nullable = False)
]
iotDataSchema = StructType(fields)

#指定监听的文件目录
dataPath = "file://home/hduser/data/spark/stream/iot2"

#读取指定目录下的源数据文件,一次一个
iotDF = spark.readStream \
    .option("mode","failFast") \
    .schema(iotDataSchema) \
    .json(dataPath)

#group by一个滑动窗口,并在temperature列上求平均值
windowAvgDF = iotDF \
    .groupBy(window("ts", "10 minutes", "5 minutes")) \
    .agg(avg("temperature").alias("avg_temp"))

windowAvgDF.printSchema()

#执行流查询,将结果输到控制台
windowAvgDF \
    .select("window.start", "window.end", "avg_temp") \
    .orderBy("start") \
    .writeStream \
    .format("console") \
    .option("truncate", "false") \
    .outputMode("complete") \
    .start()

#等待流程序执行结束(作为作业文件提交时启用)
#query.awaitTermination()         #PySpark Shell交互环境下不需要这一句
```

在上面的代码中,首先读取温度数据,然后在 ts 列上构造一个长 10min、每 5min 进行滑动的滑动窗口,并在这个窗口上执行 groupBy()转换。对于每个滑动窗口,avg()函数被应用于 temperature 列。

(3)执行流程序。

执行以上代码,输出结果如下:

```
root
 |-- window: struct (nullable = true)
 |    |-- start: timestamp (nullable = true)
 |    |-- end: timestamp (nullable = true)
 |-- avg_temp: double (nullable = true)
```

```
------------------------------------------------
Batch: 0
------------------------------------------------
+-------------------+-------------------+--------+
|start              |end                |avg_temp|
+-------------------+-------------------+--------+
|2017-06-02 07:55:00|2017-06-02 08:05:00|99.5    |
|2017-06-02 08:00:00|2017-06-02 08:10:00|101.25  |
|2017-06-02 08:05:00|2017-06-02 08:15:00|102.75  |
|2017-06-02 08:10:00|2017-06-02 08:20:00|103.75  |
|2017-06-02 08:15:00|2017-06-02 08:25:00|105.0   |
+-------------------+-------------------+--------+
```

上面的输出显示在合成数据集中有 5 个窗口。注意每个窗口的开始时间间隔为 5min，这是在 groupBy() 转换中指定的滑动间隔的长度。

在上面的分析结果中，可以看出 avg_temp 列所代表的机架平均温度在上升。那么大家思考一下，机架平均温度的上升，是因为其中某个机架的温度升高从而导致平均温度的升高，还是所有机架的温度都在升高？

（4）代码重构。

为了弄清楚到底是哪些机架在不断升温，重构上面的代码，把 rack 列添加到 groupBy() 转换中，代码如下：

```
#第7章/streaming_iot2.py

from pyspark.sql import SparkSession
from pyspark.sql.types import *
from pyspark.sql.functions import *

#创建 SparkSession 实例
spark = SparkSession \
    .builder \
    .appName("sliding window demo") \
    .getOrCreate()

#将 shuffle 后的分区数设置为 2（测试环境下）
spark.conf.set("spark.sql.shuffle.partitions",2)

#设置日志级别
#spark.sparkContext.setLogLevel("WARN")

#为 IoT 事件数据创建一个 schema
fields = [
    StructField("rack", StringType(), nullable = False),
    StructField("temperature", DoubleType(), nullable = False),
    StructField("ts", TimestampType(), nullable = False)
]
```

```
iotDataSchema = StructType(fields)

#指定监听的文件目录
dataPath = "file://home/hduser/data/spark/stream/iot2"

#读取指定目录下的源数据文件，一次一个
iotDF = spark.readStream \
    .option("mode","failFast") \
    .schema(iotDataSchema) \
    .json(dataPath)

#group by 一个滑动窗口和 rack 列，并在 temperature 列上求平均值
windowAvgDF = iotDF \
    .groupBy(window("ts", "10 minutes", "5 minutes"), "rack") \
    .agg(avg("temperature").alias("avg_temp"))

windowAvgDF.printSchema()

#分别报告每个机架随时间而变化的温度
windowAvgDF \
    .select("rack","window.start", "window.end", "avg_temp") \
    .orderBy("rack","start") \
    .writeStream \
    .format("console") \
    .option("truncate", "false") \
    .outputMode("complete") \
    .start()

#等待流程序执行结束（作为作业文件提交时启用）
#query.awaitTermination()              #PySpark Shell 交互环境下不需要这一句
```

执行以上代码，输出结果如下：

```
root
 |-- window: struct (nullable = true)
 |    |-- start: timestamp (nullable = true)
 |    |-- end: timestamp (nullable = true)
 |-- rack: string (nullable = false)
 |-- avg_temp: double (nullable = true)

-------------------------------------------
Batch: 0
-------------------------------------------
+-----+-------------------+-------------------+--------+
|rack |start              |end                |avg_temp|
+-----+-------------------+-------------------+--------+
|rack1|2017-06-02 07:55:00|2017-06-02 08:05:00|99.5    |
|rack1|2017-06-02 08:00:00|2017-06-02 08:10:00|100.0   |
```

```
|rack1|2017-06-02 08:05:00|2017-06-02 08:15:00|100.75 |
|rack1|2017-06-02 08:10:00|2017-06-02 08:20:00|101.5  |
|rack1|2017-06-02 08:15:00|2017-06-02 08:25:00|102.0  |
|rack2|2017-06-02 07:55:00|2017-06-02 08:05:00|99.5   |
|rack2|2017-06-02 08:00:00|2017-06-02 08:10:00|102.5  |
|rack2|2017-06-02 08:05:00|2017-06-02 08:15:00|104.75 |
|rack2|2017-06-02 08:10:00|2017-06-02 08:20:00|106.0  |
|rack2|2017-06-02 08:15:00|2017-06-02 08:25:00|108.0  |
+-----+-------------------+-------------------+-------+
```

从上面的输出结果表中，可以清楚地看出来，机架 1 的平均温度低于 103，而机架 2 升温的速度要远快于机架 1，所以应该关注的是机架 2。

7.2 水印

在流处理引擎中，水印是一种常用的技术，用于限制维护它所需的状态数量，以及处理延迟数据。

7.2.1 限制维护的聚合状态数量

通过应用于事件时间上的窗口聚合（固定窗口聚合或滑动窗口聚合），在 PySpark 结构化流中可以很容易地执行常见的和复杂的流处理操作。事实上，任何时候在流查询上执行聚合时，都必须维护中间聚合状态。

首先理解"维护中间聚合状态"的含义。例如，在实时单词统计流程序中，随着每行文本的到来，流引擎会进行分词和计数处理。在分组聚合中，在用户指定的分组列中为每个唯一值（每个单词）维护聚合值（例如计数），代码如下：

```
#第7章/streaming_state.py

from pyspark.sql.functions import explode
from pyspark.sql.functions import split

#将shuffle后的分区数设置为10（测试环境下）
spark.conf.set("spark.sql.shuffle.partitions",10)

#从Socket数据源读取流数据
lines = spark.readStream \
    .format("socket") \
    .option("host", "localhost") \
    .option("port", "9999") \
    .load()

#将行拆分为单词
```

```
words = lines.select(
   explode(split(lines.value, " ")).alias("word")
)

#生成运行时单词计数
wordCounts = words.groupBy("word").count()

query = wordCounts.writeStream \
   .format("console") \
   .outputMode("complete") \
   .start()

#等待流程序执行结束（作为作业文件提交时启用）
#query.awaitTermination()
```

处理过程和状态维护过程如图 7-3 所示。

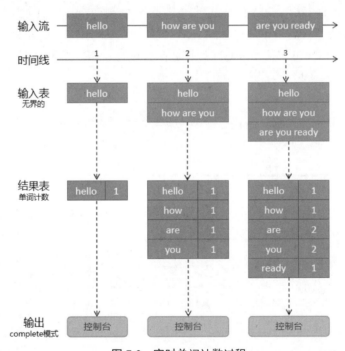

图 7-3　实时单词计数过程

在这个过程中，PySpark 结构化流并没有具体化整个表，而只是从流数据源读取最新的可用数据，对其进行增量处理以更新结果，然后丢弃源数据。它只保留更新结果所需的最小中间状态数据，即上面示例中的中间计数。

这个中间状态存储在 PySpark executors 的内存中，以及版本化的 key-value 状态存储中（其中 key 是分组名称，value 是中间聚合值），并将其写到一个 WAL（预写日志，该日志应该被配置为驻留在像 HDFS 这样的稳定存储系统中）中。在每个触发点上，该状态都在内存中的状态存储中读取和更新，然后写入预写日志中。在失败的情况下，当 PySpark 结构化的流应用程序重新启动时，状态从预写日志中还原并恢复。这种容错的状态管理显然会在结构化流引擎中产生一些资源和处理开销，因此，开销的大小与它需要维护的状态量成正比，因此，保持状态的数量在一个可以接受大小是很重要的；换句话说，状态的规模不应该无限增大，也就是流引擎不应该无限地维护状态，必须在合适时清除旧的、不再需要的状态。

对于结构化流来讲，在滑动事件时间窗口上的聚合与分组聚合非常相似。在基于窗口的聚合中，对于输入行的事件时间落入的每个窗口，该窗口的聚合值都需要被维护。用一个例子来理解这一点，对前面的示例进行修改，让输入流包含文本行及行生成的时间。现在想要统计在 10min 的窗口（每 5min 更新）中的单词计数。也就是说，统计在 12:00-12:10、12:05-12:15、12:10-12:20 等 10min 的窗口（时间间隔内）收到的单词数量。需要注意，12:00-12:10 表示在 12:00 之后和 12:10 之前到达的数据。现在，考虑 12:07 收到的一个单词。这个单词应该增加对应的两个窗口 12:00-12:10 和 12:05-12:15 的计数，因此计数将由分组键(单词)和窗口（可以从事件时间计算）两者进行索引。这个过程和结果表如图 7-4 所示。

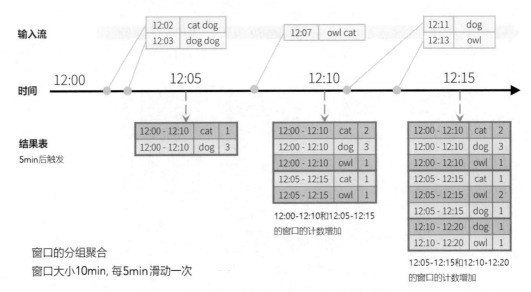

图 7-4　基于事件时间滑动窗口的单词计数过程

由于这个窗口类似于分组，在代码中，可以使用 groupBy()和 window()操作来表示窗口聚合，代码片段如下：

```
#流 DataFrame, schema { timestamp: Timestamp, word: String }
words = ...
```

```
#按窗口和单词对数据进行分组,并计算每组的单词计数
windowedCounts = words.groupBy(
    window(words.timestamp, "10 minutes", "5 minutes"),
    words.word
).count()

...
```

考虑到滑动窗口的性质,窗口的数量将会无限增多。这意味着执行滑动窗口聚合会导致中间状态无限地增长,因此,必须有一种方法可以删除不再更新的旧状态。在 PySpark 结构化流处理技术中,通过一种叫作水印(watermarking)的技术完成。

从 PySpark 2.1 开始,在结构化流 API 中引入了水印。本质上,水印是一个时间阈值,用来指定一个时间窗口结束后最多再等待多长时间就开始计算、更新状态并删除旧的状态。可以通过简单地在查询中添加 withWatermark 运算符来启用水印,代码如下:

```
DataFrame.withWatermark(eventTime: str, delayThreshold: str) →
pyspark.sql.dataframe.DataFrame
```

这种方法需要两个参数:

(1)一个事件时间列(eventTime,必须与聚合操作的时间相同)。

(2)一个阈值(delayThreshold),用于指定延迟多长时间的数据应该被处理(以事件时间作为单位,可以是秒、分钟或小时)。

聚合的状态将由 Spark 引擎进行维护,直到 max(eventTime) – delayThreshold > T,其中 max(eventTime)是引擎看到的最近事件时间,T 是窗口的结束时间。

指定水印的最大好处之一是能让结构化流引擎可以安全地删除比水印更"古老"的窗口的聚合状态。在生产环境下执行任何类型聚合的流应用程序都应该指定一个水印来避免内存不足的问题。

7.2.2 处理迟到的数据

在现实系统中,因为存在着各种可能,例如网络拥挤、网络中断或数据生成器(如移动设备等)不在线或传感器断开连接等,因此,不能保证数据会按照创建的顺序到达流处理引擎。为了容错,因此有必要处理这种乱序数据。要处理这个问题,流引擎必须保持聚合的状态。如果发生延迟事件,则当延迟事件到达时可以重新处理查询,但这意味着所有聚合的状态必须无限期地保持,这将导致内存使用量也无限期地增长。这在实际生产场景中是不现实的,除非系统有无限的资源和无限的预算,因此,作为一个实时流应用程序的开发人员,必须知道想要怎样处理延迟到达的数据。换句话说,数据延迟到达时间量是多少时才是可以接受的,或者说对这段时间量之后迟到的数据置之不理?这取决于应用场景。

处理延迟到达的事件是流处理引擎的一个关键功能。解决这个问题的方法是使用水印。水印技术是流处理引擎处理延迟的一种有效方法。水印实际上是一个时间阈值，用来指定系统等待延迟事件的时间。如果延迟到达的事件位于水印内，则用于更新查询。否则如果它的事件时间比水印的时间还要早，它就会被丢弃，而不会被流引擎进一步处理。

现在考虑一下前面基于事件时间滑动窗口的单词计数示例：如果某个事件延迟到达应用程序，会发生什么情况？例如，在12:04（事件时间）生成的单词因为某些原因，在12:11到达流应用程序，该程序接收此迟到的事件，并使用时间12:04来更新窗口12:00-12:10中的旧计数。PySpark 结构化流在基于窗口的分组中会自动这样来处理迟到的数据，因为结构化流可以长期维护部分聚合的中间状态，以便后期数据能够正确地更新旧窗口的聚合。这个过程如图7-5所示。

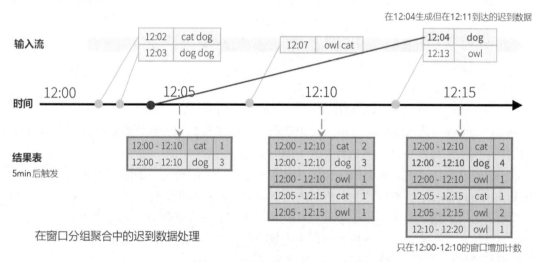

图 7-5 在窗口分组聚合中对延迟到达数据的处理

但是，如果要运行该查询很多天，系统需要绑定它在这么多天运行中累积的中间内存状态的数量。这意味着系统需要知道应用程序何时不会再接收到延迟到达的数据并从内存状态中删除一个旧的聚合。为了实现这一点，需要水印技术，它允许引擎自动跟踪数据中的当前事件时间，并尝试相应地清除旧的状态。对于在时间T结束的特定窗口，PySpark 引擎将维护其状态，并允许延迟到达的数据更新状态，直到"max(eventTime)−延迟阈值> T"。换句话说，在阈值内到达的迟到数据将被聚合，而迟于阈值时间到达的迟到数据将会被丢弃。

要在结构化流中指定水印作为流 DataFrame 的一部分，只需使用 Watermark API 并提供两个参数：事件时间列和阈值，这些数据可以以秒、分钟或小时为单位。使用 Watermark API 的模板代码如下：

```
#流 DataFrame, schema { timestamp: Timestamp, word: String }
```

```
words = ...

#通过窗口和单词分组数据,计算每个组的计数
windowedCounts = words \
    .withWatermark("timestamp", "10 minutes") \
    .groupBy(
        window(words.timestamp, "10 minutes", "5 minutes"),
        words.word) \
    .count()
...
```

在本例中，在 timestamp 列值上定义查询的水印，并定义 10 minutes 作为允许延迟到达数据的阈值。如果该查询在 update 输出模式下运行，则引擎会持续更新结果表中窗口的计数，直到窗口的水位超过水印，它落后 timestamp 列的当前事件时间 10min，如图 7-6 所示。

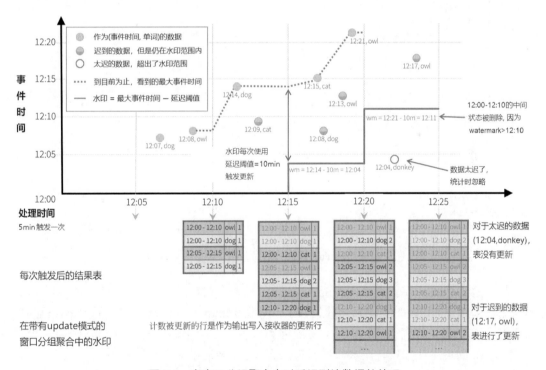

图 7-6　在窗口分组聚合中对延迟到达数据的处理

正如图 7-6 所示，引擎跟踪的最大事件时间是虚线，在每个触发器开始时设置为(最大事件时间–10min)的水印为实线。例如，当引擎观察到数据（12:14, dog）时，它为下一次触发器将水印设置为 12:04。此水印使引擎保持中间状态额外 10min，以允许统计后期数据。

例如，数据（12:09, cat）是乱序和延迟到达的，它落在窗口 12:00-12:10 和 12:05-12:15。由于它仍然在触发器中水印 12:04 之前，引擎仍然将中间计数作为状态维护，并正确更新相关窗口的计数，然而，当水印更新到 12:11 时，窗口（12:00 - 12:10）的中间状态被清除，所有后续数据（例如(12:04, donkey)）被认为太迟了而因此被忽略。注意，在每个触发器之后，更新的计数被写入 sink 作为触发器输出，这是由 update 模式决定的。

一些接收器（例如文件）可能不支持 update 模式需要的细粒度更新。为了使用它们，PySpark 结构化流还支持 append 模式。在这种模式下，只有最后的计数被写入 sink 中，如图 7-7 所示。

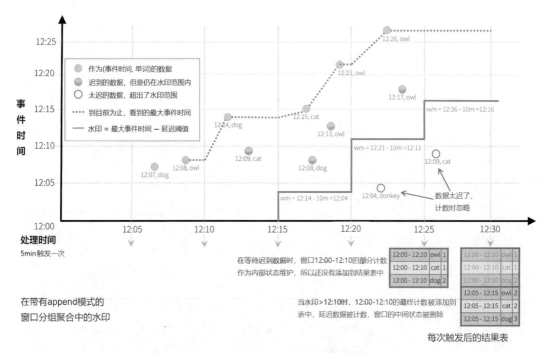

图 7-7 在 append 模式的窗口分组聚合中使用水印

与前面的 update 模式类似，PySpark 引擎维护每个窗口的中间计数，但是，部分计数不会更新到结果表，也不会写入接收器。引擎等待 10min 来统计延迟到达的数据，然后删除小于水印的窗口的中间状态，并将最终的计数添加到结果表/接收器。例如，窗口 12:00-12:10 的最终计数只有在水印更新为 12:11 后才会被添加到结果表中。

注意：在非流数据集上使用 withWatermark 是无效的。由于水印不应该以任何方式影响任何批处理查询，所以 PySpark 引擎将直接忽略它。

下面通过一个完整的应用示例来理解和掌握使用水印处理迟到数据的方法。

【例 7-3】使用水印来处理延迟到达的移动电话操作事件数据。

建议按以下步骤操作。

（1）准备数据。

在本示例中使用文件数据源，代表移动电话操作事件数据存储在两个 JSON 格式的文件中。每个事件由 3 个字段组成。

① id：表示手机的唯一 ID，字符串类型。

② action：表示用户所采取的操作。该操作的可能值是"open"或"close"。

③ ts：表示用户 action 发生时的时间戳。这是事件时间。

本示例准备了两个存储移动电话事件数据的 JSON 文件 file1.json 和 file2.json，其中第 2 个数据文件 file2.json 中包含的是迟到的数据。

file1.json 文件的内容如下：

```
{"id":"phone1","action":"open","ts":"2018-03-02T10:15:33"}
{"id":"phone2","action":"open","ts":"2018-03-02T10:22:35"}
{"id":"phone3","action":"open","ts":"2018-03-02T10:33:50"}
```

file2.json 文件的内容如下：

```
{"id":"phone4","action":"open","ts":"2018-03-02T10:29:35"}
{"id":"phone5","action":"open","ts":"2018-03-02T10:11:35"}
```

注意观察这两个数据文件中的数据。数据以这样一种方式设置，即 file1.json 文件中的每行进入了它自己的 10min 窗口，那么 file1.json 的处理会形成 3 个窗口：10:10:00-10:20:00、10:20:00-10:30:00 和 10:30:00-10:40:00。示例中将水印指定为 10min。file2.json 文件中的数据代表迟到的数据，其中第 1 行落在 10:20:00-10:30:00 窗口中，所以即使它到达的时间较晚，它的时间戳仍然在水印的阈值范围内，因此它将被处理。file2.json 文件中的最后一行数据的时间戳在 10:10:00-10:20:00 窗口中，由于它超出了水印的阈值，所以它将被忽略，而不会被处理。

为了模拟数据流的行为，应把这两个 JSON 文件复制到 HDFS 的指定目录下。

（2）代码编写。

实现的流查询代码如下：

```python
#第7章/streaming_late_date.py

from pyspark.sql import SparkSession
from pyspark.sql.types import *
from pyspark.sql.functions import *

#创建 SparkSession 实例
spark = SparkSession \
    .builder \
    .appName("late data demo") \
    .getOrCreate()
```

```python
#将shuffle后的分区数设置为2（测试环境下）
spark.conf.set("spark.sql.shuffle.partitions",2)

#设置日志级别
#spark.sparkContext.setLogLevel("WARN")

#为手机事件数据创建一个schema
fields = [
    StructField("id", StringType(), nullable = False),
    StructField("action", StringType(), nullable = False),
    StructField("ts", TimestampType(), nullable = False)
]
mobileDataSchema = StructType(fields)

#指定监听的文件目录
dataPath = "/data/spark/stream/mobile4"

#读取指定目录下的源数据文件，一次一个
mobileDF = spark.readStream \
    .option("maxFilesPerTrigger", 1) \
    .option("mode","failFast") \
    .schema(mobileDataSchema) \
    .json(dataPath)

#设置一个带有水印的流DataFrame，并按ts和action列分组
#水印，10min，必须先于groupBy调用
#指定的窗口列必须与水印中指定的列一致
windowCountDF = mobileDF \
    .withWatermark("ts", "10 minutes") \
    .groupBy(window("ts", "10 minutes"), "action") \
    .count()

windowCountDF.printSchema()

#输出到控制台
windowCountDF \
    .select("window.start", "window.end", "action", "count") \
    .writeStream \
    .format("console") \
    .option("truncate", "false") \
    .outputMode("update") \
    .start()

#等待流程序执行结束（作为作业文件提交时启用）
#query.awaitTermination()
```

执行程序，输出源数据的结构，内容如下：

```
root
 |-- window: struct (nullable = false)
 |    |-- start: timestamp (nullable = true)
 |    |-- end: timestamp (nullable = true)
 |-- action: string (nullable = false)
 |-- count: long (nullable = false)
```

当它读取第 1 个流数据文件 file1.json 时，输出结果如下：

```
-------------------------------------------
Batch: 0
-------------------------------------------
+-------------------+-------------------+------+-----+
|start              |end                |action|count|
+-------------------+-------------------+------+-----+
|2018-03-02 10:30:00|2018-03-02 10:40:00|open  |1    |
|2018-03-02 10:10:00|2018-03-02 10:20:00|open  |1    |
|2018-03-02 10:20:00|2018-03-02 10:30:00|open  |1    |
+-------------------+-------------------+------+-----+
```

正如所期望的，每行都落在它自己的窗口内。

当它读取第 2 个流数据文件 file2.json 时，输出结果如下：

```
-------------------------------------------
Batch: 1
-------------------------------------------
+-------------------+-------------------+------+-----+
|start              |end                |action|count|
+-------------------+-------------------+------+-----+
|2018-03-02 10:20:00|2018-03-02 10:30:00|open  |2    |
+-------------------+-------------------+------+-----+
```

注意到窗口 10:20:00-10:30:00 的 count 现在被更新为 2，窗口 10:10:00-10:20:00 没有变化。如前所述，因为 file2.json 文件中的最后一行的时间戳落在 10min 的水印阈值之外，因此它不会被处理。

（3）如果删除对 Watermark API 的调用，则输出结果如下：

```
-------------------------------------------
Batch: 1
-------------------------------------------
+-------------------+-------------------+------+-----+
|start              |end                |action|count|
+-------------------+-------------------+------+-----+
|2018-03-02 10:10:00|2018-03-02 10:20:00|open  |2    |
|2018-03-02 10:20:00|2018-03-02 10:30:00|open  |2    |
+-------------------+-------------------+------+-----+
```

可以看出，因为没有指定水印，所以迟到的数据也不会被删除，对窗口 10:10:00-10:20:00 的 count 计数被更新为 2。

最后，因为水印是一个有用的特性，因此理解聚合状态被正确清理的条件是很重要的：

（1）输出模式不能是 complete 模式，必须是 update 或 append 模式。原因是，complete 模式的语义规定必须维护所有聚合数据，并且不违反这些语义，水印不能删除任何中间状态。

（2）通过 groupBy 转换的聚合必须直接位于 event-time 列或 event-time 列的窗口上。

（3）在 Watermark API 和 groupBy() 转换中指定的 event-time 列必须是同一个。

（4）当设置一个流 DataFrame 时，必须在调用 groupBy() 转换之前调用 Watermark API，否则它将被忽略。

7.3 处理重复数据

当数据源多次发送相同的数据时，实时流数据中的数据就会重复。在流处理中，由于流数据的无界性，去除重复数据是一种非常具有挑战性的任务。

不过，PySpark 结构化流使流应用程序能够轻松地执行数据去重操作，因此这些应用程序可以通过在到达时删除重复的数据来保证精确一次处理。结构化流所提供的数据去重特性可以与水印一起工作，也可以不使用水印。不过，需要注意的一点是，在执行数据去重时，如果没有指定水印，在流应用程序的整个生命周期中，结构化流需要维护的状态将无限增长，这可能会导致内存不足的问题。使用水印，比水印更老的数据会被自动删除，以避免重复的可能。

在结构化流中执行重复数据删除操作的 API 很简单，它只有一个输入参数，该输入参数是用来唯一标识每行的列名的列表。这些列的值将被用于执行重复检测，并且结构化流将把它们存储为中间状态。下面通过一个示例来演示这个 API 的使用。

【例 7-4】处理重复到达的移动电话操作事件数据。

建议按以下步骤操作。

（1）准备数据。

在本例中使用文件数据源，代表移动电话操作事件数据存储在两个 JSON 格式的文件中。每个事件由 3 个字段组成。

① id：表示手机的唯一 ID，字符串类型。

② action：表示用户所采取的操作。该操作的可能值是 open 或 close。

③ ts：表示用户 action 发生时的时间戳，这是事件时间。

本示例准备了两个存储移动电话操作事件数据的 JSON 文件 file1.json 和 file2.json，其中 file2.json 文件中包含了与 file1.json 文件中重复的记录。

file1.json 文件的内容如下:

```
{"id":"phone1","action":"open","ts":"2018-03-02T10:15:33"}
{"id":"phone2","action":"open","ts":"2018-03-02T10:22:35"}
{"id":"phone3","action":"open","ts":"2018-03-02T10:23:50"}
```

观察上面的数据,每行都是唯一的 id 和 ts 列。

file2.json 文件的内容如下:

```
{"id":"phone1","action":"open","ts":"2018-03-02T10:15:33"}
{"id":"phone2","action":"open","ts":"2018-03-02T10:22:35"}
{"id":"phone4","action":"open","ts":"2018-03-02T10:29:35"}
{"id":"phone5","action":"open","ts":"2018-03-02T10:01:35"}
```

观察上面的数据,前两行与 file1.json 文件中的前两行数据重复,第 3 行是唯一的,第 4 行也是唯一的,但延迟到达(所以在后面的代码中应该不被处理)。

为了模拟数据流的行为,应把这两个 JSON 文件复制到 HDFS 的指定目录下。

(2) 编写流处理代码。

编写流处理代码逻辑,在代码中基于 id 列进行分组 count() 聚合。id 列和 ts 列共同定义为 key,代码如下:

```python
#第7章/streaming_duplicates.py
from pyspark.sql import SparkSession
from pyspark.sql.types import *
from pyspark.sql.functions import *

#创建 SparkSession 实例
spark = SparkSession \
    .builder \
    .appName("late data demo") \
    .getOrCreate()

#将 shuffle 后的分区数设置为 2(测试环境下)
spark.conf.set("spark.sql.shuffle.partitions",2)

#设置日志级别
#spark.sparkContext.setLogLevel("WARN")

#为手机事件数据创建一个 schema
fields = [
    StructField("id", StringType(), nullable = False),
    StructField("action", StringType(), nullable = False),
    StructField("ts", TimestampType(), nullable = False)
]
```

```
mobileDataSchema = StructType(fields)

#指定监听的文件目录
dataPath = "/data/spark/stream/mobile4"

#读取指定目录下的源数据文件,一次一个
mobileDF = spark.readStream \
    .option("maxFilesPerTrigger", 1) \
    .option("mode","failFast") \
    .schema(mobileDataSchema) \
    .json(dataPath)

#添加水印,去重,分组聚合操作
windowCountDupDF = mobileDF \
    .withWatermark("ts", "10 minutes") \
    .dropDuplicates(["id", "ts"]) \
    .groupBy("id") \
    .count()

#将结果输到控制台
windowCountDupDF.writeStream \
    .format("console") \
    .option("truncate", "false") \
    .outputMode("update") \
    .start()

#等待流程序执行结束(作为作业文件提交时启用)
#query.awaitTermination()
```

(3) 执行流程序。

当读取 file1.json 源数据文件时,输出结果如下:

```
+------+-----+
|id    |count|
+------+-----+
|phone3|1    |
|phone1|1    |
|phone2|1    |
+------+-----+
```

当读取 file2.json 源数据文件时,输出结果如下:

```
+------+-----+
|id    |count|
+------+-----+
|phone4|1    |
+------+-----+
```

如预期所料，当读取 file2.json 源数据文件时，输出结果中只有一行显示在控制台中。原因是前两行是 file1.json 文件中前两行的重复，因此它们被过滤掉了（去重）。最后一行的时间戳是 10:10:00，这被认为是迟到数据，因为时间戳比 10min 的水印阈值更迟，因此，最后一行没有被处理，也被删除了。

7.4 容错

5min

当开发重要的流应用程序并将其部署到生产环境中时，最重要的考虑之一就是故障恢复。根据墨菲定律，任何可能出错的地方都会出错。机器将会有故障，软件将会有缺陷。当有故障时，结构化流提供了一种方法来重新启动或恢复流应用程序，并从停止的地方继续执行。

要利用这种恢复机制，需要配置流应用程序使用检查点和预写日志 WAL（Write Ahead Logs，预写日志）。理想情况下，检查点的位置应该是一个可靠的、容错的文件系统的路径，例如 HDFS 或 S3。结构化流将所有的进度信息定期保存到检查点位置，例如正在处理的数据的偏移细节和中间状态值，因此，在任何失败的情况下，它可以完全重新处理相同的数据，确保得到端到端的一次保证。这些存储已处理偏移量信息的 WAL 文件以 JSON 格式保存到 HDFS 或 S3 Bucket 或 Azure Blob 等容错存储中，以实现前向兼容性，如图 7-8 所示。

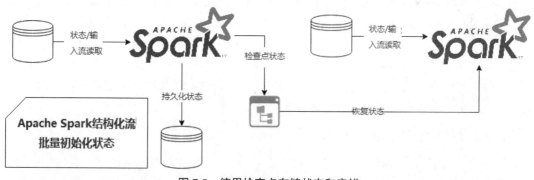

图 7-8　使用检查点存储状态和容错

向流查询添加检查点位置非常简单，只需在流查询中添加一个选项，将 checkpointLocation 作为名称并将路径作为值，代码如下：

```
#向一个流查询添加 checkpointLocation 选项
query = df.writeStream \
    .format("console") \
    .option("truncate",false) \
```

```
.option("checkpointLocation","/ck/location") \ #设置检查点
.outputMode("append") \
.start()
```

如果查看指定的检查点位置,则应该看到以下子目录:commits、metadata、offsets、sources 和 stats。这些目录中的信息专用于特定流查询,因此,每个流查询都必须使用不同的检查点位置,如图 7-9 所示。

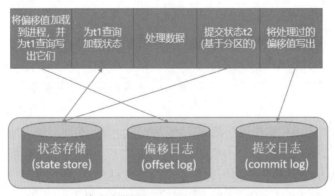

图 7-9 检查点位置

通过检查点,得到了完全的容错保证,从而确保从故障中恢复。结构化流可以结合特定的接收器/数据源和 WAL 文件提供恰好一次的交付语义。接收器和数据源必须是可重放的,就像 Kafka 和 Kinesis 等。否则它将是 at-least once(精确一次)交付语义。

就像大多数软件应用程序一样,流应用程序将随着时间的推移而不断重构,因为需要改进处理逻辑、性能或者修复 Bug。重要的是要记住,这可能会影响在检查点位置保存的信息,并知道哪些更改被认为是安全的。概括地说,有两类变化:一类是对流应用程序代码的更改,另一类是对 PySpark 运行时的更改。

1. 流应用程序代码更改

检查点位置的信息被设计为对流应用程序的变化有一定的弹性。有一些变化将被认为是不相容的变化。第 1 个是通过改变 key 列、添加更多的 key 列或者删除一个现存的 key 列来改变聚合的方式。第 2 种方法是改变用于存储中间状态的类结构,例如,当一个字段被移除或者字段的类型从字符串转换为整数时。当在重新启动期间检测到不兼容的更改时,结构化流将通过一个异常进行通知。在这种情况下,必须使用新的检查点位置,或者删除先前检查点位置的内容。

2. 运行时更改

检查点格式被设计为向前兼容,这样当 PySpark 跨越补丁版本或小版本的更新时(例

如从 PySpark 2.2.0 升级到 2.2.1 或从 PySpark 2.2.x 升级到 2.3.x），流应用程序应该能够从一个旧的检查点重新启动，甚至可以通过在流转换中做有限的更改来恢复处理（例如，添加新的过滤器子句来删除损坏的数据等），并且 WAL 文件可以无缝地工作。唯一的例外是当有严重的 Bug 修复时。不过通常不需要担心，当 Spark/PySpark 引入不兼容的变更时，它会在发行说明中清楚地进行说明。

如果由于不兼容的问题，无法启动一个使用现有检查点位置的流应用程序，就必须使用一个新的检查点位置，并且可能还需要为流应用程序提供一些关于偏移量的信息来读取数据。

7.5 流查询度量指标

与其他长时间运行的应用程序（如在线服务）类似，有必要了解流应用程序正在进行的进展、传入的数据速率或者中间状态所消耗的内存数量等信息。结构化流提供了一些 API 来提取关于最近执行进度的信息，以及在流应用程序中监控所有流媒体查询的异步方式。

流查询最有用的信息是其当前状态。通过对 pyspark.sql.streaming.StreamingQuery.status 属性的调用，可以以可读的格式检索和显示这些信息，返回的是 JSON 格式的状态信息。例如，调用 StreamingQuery.status 属性，代码如下：

```
#以 JSON 格式查询状态信息
#从上面的示例中使用一个流查询
query.status
```

返回内容如下：

```
{
"message" : "Waiting for data to arrive",
"isDataAvailable" : false,
"isTriggerActive" : false
}
```

很明显，在当前状态属性被调用时，前面的状态提供了关于流查询的基本信息。为了从最近的进展中获得更多的细节，例如传入的数据速率、处理速率、水印、数据源的偏移量，以及一些关于中间状态的信息，可以调用 StreamingQuery.recentProgress 属性。这个属性返回 StreamingQueryProgress 类的实例的一个数组，它将细节转换为 JSON 格式。默认情况下，每个流查询都被配置为保持 100 个进度更新，这个数字可以通过更新 PySpark 配置 spark.sql.streaming.numRecentProgressUpdates 来改变。要查看最新的流查询进度，可以调用属性 StreamingQuery.lastProgress。关于流查询进度的示例，内容如下：

```
#流查询进度结点
{
  "id" : "9ba6691d-7612-4906-b64d-9153544d81e9",
  "runId" : "c6d79bee-a691-4d2f-9be2-c93f3a88eb0c",
  "name" : null,
  "timestamp" : "2018-04-23T17:20:12.023Z",
  "batchId" : 0,
  "numInputRows" : 3,
  "inputRowsPerSecond" : 250.0,
  "processedRowsPerSecond" : 1.728110599078341,
  "durationMs" : {
    "addBatch" : 1548,
    "getBatch" : 8,
    "getOffset" : 36,
    "queryPlanning" : 110,
    "triggerExecution" : 1736,
    "walCommit" : 26
  },
  "eventTime" : {
    "avg" : "2017-09-06T15:10:04.666Z",
    "max" : "2017-09-06T15:11:10.000Z",
    "min" : "2017-09-06T15:08:53.000Z",
    "watermark" : "1970-01-01T00:00:00.000Z"
  },
  "stateOperators" : [ {
    "numRowsTotal" : 1,
    "numRowsUpdated" : 1,
    "memoryUsedBytes" : 16127
  } ],
  "sources" : [ {
    "description" : "FileStreamSource[file:<path>/data/input]",
    "startOffset" : null,
    "endOffset" : {
      "logOffset" : 0
    },
    "numInputRows" : 3,
    "inputRowsPerSecond" : 250.0,
    "processedRowsPerSecond" : 1.728110599078341
  } ],
  "sink" : {
    "description" : "org.apache.spark.sql.execution.streaming.
    ConsoleSinkProvider@37dc4031"
  }
}
```

查看上面显示的流进度的详细信息，有一些重要的关键指标值得注意。输入率表示从输入源流入流应用程序的传入数据量。处理速率代表流应用程序处理传入数据的速度有多快。在理想状态下，处理速率应该高于输入速率，如果不是这样，则需要考虑在 PySpark 集群中增加节点的数量。如果流应用程序通过隐式 groupBy 转换或显式地通过任意状态处理 APIs 来保持状态，则关注 stateOperators 部分中的指标是很重要的。

PySpark UI 在 job、stages 和 task 级别上提供了丰富的度量标准。流应用程序中的每个触发器都被映射到 PySpark UI 中的一个作业，在那里查询计划和任务持续时间可以很容易地检查。

7.6 结构化流案例：运输公司车辆超速实时监测

7min

想象一个物流公司车队管理解决方案，其中车队中的车辆启用了无线网络功能。每辆车定期报告其地理位置和许多操作参数，如燃油水平、速度、加速度、轴承和发动机温度等。物流公司希望利用这一遥测数据流实现一系列应用程序，以帮助他们管理业务的运营和财务方面。

假设现在需要开发一个流应用程序，计算车辆每几秒的平均速度，用来检查车辆是否超速。程序整体架构如图 7-10 所示。

图 7-10　运输公司车辆超速实时监测程序架构

在本案例中，使用 Kafka 作为流数据源，将从 Kafka 的 cars 主题来读取这些事件。同时也将 Kafka 作为流的 Data Sink，检测出的超速事件将写入 Kafka 的 fastcars 主题。

7.6.1　实现技术剖析

PySpark 在处理实时数据流时，它会等待一个非常小的间隔，例如 1s（或者甚至 0s，即尽可能地快）。将在此间隔期间收到的所有事件合并到一个微批处理中。在结构化流中，微批数据被构造为一个流式的 DataFrame(无界表)，如图 7-11 所示。

图 7-11　PySpark 结构化流以微批方式处理流数据

这个微批处理由驱动程序调度，作为任务在执行器（executor）中执行。完成微批处理执行后，将再次收集并调度下一批。这种调度是经常进行的，给人以流执行的印象，如图 7-12 所示。

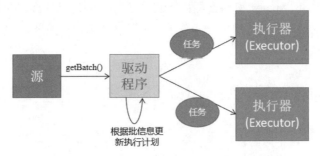

图 7-12　微批处理作为任务在 Executor 中执行

在本案例中使用 Kafka 作为流数据源，流处理程序将从 Kafka 的 cars 主题来读取这些事件，代码如下：

```
df = spark.readStream \
    .format("kafka") \
    .option("kafka.Bootstrap.servers", "localhost:9092") \
    .option("subscribe", "cars") \
    .option("startingOffsets", "earliest") \
    .load()
```

为了模拟车辆发送传感器数据，在本案例的配套源码中提供了一个已经创建好的 Kafka 生产者程序 fastcars_fat.jar（这是一个 Java 程序，执行 JAR 包），它将 id、speed（速度）、acceleration（加速度）和 timestamp（时间戳）写入 Kafka 的 cars 主题。注意，这里的 timestamp 时间戳是在事件源处生成事件（消息）的时间，即代表事件时间。这个数据源的产生过程如图 7-13 所示。

图 7-13　模拟车辆将传感器数据发送到 Kafka

接下来，在流处理程序中对从 Kafka cars 主题拉取的原始数据进行解析，这样就有了一个可以使用的结构，代码如下：

```
#解析从 Kafka 拉取的值 value
import pyspark.sql.functions as F
```

```
convertedDF = df \
   .selectExpr("CAST(value as string)") \
   .select(F.split("value",",").alias("data")) \
   .select(
      F.col("data")[0].alias("car_id"),
      F.col("data")[1].alias("speed"),
      F.col("data")[2].alias("acceleration"),
      F.substring(F.col("data")[3],0,10).alias("ts")
   ) \
   .selectExpr(
      "car_id",
      "CAST(speed as int)",
      "CAST(acceleration as double)",
      "CAST(ts as bigint)"
   ) \
   .withColumn("ts",F.to_timestamp(F.from_unixtime("ts")))
convertedDF.printSchema()
```

这时产生的流 DataFrame 的结构如图 7-14 所示。

```
>>> convertedDF.printSchema()
root
 |-- car_id: string (nullable = true)
 |-- speed: integer (nullable = true)
 |-- acceleration: double (nullable = true)
 |-- ts: timestamp (nullable = true)
```

图 7-14　从 Kafka 的 cars 主题摘取的数据格式

注意，在上面的代码中，因为从 Kafka 中拉取的数据 value 值为二进制，所以需要先把它转换为字符串，然后调用 split()方法将其分割为多个字段，其中模拟数据源生成的日期是 13 位数字的毫秒值，所以在转换为时间戳之前，需要先调用 substring()方法截掉最后三位数字（变成 10 位数字的秒值）。最后，将各个字段由字符串类型转换为相应的数据类型。

接下来执行聚合操作，这从求每辆车的平均速度开始。可以通过对 car_id 执行 groupBy()并应用 avg()聚合函数实现，代码如下：

```
aggregates = convertedDF \
   .groupBy("car_id") \
   .agg(F.avg("speed"))
```

在结构化流程序中，可以使用触发器控制微批处理的时间间隔。在 PySpark 中，触发器被设置为指定在检查新数据是否可用之前等待多长时间。如果没有设置触发器，一旦完成前一个微批处理，PySpark 将立即检查新数据的可用性。

这里需要计算车辆在过去 5s 内的平均速度。为此，需要根据事件时间将事件分组为 5s 间隔时间组，这种分组称为窗口。

在 PySpark 中，窗口是通过在 groupBy()子句中添加额外的 key 实现的。对于每条消息，它的事件时间（传感器生成的时间戳）用于标识消息属于哪个窗口。基于窗口的类型（滚动/滑动），一个事件可能属于一个或多个窗口。

为了实现窗口化，PySpark 添加了一个名为 window 的新列，并将提供的 ts 列分解为一个或多行（基于它的值、窗口的大小和滑动），并在该列上执行 groupBy()操作。这将隐式地将属于一个时间间隔的所有事件拉到同一个"窗口"中。

这里根据 window 和 car_id 对 convertedDF 数据进行分组。注意，window()是 PySpark 中的一个函数，它返回一个列，代码如下：

```
#4s 大小的滚动窗口
aggregates = convertedDF
    ..groupBy(F.window("ts","4 seconds"), "car_id")
    .agg(F.avg("speed").alias("speed"))
    .where("speed > 100")
```

定义大小为 4s、滑动为 2s 的滑动窗口，代码如下：

```
convertedDF.groupBy(F.window("ts", "4 seconds", "2 seconds"), "car_id")
```

这将产生一个 car_id、平均速度和相应时间窗口的 DataFrame，输出如图 7-15 所示。

```
Batch: 17
-------------------------------------------------+------+-----+
|window                                          |car_id|speed|
+------------------------------------------------+------+-----+
|{2022-02-27 11:44:24, 2022-02-27 11:44:28}|car3  |108.0|
|{2022-02-27 11:44:24, 2022-02-27 11:44:28}|car6  |112.0|
```

图 7-15 产生一个 car_id、平均速度和相应时间窗口的 DataFrame

PySpark 提供了 3 种输出模式：complete、update 和 append。在处理微批处理后，PySpark 更新状态和输出结果的方式各不相同，如图 7-16 所示。

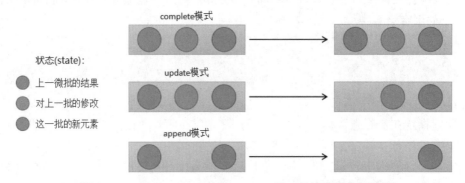

图 7-16 complete、update 和 append 这 3 种输出模式的区别

在每个微批处理期间,PySpark 会更新前批处理中的一些 key 的值,有些是新的,有些保持不变。在 complete 模式下,会输出所有的行,而在 update 模式下,只输出新的和更新的行。append 模式略有不同,在 append 模式中,不会有任何更新的行,它只输出新增的行。在流程序中指定输出模式,代码如下:

```
writeToKafka = aggregates \
    .selectExpr("CAST(car_id as string) as key", "warn as value") \
    .writeStream \
    .format("kafka") \
    .option("kafka.Bootstrap.servers","xueai8:9092") \
    .option("topic", "fastcars") \
    .option("startingOffsets", "earliest") \
    .option("checkpointLocation", "/tmp/carsck/") \
    .outputMode("update") \
    .start()
```

在使用 Kafka 接收器时,检查点位置是必需的,它支持故障恢复和精确的一次处理。运行应用程序,输出如图 7-17 所示。

图 7-17 update 模式下流程序的输出结果

需要注意,结构化流 API 隐式地跨批维护聚合函数的状态。例如,在上面的例子中,第 2 个微批计算的平均速度将是第 1 批和第 2 批接收的事件的平均速度。作为用户,我们不需要自定义状态管理,但随着时间的推移,维护一个庞大的状态也会带来成本。这可以通过水印实现控制。

在 PySpark 中,水印用于根据当前最大事件时间决定何时清除状态。基于用户所指定的延迟,水印滞后于目前所看到的最大事件时间。例如,如果 delay 是 3s,当前最大事件时间是 10:00:45,则水印是在 10:00:42。这意味着 Spark 将保持结束时间小于 10:00:42 的窗口的状态。在程序中指定水印,代码如下:

```
#使用 ts 字段设置水印,最大延迟为 3s
aggregates = convertedDF \
    .withWatermark("ts", "3 seconds") \
    .groupBy(F.window("ts","4 seconds"), "car_id") \
    .agg(F.avg("speed").alias("speed")) \
    .where("speed > 100") \
    .withColumn("warn",F.concat_ws(", ","car_id","speed"))
```

需要理解的一个细微但重要的细节是，当使用基于 EventTime 的处理时，只有当接收到具有更高时间戳值的消息/事件时，时间才会前进。可以将它看作 PySpark 内部的时钟，但与每秒滴滴计时（基于处理时间）的普通时钟不同，该时钟只在接收到具有更高时间戳的事件时才移动。

参看下面这个示例，以深入理解在消息到达较迟时 PySpark 引擎会如何去处理。这里将集中于 10:00~10:10 的单个窗口和 5s 的最大延迟，代码如下：

```
.withWatermark("ts", "5 seconds")
```

引擎跟踪的最大事件时间，在每个触发器开始时将水印设置为最大事件时间–5s，如图 7-18 所示。

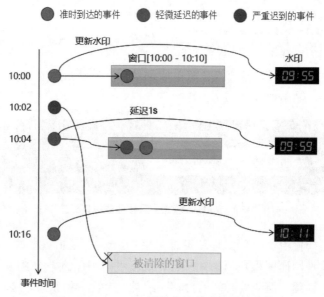

图 7-18 PySpark 结构化流处理迟到的事件

对于图 7-18 的说明如下：

（1）一个时间戳为 10:00 的事件到达并落在窗口[10:00,10:10)，水印被更新为 timestamp –5。

（2）在数据源处生成时间戳为 10:02 的事件，但该事件会有很大的延迟。这个事件应该落在窗口[10:00,10:10)中。

（3）时间戳为 10:04 的事件在 10:05 时到达，稍微有些延迟，但它仍然落在窗口[10:00,10:10)中，因为当前水印为 09:55，即小于窗口结束时间。水印更新到 10:04 – 00:05 = 09:59。

（4）一个时间戳为 10:16 的事件到达，它将水印更新为 10:11（此事件将落在窗口[10:10,10:20)，但与这里讨论的内容不相关）。

（5）带有时间戳 10:02 的延迟事件姗姗来迟，但是此时窗口[10:00,10:10]已经被清除，因此该事件将被删除。

设置水印将确保状态不会永远增长。另外，需要注意本示例中一个延迟事件是如何处理的，而另一个是如何被忽略的（因为已经太迟了）。

7.6.2 完整实现代码

下面给出本案例的完整代码。

【例 7-5】运输公司车辆超速实时监测完整实现。

实现代码如下：

```
#第7章/streaming_car.py

from pyspark.sql import SparkSession
from pyspark.sql.types import *
import pyspark.sql.functions as F

#创建 SparkSession 实例
spark = SparkSession.builder.appName("Cars Demo").getOrCreate()

#将 shuffle 后的分区数设置为 10（测试环境下）
spark.conf.set("spark.sql.shuffle.partitions",10)

#设置日志级别
#spark.sparkContext.setLogLevel("WARN")

#创建一个流来监听 cars topic 的消息
df = spark.readStream \
    .format("kafka") \
    .option("kafka.Bootstrap.servers", "xueai8:9092") \
    .option("subscribe", "cars") \
    .option("startingOffsets", "earliest") \
    .load()

#查看 schema
df.printSchema()

#解析从 Kafka 拉取的值 value
convertedDF = df \
    .selectExpr("CAST(value as string)") \
    .select(F.split("value",",").alias("data")) \
    .select(
        F.col("data")[0].alias("car_id"),
        F.col("data")[1].alias("speed"),
        F.col("data")[2].alias("acceleration"),
```

```python
        F.substring(F.col("data")[3],0,10).alias("ts")
    ) \
    .selectExpr(
        "car_id",
        "CAST(speed as int)",
        "CAST(acceleration as double)",
        "CAST(ts as bigint)"
    ) \
    .withColumn("ts",F.to_timestamp(F.from_unixtime("ts")))

convertedDF.printSchema()

#执行不带窗口的聚合
#aggregates = convertedDF.groupBy("car_id").agg(F.avg("speed"))

#执行滑动窗口聚合,窗口定义:大小为4s、滑动为1s
#convertedDF.groupBy(F.window("ts","4 seconds","1 seconds"), "car_id")

#如果使用处理时间
#convertedDF.groupBy(window(current_timestamp(),"4 seconds"), "car_id")

#执行带有水印的窗口聚合
aggregates = convertedDF \
    .withWatermark("ts", "3 seconds") \
    .groupBy(F.window("ts","4 seconds"), "car_id") \
    .agg(F.avg("speed").alias("speed")) \
    .where("speed > 100") \
    .withColumn("warn",F.concat_ws(", ","car_id","speed"))

aggregates.printSchema()

#将结果写到控制台
writeToConsole = aggregates \
    .writeStream \
    .format("console") \
    .option("truncate", "false") \
    .outputMode("append") \
    .start()

#将结果写到Kafka
writeToKafka = aggregates \
    .selectExpr("CAST(car_id as string) as key", "warn as value") \
    .writeStream \
    .format("kafka") \
    .option("kafka.Bootstrap.servers","xueai8:9092") \
    .option("topic", "fastcars") \
    .option("startingOffsets", "earliest") \
```

```
            .option("checkpointLocation", "/tmp/carsck/") \
            .outputMode("update") \
            .start()

#等待多个流结束（作为作业文件提交集群上运行时启用）
#在 PySpark Shell 命令行，在查询对象上调用.stop()方法停止查询
#spark.streams.awaitAnyTermination()
```

7.6.3 执行步骤演示

本案例涉及与 Kafka 的集成，以及使用自定义的数据源程序，所以执行步骤稍稍有点复杂，建议按以下说明操作。

1. 测试数据源

本案例用到一个 Kafka 生产者程序。该程序是以 Java 开发、编译和打包的 JAR 包，它位于配套的源码包中，名为 fastcars_fat.jar，可使用 JAR 命令来运行它。建议按以下步骤测试和掌握该数据源程序的使用。

（1）启动 ZooKeeper。在第 1 个终端窗口，执行的命令如下：

```
$ cd ~/bigdata/kafka_2.12-2.4.1
$ ./bin/zookeeper-server-start.sh config/zookeeper.properties
```

这将在 2181 端口启动 ZooKeeper 进程，并让 ZooKeeper 在后台工作。

（2）接下来，启动 Kafka 服务器。在第 2 个终端窗口，执行的命令如下：

```
$ cd ~/bigdata/kafka_2.12-2.4.1
$ ./bin/kafka-server-start.sh config/server.properties
```

（3）创建主题 cars。在第 3 个终端窗口，执行的命令如下：

```
$ cd ~/bigdata/kafka_2.12-2.4.1
$ ./bin/kafka-topics.sh --create --zookeeper localhost:2181
--replication-factor 1 --partitions 1 --topic cars
```

```
#查看已有的主题
$ ./bin/kafka-topics.sh --list --zookeeper localhost:2181
```

（4）运行 Kafka 消费者脚本，指定要连接的 Kafka 服务器名称和端口号，以及要连接的主题。继续在第 3 个终端窗口，执行的命令如下：

```
$ ./bin/kafka-console-consumer.sh --topic cars --Bootstrap-server
localhost:9092
```

（5）运行模拟数据源程序 JAR 包。打开第 4 个终端窗口，执行的命令如下：

```
$ Java -jar ~/jars/fastcars_fat.jar localhost:9092 cars
```

其中参数 localhost:9092 是要连接的 Kafka 服务器名称和端口号，参数 cars 是连接的主题。这两个参数可以根据自己的需要修改，但是要和消费者监听的服务器和主题一致。

执行过程如图 7-19 所示。

图 7-19　执行模拟数据源程序

（6）回到第 3 个终端窗口（运行消费者脚本的窗口），可以看到输出了接收的数据信息，如图 7-20 所示。

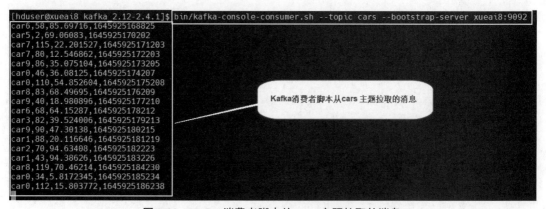

图 7-20　Kafka 消费者脚本从 cars 主题拉取的消息

（7）如果要终止数据源程序或消费者脚本的运行，同时按下 Ctrl + C 快捷键即可。

2. 执行流程序

接下来，就可以运行 7.6.2 节开发的 PySpark 结构化流程序，来实时监测运输车辆的车速,并及时将监测到的超速车辆信息发送到 Kafka 的 fastcars 主题。建议按以下步骤操作。

(1)启动 ZooKeeper(如果已经启动,略过此步骤)。在第 1 个终端窗口,执行的命令如下:

```
$ cd ~/bigdata/kafka_2.12-2.4.1
$ ./bin/zookeeper-server-start.sh config/zookeeper.properties
```

这将在 2181 端口启动 ZooKeeper 进程,并让 ZooKeeper 在后台工作。

(2)接下来,启动 Kafka 服务器(如果已经启动,略过此步骤)。在第 2 个终端窗口,执行的命令如下:

```
$ cd ~/bigdata/kafka_2.12-2.4.1
$ ./bin/kafka-server-start.sh config/server.properties
```

(3)创建主题 fastcars(如果已经创建,略过此步骤)。在第 3 个终端窗口,执行的命令如下:

```
$ cd ~/bigdata/kafka_2.12-2.4.1
$ ./bin/kafka-topics.sh --create --zookeeper localhost:2181
--replication-factor 1 --partitions 1 --topic fastcars

#查看已有的主题
$ ./bin/kafka-topics.sh --list --zookeeper localhost:2181
```

(4)运行 Kafka 消费者脚本,监听超速车辆信息。继续在第 3 个终端窗口,执行的命令如下:

```
$ ./bin/kafka-console-consumer.sh --topic fastcars --Bootstrap-server xueai8:9092
```

(5)运行流程序。打开第 4 个终端窗口,启动 PySpark Shell,运行 7.6.2 节开发的流程序代码(也可以保存到.ipynb 文件中,使用 spark-submit 命令提交执行)。

(6)运行模拟数据源程序 JAR 包。打开第 5 个终端窗口,执行的命令如下:

```
$ Java -jar ~/jars/fastcars_fat.jar localhost:9092 cars
```

(7)回到第 3 个终端窗口,如果一切正常,则可以看到接收的超速信息,如图 7-21 所示。

图 7-21 Kafka 消费者脚本从 fastcars 主题拉取的超速消息

(8) 如果要结束程序的运行，则可按以下顺序关闭：

第 1 步，先关闭数据源程序，同时按下 Ctrl + C 快捷键。

第 2 步，再关闭流处理程序，键入 exit() 函数，然后按 Enter 键。

第 3 步，然后关闭消费者脚本程序，同时按下 Ctrl + C 快捷键。

第 4 步，先关闭 Kafka 集群，同时按下 Ctrl + C 快捷键。

第 5 步，最后关闭 ZooKeeper 集群，同时按下 Ctrl + C 快捷键。

第 8 章 PySpark 大数据分析综合案例

CHAPTER 8

本章的综合案例涉及数据的采集（使用爬虫程序）、数据集成、数据预处理、大数据存储、Hive 数据仓库应用、大数据 ELT 实现和大数据结果展现等全流程所涉及的各种典型操作，涵盖 Linux、MySQL、Hadoop 3、Flume、PySpark 3.x.x、Hive、Flask Web 框架、ECharts 组件和 PyCharm、Zeppelin Notebook 等系统和软件的使用方法。通过本项目，将有助于读者综合运用主流大数据技术及各种工具软件，掌握大数据离线批处理的全流程操作。

21min

8.1 项目需求说明

就业是最大的民生，是社会稳定的主要支撑。在经历新冠肺炎疫情的两年多里，就业压力越发明显，毋庸置疑，2022 年就业形势会更严峻。

从新增数据看，2022 年，全国大学生毕业人数将首次突破 1000 万人，达到惊人的 1020 万人，其中高学历毕业生人数，985、211 高校毕业生达到 75 万人。其他一本毕业生人数超过 100 万人，二本毕业生人数高达 370 万人。专科大学生毕业人数近 460 万人之多。2022 年，全国各大高校将近有 130 万名研究生毕业。每年约 85% 的应届毕业生会选择直接工作。

据 2020 年年底教育部统计，2019 年度我国出国留学人员总数约为 70.35 万人，较上一年度增长 6.25%；各类留学回国人员总数为 58.03 万人。基于该数据推测，大概会有 100 多万国外留学人员和其他人员回国求职。

综合测算，新增就业大学生人数将超过 1100 万人。

面对如此严峻的就业形势，求职乃至职业选择，对于求职者来讲就显得至关重要，所以招聘需求分析应成为每个职场人和将要走入职场者关注的焦点，要不断思考和回答：在特定的发展阶段及特定的文化背景下，面对变动的市场环境和弹性的岗位要求，企业到底需要什么样的求职者？不同城市、不同行业、不同岗位对求职者的要求及薪酬待遇是什么样的？国企和民企、内资和外资企业该怎么选？

因此，广大求职者及至整个社会都需要及时地分析行业企业招聘信息，为求学、求职提供指导性建议。本案例综合运用大数据分析和可视化技术，对使用爬虫程序从互联网上采集到的某头部招聘网站招聘岗位数据进行多维度分析，并可视化展示分析结果，以供有需要的用户参考。

8.2 项目架构设计

本项目涵盖了大数据处理的完整流程，包括数据采集（爬取）、数据集成、数据 ETL、Hive 数据仓库、数据清洗和预处理、数据分析、数据分析结果可视化等步骤，涉及的技术包括 Python 爬虫程序开发、Flume 数据采集技术、PySpark SQL 实现 ETL 数据抽取、Hive 数据仓库技术、Flask Web 开发及 Apache ECharts 图表应用技术等。通过本项目，读者可以完整地掌握一个大数据项目的处理流程。

本项目架构流程如图 8-1 所示。

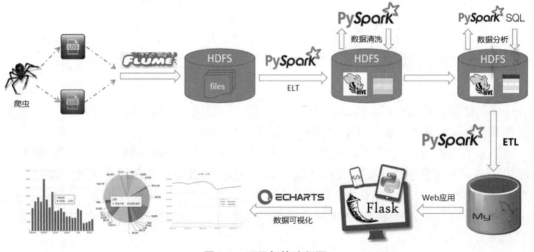

图 8-1 项目架构流程图

项目开发步骤如下：

（1）数据采集：编写 Python 爬虫程序，爬取 51job 网站的 5 个热门城市（北京、上海、广州、深圳、杭州）的热门岗位（大数据、数据分析、Java、Python）的招聘数据，并保存到文件中。

（2）数据集成：使用 Flume 自动监测并将采集到的数据文件导入 HDFS 中存储。

（3）数据 ELT：使用 PySpark 建立 ELT 管道，将集成的数据文件导入 Hive 数据仓库中的 ODS 层。

（4）数据清洗：使用 PySpark + Hive 进行数据清洗和整理。

（5）数据分析：使用 PySpark SQL + Hive 进行数据多维度分析。

（6）数据导出：使用 PySpark 建立 ETL 管道，将分析结果导入 MySQL 数据库。
（7）数据可视化：使用 Python Flask + ECharts 将分析结果通过网页可视化展示。

8.3 项目实现：数据采集

编写 Python 爬虫程序，从 51job 招聘网站上爬取 5 个热门城市（北京、上海、广州、深圳、杭州）的热门岗位（大数据、数据分析、Java、Python）的招聘信息。

本案例提供了两套实现方案。一个是使用 requests 库编写爬虫程序，另一个是使用 Scrapy 框架编写爬虫程序。大家根据自己的需要选择其中一个即可。

8.3.1 爬虫程序实现：使用 requests 库

Requests 是一个 Python 模块，可以使用它发送各种 HTTP 请求。它是一个易于使用的库，具有许多特性，如从在 url 中传递参数到发送自定义头部和 SSL 验证。

使用 requests 可以发送 HTTP/1.1 请求。可以使用简单的 Python 字典添加头、表单数据、multi-part 文件和参数，并以相同的方式访问响应数据。

在使用 requests 库之前，首先要安装它。在命令行窗口下，执行的安装命令如下：

```
pip install -U requests
```

使用 requests 库访问网络数据非常简单。例如，要使用 requests 库请求网易的网站首页内容，代码如下：

```
import requests

url = "http://www.163.com"
r = requests.get(url)
print(type(r))                  #返回 requests.models.Response
#r.encoding = "utf8"            #设置编码
print(r.text)
```

对以上代码的说明如下：

（1）requests.get()方法返回一个 requests.models.Response 类型的 Python 对象，包含从服务器返回的 HTTP 响应信息，requests 负责解析。

（2）r.text 包含 HTTP response 内容体，为文本格式。

除了 requests.get()方法，还有 request.post()方法，以及通用的 requests.request('GET', url)方法。

【例 8-1】使用 Python 的 requests 库编写爬虫程序，爬取 51job 网站最新的招聘信息，并将爬取到的招聘信息保存到 CSV 格式的文件中。招聘信息限定为热门城市（北京、上海、广州、深圳和杭州）的热门岗位（大数据、数据分析、Java、Python）。

实现代码如下：

```python
#第8章/job51_crawler.py

import requests
from bs4 import BeautifulSoup
import pandas as pd
import time
import random
import re

"""
    通过循环遍历分页爬取多个岗位、多个城市（一线城市）的岗位需求信息，
    并将爬取的数据写入 CSV 文件中存储。
"""

#构造请求头，假装自己是浏览器
headers = {
        "user-agent": "Mozilla/5.0 (Windows NT 10.0; Win64; x64; rv:89.0) Gecko/20100101 Firefox/89.0",
        "Host": "search.51job.com",
        "Referer": "https://search.51job.com",
        "Cookie": "替换为自己浏览器的Cookie",
        "Accept": "application/json, text/javascript, */*; q=0.01",
        "Accept-Encoding": "gzip, deflate, br"
    }

#城市编码
#city_list = ["010000", "020000", "030200", "040000", "080200"]
#城市编码与城市名称的映射
#city_dict = {"010000": "北京", "020000": "上海", "030200": "广州", "040000": "深圳", "080200": "杭州"}

#url
#动态指定查询的岗位关键字和分页数
url = "https://search.51job.com/list/010000%252c020000%252c030200%252c040000%252c080200,000000,0000,00,9,99,{},2,{}.html"

#查询字符串参数
params = {
    "lang": "c",
    "postchannel": "0000",
    "workyear": 99,
    "cotype": 99,
    "degreefrom": 99,
```

```python
    "jobterm": 99,
    "companysize": 99,
    "ord_field": 0,
    "dibiaoid": 0,
    'line': '',
    'welfare': '',
}

#需要查询的岗位关键字
job_list = ["Java", "Python", "数据分析", "大数据"]

#指定存储文件:使用制表位(Tab)作为分隔符
result_file = "zhaopin.tsv"

#先写入标题
file_header = 
['jobid','job_name','job_title','job_href','company_name','company_href',
'providesalary_text','workarea','workarea_text','companytype_text',
'degreefrom','workyear','issuedate','jobwelf','jobwelf_list','attribute_
text', 'companysize_text','companyind_text','label']

with open(result_file, mode='a', encoding='utf-8') as f:
    f.write('\t'.join(file_header))        #先写入标题行
    f.write('\n')
    time_start = time.time()
    try:
        for job_name in job_list:
            #向服务器发起请求,并保存返回的响应对象
            response_job = requests.get(url.format(job_name,1),
headers=headers, params=params)

            #将返回的 JSON 数据转换为 dict
            result = response_job.json()

            ##目标1:获取总页数
            total_page = int(result["total_page"])
            print("#################{}岗位的总页数为{}
#################".format(job_name,total_page))

            ###分页搜索
            for page_index in range(total_page):
                print("#########岗位:{},页码:{}#########".format(job_name,
page_index))
                #向服务器发起请求,并保存返回的响应对象
                response_page = requests.get(url.format(job_name,
page_index), headers=headers, params=params)
```

```
            #将返回的 JSON 数据转换为 dict
            result_page = response_page.json()

            #提取搜索引擎搜索的结果
            engine_search_result = result_page["engine_jds"]

            #遍历一页中的每个岗位招聘信息（每个岗位是个 dict）
            for job_one in engine_search_result:
                try:
                    #取其中 18 个有用的字段
f.write('{}\t{}\t{}\t{}\t{}\t{}\t{}\t{}\t{}\t{}\t{}\t{}\t{}\t{}\
\t{}\t{}\t{}\n'.format(job_one['jobid'], job_one['job_name'],
job_one['job_title'], job_one['job_href'], job_one['company_name'],
job_one['company_href'], job_one['providesalary_text'],
job_one['workarea'], job_one['workarea_text'],
job_one['companytype_text'], job_one['degreefrom'], job_one['workyear'],
job_one['issuedate'], job_one['jobwelf'], job_one['jobwelf_list'],
ob_one['attribute_text'], job_one['companysize_text'],
job_one['companyind_text'], job_name))
                except Exception as e2:
                    print(e2)
                    pass

        time_end = time.time()
        print("共用时{}".format(time_end - time_start))
    except Exception as e:
        print("异常")
        print(e)
```

运行以上的爬虫程序，可以获得一个名为 zhaopin.tsv 的招聘数据存储文件。

8.3.2 爬虫程序实现：使用 Scrapy 框架

Scrapy 是一个异步处理爬虫框架，该框架使用纯 Python 语言编写。Scrapy 常应用在包括数据挖掘、信息处理或存储历史数据等一系列的程序中。通常用户可以很简单地通过 Scrapy 框架实现一个爬取网站数据、提取结构性数据的网络爬虫程序。

Scrapy 框架的架构如图 8-2 所示。

其中各组件的作用说明如下。

（1）Scrapy Engine（引擎）：负责 Spider、Item Pipeline、Downloader、Scheduler 中间的通信，以及信号、数据传递等。

（2）Scheduler（调度器）：它负责接受引擎发送过来的 Request 请求，并按照一定的方式进行整理排列、入队，当引擎需要时，交还给引擎。

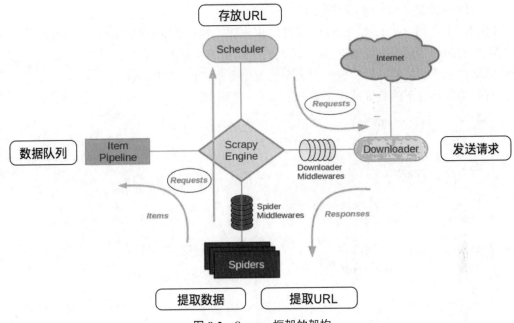

图 8-2　Scrapy 框架的架构

（3）Downloader（下载器）：负责下载 Scrapy Engine 发送的所有 Requests 请求，并将其获取的 Responses 交还给 Scrapy Engine，由引擎交给 Spider 爬虫程序来处理。

（4）Spider（爬虫）：它负责处理所有 Responses，从中分析及提取数据，以便获取 Item 字段需要的数据，并将需要跟进的 URL 提交给引擎，再次进入 Scheduler（调度器）。

（5）Item Pipeline（管道）：它负责处理 Spider 中获取的 Item，并进行后期处理（详细分析、过滤、存储等）的地方。

（6）Downloader Middlewares（下载中间件）：可以将它当作一个可以自定义扩展下载功能的组件。

（7）Spider Middlewares（Spider 中间件）：可以将它理解为一个可以自定义扩展的功能组件（例如进入 Spider 的 Responses，以及从 Spider 出去的 Requests）。

在使用 Scrapy 框架之前，首先要安装它。在命令行窗口下，执行的安装命令如下：

```
pip install -U scrapy
```

开发 Scrapy 爬虫程序的过程共分为 4 步，分别说明如下。

（1）新建项目（scrapy startproject xxx）：新建一个新的爬虫项目。

（2）明确目标（编写 items.ipynb）：明确想要抓取的目标，可理解为爬取数据的数据结构。

（3）制作爬虫（spiders/xxspider.ipynb）：制作爬虫程序，开始爬取网页。

（4）存储内容（pipelines.ipynb）：设计管道存储爬取到的内容。

下面的示例使用 Scrapy 编写爬虫程序。

【例 8-2】使用 Python 的 Scrapy 编写爬虫程序，爬取 51job 网站最新的招聘信息，并将爬取到的招聘信息保存到 CSV 格式的文件中。招聘信息限定为热门城市（北京、上海、广州、深圳和杭州）的热门岗位（大数据、数据分析、Java、Python）。

首先编辑项目下的数据结构文件 items.py，代码如下：

```python
#第8章/scrapy_spider/items.py

import scrapy
from scrapy import Item, Field

class Job51Item(scrapy.Item):
    #define the fields for your item here like:
    jobid = scrapy.Field()
    job_name = scrapy.Field()
    job_title = scrapy.Field()
    job_href = scrapy.Field()
    company_name = scrapy.Field()
    company_href = scrapy.Field()
    providesalary_text = scrapy.Field()
    workarea = scrapy.Field()
    workarea_text = scrapy.Field()
    companytype_text = scrapy.Field()
    degreefrom = scrapy.Field()
    workyear = scrapy.Field()
    issuedate = scrapy.Field()
    jobwelf = scrapy.Field()
    jobwelf_list = scrapy.Field()
    attribute_text = scrapy.Field()
    companysize_text = scrapy.Field()
    companyind_text = scrapy.Field()
    label = scrapy.Field()
```

然后编辑项目下的爬虫程序文件 spiders/jobspider.py 的处理逻辑，代码如下：

```python
#第8章/scrapy_spider/spiders/jobspider.py

import json
from urllib.parse import urlencode

import scrapy
from scrapy import Request
import time
from job51.items import Job51Item
```

```python
class JobspiderSpider(scrapy.Spider):
    #代表爬虫项目名称
    name = 'jobspider'
    #allowed_domains 代表允许爬取的域名
    allowed_domains = ['https://search.51job.com']
    #URL 列表，包含 spider 在启动时爬取的 URL 列表，用于定义初始请求
    url = "https://search.51job.com/list/{},000000,0000,00,9,99,{},2,{}.html?{}"
    city = ""                    #城市参数
    job_name = ""                #岗位参数
    page_index = 1               #分页索引
    params = ""                  #查询字符串参数

    #构造 URL 列表
    def start_requests(self):
        #查询字符串参数
        job_params = {
            "lang": "c",
            "postchannel": "0000",
            "workyear": 99,
            "cotype": 99,
            "degreefrom": 99,
            "jobterm": 99,
            "companysize": 99,
            "ord_field": 0,
            "dibiaoid": 0,
            'line': '',
            'welfare': '',
        }
        self.params = urlencode(job_params)
        #需要查询的热门城市（一线城市）
        city_list = ["010000", "020000", "030200", "040000", "080200"]
        #需要查询的岗位关键字
        job_list = ["Java", "Python", "数据分析", "大数据"]
        for city in city_list:
            self.city = city
            for job_name in job_list:
                self.job_name = job_name
                yield \
                    scrapy.Request(url=self.url.format(self.city,self.job_name,1,self.params),dont_filter=True, callback=self.parse)

    #parse 方法是每个爬行器的核心
    #每次 Scrapy 下载一个 URL 时都会调用这种方法，大多数情况下是用这种方法编写提取代码的
```

```python
#当启动爬虫时,会自动执行parse函数,返回start_urls的response
def parse(self, response):
    #print(response.text)
    result = json.loads(response.text)
    #print(result)
    ##目标1:获取总页数
    total_page = int(result["total_page"])
    print("总页数: ", total_page)
    ##分页搜索
    for page_index in range(total_page):
        yield Request(url=self.url.format(self.city, self.job_name, page_index, self.params), dont_filter=True, callback=self.parseJSON)

#解析返回的JSON数据
def parseJSON(self, response):
    #加载为json
    result_page = json.loads(response.text)
    #提取搜索引擎搜索的结果
    #engine_search_result = result_page["engine_search_result"]
    engine_search_result = result_page["engine_jds"]
    print(engine_search_result)

    #遍历一页中的每个岗位招聘信息(每个岗位是个dict)
    for job_one in engine_search_result:
        item = Job51Item()
        try:
            #取其中18个有用的字段
            item["jobid"] = job_one['jobid']
            item["job_name"] = job_one['job_name']
            item["job_title"] = job_one['job_title']
            item["job_href"] = job_one['job_href']
            item["company_name"] = job_one['company_name']
            item["company_href"] = job_one['company_href']
            item["providesalary_text"] = job_one['providesalary_text']
            item["workarea"] = job_one['workarea']
            item["workarea_text"] = job_one['workarea_text']
            item["companytype_text"] = job_one['companytype_text']
            item["degreefrom"] = job_one['degreefrom']
            item["workyear"] = job_one['workyear']
            item["issuedate"] = job_one['issuedate']
            item["jobwelf"] = job_one['jobwelf']
            item["jobwelf_list"] = job_one['jobwelf_list']
            item["attribute_text"] = job_one['attribute_text']
            item["companysize_text"] = job_one['companysize_text']
            item["companyind_text"] = job_one['companyind_text']
```

```
            item["label"] = self.job_name
        except Exception as e2:
            print(e2)
            pass
    print(item)
    #将结果内容以Item的形式返回
    yield item
```

接下来编辑项目下的数据处理管道文件pipelines.py的处理逻辑,代码如下:

```
#第8章/scrapy_spider/pipelines.py

from itemadapter import ItemAdapter

#pipelines是数据的输出模块
#爬虫生成的数据,需要通过pipelines输到文件或数据库等
#运行爬虫时,会自动执行process_item函数,将Item输到不同的文件中
#在close_spider函数中记得对资源进行close
class Job51Pipeline:
    def __init__(self):
        self.csvFileName = "zhaopin.tsv"
        #先写入标题
        self.file_header = ['jobid', 'job_name', 'job_title', 'job_href',
'company_name', 'company_href', 'providesalary_text', 'workarea',
'workarea_text', 'companytype_text', 'degreefrom', 'workyear', 'issuedate',
'jobwelf', 'jobwelf_list', 'attribute_text', 'companysize_text',
'companyind_text', 'label']

    def open_spider(self, spider):
        #打开文件
        self.file = open(self.csvFileName, 'w', encoding='utf-8')
        self.file.write('\t'.join(self.file_header))   #先写入标题行
        self.file.write('\n')

    def process_item(self, item, spider):
        #读取item中的数据,一个item写一行
        #写入文件
        try:
            #取其中18个有用的字段
            self.file \
                .write('{}\t{}\t{}\t{}\t{}\t{}\t{}\t{}\t{}\t{}\t{}\t{}\t{}\t{}\t{}\t{}\t{}\t{}\n'.format(item['jobid'], item['job_name'],
item['job_title'],item['job_href'],item['company_name'],item['company_href'], item['providesalary_text'],item['workarea'], item['workarea_text'], item['companytype_text'], item['degreefrom'], item['workyear'],
```

```
        item['issuedate'], item['jobwelf'], item['jobwelf_list'], item['attribute_
text'], item['companysize_text'], item['companyind_text'], item['label']))
        except Exception as e2:
            print(e2)
            pass
    #返回item
    return item

def close_spider(self, spider):
    self.file.close()
```

运行以上的爬虫程序,可以获得一个名为 zhaopin.tsv 的招聘数据存储文件。

8.4 项目实现:数据集成

本步骤使用 Flume 自动监测并将采集到的数据文件导入 HDFS 中存储。这一步是可选的,用户可根据自己的要求决定是否采用。如果没有要求,则可以直接采用 HDFS Shell 命令将采集到的数据上传到 HDFS 上存储。

8.4.1 Flume 简介

Flume 是一个高可用的,高可靠的,分布式的海量日志采集、聚合和传输的系统,Flume 支持在日志系统中定制各类数据发送方,用于收集数据;同时,Flume 提供对数据进行简单处理并写到各种数据接受方(可定制)的能力。

Flume 主要由以下 3 个重要的组件构成。

(1) Source:完成对日志数据的收集,分成 transition 和 event 打入 channel 之中。

(2) Channel:主要提供一个队列功能,对 source 提供的数据进行简单的缓存。

(3) Sink:取出 Channel 中的数据,存储到相应的存储文件系统、数据库或者提交到远程服务器。

Flume 架构如图 8-3 所示。

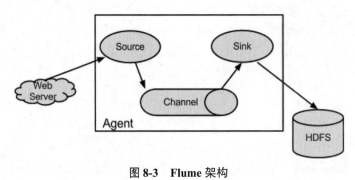

图 8-3 Flume 架构

8.4.2 安装和配置 Flume

首先从 Flume 的官网下载并安装包（本书使用的是 apache-flume-1.9.0-bin.tar.gz），并将安装包上传到 Linux 系统中，然后按以下步骤安装和配置 Flume。

（1）切换到安装包所在目录，解压下载的安装包，命令如下：

```
$ tar -zxvf apache-flume-1.9.0-bin.tar.gz -C /home/hduser/bigdata
```

上面将 apache-flume-1.9.0-bin.tar.gz 解压到了 /home/hduser/bigdata 目录下，然后重命名文件夹，以简化引用，命令如下：

```
$ cd /home/hduser/bigdata/
$ mv apache-flume-1.9.0-bin flume-1.9.0
```

（2）现在编辑/etc/profile 文件（或.bashrc 文件）以更新 Apache Flume 的环境变量，以便可以从任何目录访问它。例如，在 nano 编辑器中打开 profile 文件，命令如下：

```
$ nano /etc/profile
```

在打开的文件中，添加如下几行内容：

```
export FLUME_HOME=/home/hduser/bigdata/flume-1.9.0
export FLUME_CONF_DIR=$FLUME_HOME/conf
export PATH=$PATH:$FLUME_HOME/bin
```

然后按下 Ctrl + O 快捷键保存修改，按下 Ctrl + X 快捷键退出 nano 编辑器，并执行 source 命令使环境变量生效，命令如下：

```
$ source /etc/profile
```

（3）修改 flume-env.sh 配置文件。

Flume 安装包中默认没有 flume-env.sh 配置文件，用户需要复制 flume-env.sh.template 文件并去掉后缀.template，命令如下：

```
$ cd /home/hduser/bigdata/flume-1.9.0/conf
$ cp ./flume-env.sh.template ./flume-env.sh
$ nano ./flume-env.sh
```

使用编辑器打开 flume-env.sh 文件后，在文件的最后增加下面这行内容，用于设置 JAVA_HOME 变量：

```
export JAVA_HOME=/opt/Java/jdk1.8.0_281
```

注意：这里应修改为读者自己的 JDK 安装目录。

然后按下 Ctrl + O 快捷键保存修改，按下 Ctrl + X 快捷键退出 nano 编辑器。

（4）验证 Flume 的安装是否成功。在命令行执行的命令如下：

```
$ flume-ng version
```

如果安装成功，则应该出现 Flume 的版本信息，类似下面的内容：

```
Flume 1.9.0
Source code repository: https://git-wip-us.apache.org/repos/asf/flume.git
Revision: d4fcab4f501d41597bc616921329a4339f73585e
Compiled by fszabo on Mon Dec 17 20:45:25 CET 2018
From source with checksum 35db629a3bda49d23e9b3690c80737f9
```

如果看到 Flume 的版本信息，则说明 Apache Flume 已经安装成功了。

8.4.3 实现数据集成

接下来，将 Flume 集成到项目中，使用 Flume 监视招聘数据文件所在目录，并将新增加的招聘数据文件自动上传到指定的 HDFS 目录中。

【例 8-3】编写 Flume 配置文件，将 source 类型指定为 spoolDir，并将 sink 类型指定为 hdfs。

首先，在 $FLUME_HOME/conf/ 目录下，创建一个名为 zhaopin.conf 的纯文本文件，并编辑内容如下：

```
#第 8 章/zhaopin.conf

#列出 Agent 包含的 sources、sinks 和 channels
LogAgent.sources = mysource
LogAgent.channels = mychannel
LogAgent.sinks = mysink

#为 source 设置 channel
LogAgent.sources.mysource.channels = mychannel
#配置名为"mysource"的 source，采用 spooldir 类型，从本地目录~/job51 获取数据
LogAgent.sources.mysource.type = spooldir
LogAgent.sources.mysource.spoolDir = /home/hduser/job51

#为 sink 设置 channel
LogAgent.sinks.mysink.channel = mychannel
#配置名为"mysink"的 sink，将结果写入 HDFS 中，每个文件 10 000 行数据
LogAgent.sinks.mysink.type = hdfs
LogAgent.sinks.mysink.hdfs.fileType = DataStream
LogAgent.sinks.mysink.hdfs.path = hdfs://localhost:8020/job51
LogAgent.sinks.mysink.hdfs.batchSize = 1000
LogAgent.sinks.mysink.hdfs.rollSize = 0
LogAgent.sinks.mysink.hdfs.rollCount = 10000
LogAgent.sinks.mysink.hdfs.useLocalTimeStamp = true

#配置名为"mychannel"的 channel，采用 memory 类型
LogAgent.channels.mychannel.type = memory
LogAgent.channels.mychannel.capacity = 10 000
```

上面配置的含义是，将/home/hduser/job51/目录下新增的数据文件自动上传到 HDFS 文件系统的/job51/目录下。

然后启动 Flume Agent。将目录切换到 Flume 的安装目录，在命令行执行的命令如下：

```
$ ./bin/flume-ng agent -c ./conf -f ./conf/zhaopin.conf -n LogAgent
-Dflume.root.logger=INFO,console
```

最后，将爬虫程序爬取的 zhaopin.tsv 文件复制到/home/hduser/job51/目录下，然后 Flume 会自动将其复制/上传到 hdfs://localhost:8020/job51/目录下(Flume Sink 指定的位置)。

8.5 项目实现：数据 ELT

到此为止，已经爬取了热门城市的热门岗位数据，并保存到了 HDFS 分布式文件系统中。接下来使用 PySpark 编写批处理程序，建立 ELT 数据处理管道，将集成到 HDFS 中的数据文件导入 Hive 数据仓库中的 ODS 层。

注意：ETL（Extract-Load-Transform）用来描述将数据从来源端经过抽取（extract）、加载（load）、转换（transform）至目的端的过程。

PySpark SQL 内置了对众多常用数据源的支持，可以很容易地使用它来构建 ELT 管道。

【例 8-4】编写 PySpark 批处理程序，建立 ELT 数据处理管道，将集成到 HDFS 中的数据文件导入 Hive 数据仓库中的 ODS 层。

首先定义一个用于数据处理的工具类，代码如下：

```python
#第8章/elt_util.py

from pyspark.sql import *
from pyspark.sql.types import *

#ETL 工具类
class EltUtil(object):
    """ELT 工具类"""

    def extractFileToDataFrame(self, spark, fileMap):
        """读取文件数据源，创建 DataFrame
        Args:
            spark:    SparkSession 实例
            fileMap: 要加载的文件选项，字典类型

        Returns:
            返回一个 Data Frame
        """
```

```python
        df = spark \
            .read \
            .option("header", fileMap["header"]) \
            .option("inferSchema", True) \
            .option("sep", fileMap["sep"]) \
            .csv(fileMap["filePath"])

        #返回
        return df

    def extractJDBCToDataFrame(self, spark, jdbcMap):
        """读取 JDBC 数据源,创建 DataFrame
        Args:
            spark:    SparkSession 实例
            jdbcMap:  要加载的 JDBC 配置项,字典类型

        Returns:
            返回一个 Data Frame
        """
        df = spark.read.format("jdbc").options(jdbcMap).load()
        return df

    def loadToHive(self, spark, dataframe, hiveMap):
        """将 DataFrame 导入 Hive 的 ODS 层
        Args:
            spark:     SparkSession 实例
            dataframe: 要导出的 DataFrame
            hiveMap:   要加载的 Hive 配置项,字典类型

        Returns:
            None
        """
        database = hiveMap["db"]     #要写入的数据库
        table = hiveMap["tb"]        #要写入的数据表
        column = hiveMap.get("partitionColumn")   #分区列

        #切换数据库
        spark.sql(f"create database if not EXISTS {database}")
        spark.sql(f"use {database}")

        #因为有的表需要分区,有的表不需要,所以要进行判断
        #这里使用 saveAsTable() 方法将 DataFrame 数据保存到 Hive 表中
        if column is None:
            dataframe.write \
                .format("parquet") \
                .mode("overwrite") \
                .saveAsTable(table)
```

```
            else:
                dataframe.write \
                    .format("parquet") \
                    .mode("overwrite") \
                    .partitionBy(column) \
                    .saveAsTable(table)

    def eltFromFileToHive(self, spark, fileMap, hiveMap):
        """定义一个ELT方法，包含extract from file + load to Hive
        Args:
            spark:      SparkSession 实例
            fileMap:    要加载的文件选项，字典类型
            hiveMap:    要加载的Hive配置项，字典类型

        Returns:
            None
        """
        #调用抽取文件数据源的方法，返回一个DataFrame
        df = self.extractFileToDataFrame(spark, fileMap)
        #装载到数据仓库
        self.loadToHive(spark, df, hiveMap)

    def eltFromJDBCToHive(self, spark, jdbcMap, hiveMap):
        """定义一个ELT方法，包含extract from jdbc + load to Hive
        Args:
            spark:      SparkSession 实例
            jdbcMap:    要加载的JDBC选项，字典类型
            hiveMap:    要加载的Hive配置项，字典类型

        Returns:
            None
        """
        #调用抽取文件数据源的方法，返回一个DataFrame
        df = self.extractJDBCToDataFrame(spark, jdbcMap)
        #装载到数据仓库
        self.loadToHive(spark, df, hiveMap)
```

然后定义主业务逻辑，调用上面的工具方法实现ELT任务，代码如下：

```
#第8章/job51_elt.py

from elt_util import EltUtil
from pyspark.sql import SparkSession

#创建连接Spark环境的对象SparkSession
def createSparkSession():
    spark = SparkSession.builder.appName("51job"). \
```

```
            config('hive.exec.dynamici.partition', True). \
            config('hive.exec.dynamic.partition.mode', 'nonstrict'). \
            enableHiveSupport(). \
            getOrCreate()
    return spark

#执行 ELT,将数据从文件装载到 Hive 的 ODS 层
def eltToODS(spark):
    #指定要抽取的文件数据源
    filePath = "zhaopin.tsv"
    fileOptions = {"filePath": filePath, "header": "true", "sep": "\t"}
    
    #定义要装载的 Hive 配置项
    hiveOptions = {"db": "job51_db",
                   "tb": "ods_job",
                   "partitionColumn": "label"}
    
    #ELT
    EltUtil().eltFromFileToHive(spark, fileOptions, hiveOptions)

#测试 ELT 结果
def test_eltToODS(spark):
    spark.table("job51_db.ods_job").show(5,truncate=False)

if __name__ == "__main__":
    #创建连接 Spark 环境的对象 SparkSession
    spark = createSparkSession()
    print(spark.version)
    #将招聘数据加载到 Hive 的 ODS 层
    eltToODS(spark)
    test_eltToODS(spark)
```

将 job51_elt.py 提交到 PySpark 集群上运行,它将把 zhaopin.tsv 文件中的数据抽取到 Hive 的 job51_db.ods_job 表中并分区存储。在运行这个任务时,有两点需要注意:

(1) 要确保已经启动了 Hive Metastore 服务。启动该服务的命令如下:

```
$ hive --service metastore
```

(2) 提前在 Hive 中创建数据库 job51_db。

8.6 项目实现:数据清洗与整理

现在招聘数据已经抽取到了 Hive 的 ODS 层,但是还不能直接对数据进行分析,因为原始爬取的数据总是存在着各种各样的问题,如有可能包含重复的招聘信息,有可能包含

错误的招聘信息。另外，类似薪资这样的数据，还必须先提取出来，才能进一步分析，因此有必要对数据进行清洗和整理。

【例 8-5】编写 PySpark 批处理程序，使用 PySpark 对大数据进行清洗，包括去重、错误数据处理、空值处理、属性转换、属性提取等数据预处理任务。

在项目的这一步骤中，需要完成了以下几项预处理任务：

（1）薪资数据的抽取和转换。在爬虫程序爬取的原始招聘数据中，薪资的表示形式和单位都不统一，有日薪、月薪、年薪，有的单位为千元而有的单位为万元，有的是固定薪资，有的是给出薪资范围，因此，要将这些不统一的薪资表示进行抽取和转换，统一到以元为单位，包含最低月薪和最高月薪。

（2）数据去重。在爬虫程序爬取的原始招聘数据中，有的招聘信息是重复的，为此需要对数据进行去重处理。这里判断是否重复的依据是 jobid 字段。

（3）工作城市抽取。在有的招聘信息中，工作地点细分为区，例如北京市-海淀区、北京市-朝阳区等，需要统一到城市维度，所以对这个字段（workarea_city）只取城市名。

（4）删除无意义的数据。在对招聘信息进行分析时，薪资是非常重要的一个分析指标，但是在爬虫程序爬取的原始招聘数据中，有的招聘信息并没有给出薪资信息。这里认为这些无薪资信息的招聘信息为无效数据，予以删除处理。

（5）删除非关注城市的招聘信息。本案例只分析一线热门城市"北京""上海""广州""深圳""杭州"的招聘信息，但是在爬虫程序爬取的原始招聘数据中，有的招聘信息的招聘地点不属于这 5 个热门一线城市，因此需要予以去除。

（6）最后，是属性选择。根据项目的需求，只选择保留其中对于后续分析有意义的 11 个属性，分别是 jobid、job_name、company_name、providesalary_text、workarea_city、companytype_text、jobwelf、attribute_text、companysize_text、companyind_text、label 字段。

在些预处理任务中，最复杂的是薪资的抽取和转换，因此，在下面的代码中，将这个处理封装到了一个单独的方法 processSalaryText(wage)中，并将其注册为一个 UDF 函数。

最终实现的代码如下：

```
#第8章/job51_clean.py

from elt_util import EltUtil
from pyspark.sql import SparkSession
from pyspark.sql.functions import *
from pyspark.sql.types import *

#创建连接Spark环境的对象SparkSession
def createSparkSession():
    spark = SparkSession.builder.appName("51job"). \
        config('spark.driver.maxResultSize', '512m'). \
        config("spark.driver.memory", "2g"). \
        config("spark.executor.memory", "2g"). \
```

```
            config('hive.exec.dynamici.partition', True). \
            config('hive.exec.dynamic.partition.mode', 'nonstrict'). \
            enableHiveSupport(). \
            getOrCreate()
    sc = spark.sparkContext
    sc._jsc.hadoopConfiguration().setInt("parquet.block.size", 512 * 1024)
    return spark


#自定义薪资转换函数(UDF),将薪资字段转换为最低月薪和最高月薪
def processSalaryText(wage):
    min_wage = 0.0    #最低薪资
    max_wage = 0.0    #最高薪资

    #如果薪资不为空且包含中画线(-)
    if wage.strip() and "-" in wage:
        #如果是月薪
        if wage.endswith("月"):
            if "万" in wage:
                min_wage = float(wage[0:wage.index("-")]) * 10000
                max_wage = float(wage[wage.index("-") + 1:wage.index("万")]) * 10000
            else:
                min_wage = float(wage[0:wage.index("-")]) * 1000
                max_wage = float(wage[wage.index("-") + 1:wage.index("千")]) * 1000
        #如果是年薪
        else:
            if "万" in wage:
                min_wage = float(wage[0:wage.index("-")]) * 10000 / 12
                max_wage = float(wage[wage.index("-") + 1:wage.index("万")]) * 10000 / 12
            else:
                min_wage = float(wage[0:wage.index("-")]) * 1000 / 12
                max_wage = float(wage[wage.index("-") + 1:wage.index("千")]) * 1000 / 12

    #如果薪资不为空且不包含中画线(-)
    elif wage.strip() and "-" not in wage:
        #如果是月薪
        if wage.endswith("月"):
            if "万" in wage:
                min_wage = float(wage[0:wage.index("万")]) * 10000
            else:
                min_wage = float(wage[0:wage.index("千")]) * 1000
            max_wage = min_wage
```

```python
        #如果是年薪
        elif wage.endswith("年"):
            if "万" in wage:
                min_wage = float(wage[0:wage.index("万")]) * 10000 / 12
            else:
                min_wage = float(wage[0:wage.index("千")]) * 1000 / 12
            max_wage = min_wage
        #如果是日薪
        elif wage.endswith("天"):
            min_wage = float(wage[0:wage.index("元")]) * 22.5
            max_wage = min_wage
        #如果是时薪
        elif wage.endswith("小时"):
            min_wage = float(wage[0:wage.index("元")]) * 22.5 * 6
            max_wage = min_wage

    return (min_wage, max_wage)


#将数据写入Hive数据仓库
def writeToHiveDW(spark,dataframe):
    #定义要装载的Hive配置项
    hiveOptions = {
        "db": "job51_db",
        "tb": "f_job_cleaned",
        "partitionColumn": "label"
    }
    #将清洗过后的数据传到Hive数据仓库中
    EltUtil().loadToHive(spark, dataframe, hiveOptions)


#测试ELT结果
def test_writeToHiveDW(spark):
    spark.table("job51_db.f_job_cleaned").show(5,truncate=False)


if __name__ == "__main__":
    #创建连接Spark环境的对象SparkSession
    spark = createSparkSession()
    #从Hive的ODS层将招聘数据加载到DataFrame
    df1 = spark.table("job51_db.ods_job")
    #数据去重(根据jobid去重)
    df2 = df1.dropDuplicates(["jobid"])
    #预处理 - 过滤掉错误（或无意义）的数据
    df3 = df2.filter(col("providesalary_text").isNotNull())
    #处理workarea_text字段只取城市名
```

```python
    df4 = df3.withColumn("workarea_city", split(col("workarea_text"),
"-")).getItem(0)
    hot_cities = ["北京", "上海", "广州", "深圳", "杭州"]
    df5 = df4.filter(col("workarea_city").isin(hot_cities))
    #属性选择
    cols = ["jobid", "job_name", "company_name", "providesalary_text",
"workarea_city", "companytype_text", "jobwelf", "attribute_text",
"companysize_text", "companyind_text", "label"]
    df6 = df5.select(cols)   #只留 11 个属性字段
    #注册 UDF
    convert_salary_udf = udf(processSalaryText, ArrayType(DoubleType()))
    df7 = df6 \
        .withColumn("salary_min",(convert_salary_udf("providesalary_text"))
[0]) \
        .withColumn("salary_max",(convert_salary_udf("providesalary_text"))
[1]) \
        .drop("providesalary_text")
    #将清洗和整理后的数据写入数据仓库
    writeToHiveDW(spark, df7)
    #测试写入
test_writeToHiveDW(spark)
```

将 job51_clean.py 提交到 PySpark 集群上运行，它将从 Hive 中把 job51_db.ods_job 表数据加载到内存中进行处理，并把清洗之后的分区数据存储到 Hive 的 job51_db.f_job_cleaned 表中。在运行这个任务时，要确保已经启动了 Hive Metastore 服务。启动该服务的命令如下：

```
$ hive --service metastore
```

8.7 项目实现：数据分析

经过 8.6 节的清洗和整理，现在数据已经准备好，可以用于业务分析了。

使用 PySpark SQL + Hive 从多个维度对整理后的数据集进行分析，并将分析结果存入数据集，其中部分分析维度如下：

（1）按岗位类别（Java、Python、数据分析、大数据）统计招聘岗位数。
（2）按城市（北京、上海、广州、深圳、杭州）统计招聘岗位数。
（3）按城市和岗位类别统计招聘岗位数。
（4）按公司类型（民营公司、上市公司等）统计招聘岗位数。
（5）按公司类型和岗位类别统计招聘岗位数。
（6）统计每个城市各种类型公司的数量。
（7）查看公司规模。
（8）统计数据分析相关岗位在不同行业的招聘数量。

(9)统计大数据相关岗位在不同行业的招聘数量。
(10)统计 Python 相关岗位在不同行业的招聘数量。
(11)统计 Java 相关岗位在不同行业的招聘数量。
(12)以各种方式统计平均薪资,如按城市,按岗位类别,按城市和岗位类别,按公司类型,按行业类型,等等。

【例 8-6】 编写 PySpark 批处理程序,从多个维度对整理后的数据集进行分析,并将分析结果存入数据集。

实现代码如下:

```python
#第8章/job51_analysis.py

from elt_util import EltUtil
from pyspark.sql import SparkSession
from pyspark.sql.functions import *
from pyspark.sql.types import *

#创建连接 Spark 环境的对象 SparkSession
def createSparkSession():
    spark = SparkSession.builder.appName("51job"). \
        config('spark.driver.maxResultSize', '512m'). \
        config("spark.driver.memory", "2g"). \
        config("spark.executor.memory", "2g"). \
        enableHiveSupport(). \
        getOrCreate()
    sc = spark.sparkContext
    sc._jsc.hadoopConfiguration().setInt("parquet.block.size", 512 * 1024)
    return spark

#按岗位类别(Java、Python、数据分析、大数据)统计招聘岗位数
def post_by_job_type(spark):
    #按岗位类别(Java、Python、数据分析、大数据)统计招聘岗位数
    sql = """
        select label as post_type, count(*) as post_number
        from job_tb
        group by label
    """
    postByTypeDF = spark.sql(sql)
    postByTypeDF.show()
    return postByTypeDF

#按企业类型统计招聘岗位数
def post_by_companyind(spark):
    sql = """
```

```
            select label as post_type, companyind_text as companyind, count(*) 
as post_num
                from job_tb
                group by label, companyind_text;
        """
        postByCompanyindDF = spark.sql(sql)
        postByCompanyindDF.show()
        return postByCompanyindDF

#将分析结果写入数据集
def writeToHiveMart(spark,dataframe,table):
    #定义要装载的Hive配置项
    hiveOptions = {
        "db": "job51_db",
        "tb": table
    }
    #将清洗过后的数据传到Hive数据仓库中
    EltUtil().loadToHive(spark, dataframe, hiveOptions)

#测试ELT结果
def test_writeToHiveMart(spark, table):
    spark.table(f"job51_db.{table}").show(5,truncate=False)

if __name__ == "__main__":
    #创建连接Spark环境的对象SparkSession
    spark = createSparkSession()
    #加载清洗后的数据集
    data = spark.table("job51_db.f_job_cleaned")
    #注册临时表
    data.createOrReplaceTempView("job_tb")
    #按岗位类别（Java、Python、数据分析、大数据）统计招聘岗位数
    post_by_job_type_df = post_by_job_type(spark)
    #将分析结果1写入数据集
    writeToHiveMart(spark, post_by_job_type_df, "m_job_post_tb")
    #测试写入
    test_writeToHiveMart(spark, "m_job_post_tb")
    #所有类型招聘的行业分布数据
    post_by_companyind_df = post_by_companyind(spark)
    #将分析结果2写入数据集
    writeToHiveMart(spark, post_by_companyind_df, "m_post_by_companyind")
    #测试写入
    test_writeToHiveMart(spark, "m_post_by_companyind")
```

因为这些分析任务很相似，为了清晰起见，在上面的代码中只提供了两个实现：

(1) 按岗位类别（Java、Python、数据分析、大数据）统计招聘岗位数，并将分析结果分别写入 Hive 的 job51_db.m_job_post_tb 表中。

（2）分析所有类型招聘的行业分布数据，并将分析结果分别写入 Hive 的 job51_db.m_post_by_companyind 表中。

其余的分析任务，读者自行参考实现。

在运行这个任务时，要确保已经启动了 Hive Metastore 服务。启动该服务的命令如下：

```
$ hive --service metastore
```

8.8 项目实现：分析结果导出

作为分析的最后一步，同样需要使用 PySpark 建立 ELT 管道，将 Hive 数据集中的分析结果导到 MySQL 数据库中，以便前端使用数据进行可视化展示。

PySpark SQL 内置了对众多常用数据源的支持，可以很容易地使用它来构建 ELT 管道。

【例 8-7】编写 PySpark 批处理程序，建立 ELT 管道，将 Hive 数据集中的分析结果导到 MySQL 数据库中。

实现代码如下：

```python
#第 8 章/job51_export.py

from pyspark.sql import SparkSession
"""
分析结果导出
"""

#创建连接 Spark 环境的对象 SparkSession
def createSparkSession():
    spark = SparkSession.builder.appName("51job"). \
        config('spark.driver.maxResultSize', '512m'). \
        config("spark.driver.memory", "2g"). \
        config("spark.executor.memory", "2g"). \
        config('spark.driver.extraClassPath',
               'mysql-connector-Java-5.1.47.jar'). \
        enableHiveSupport(). \
        getOrCreate()
    return spark

#将分析结果从 Hive Mart 导到 MySQL
def export_to_mysql(dataframe, table):
    #下面创建一个 prop 变量，用来保存 JDBC 连接参数
    props = {
        "driver": "com.mysql.jdbc.Driver",
        "user": "root",
```

```
        "password": "123456"
    }
    #连接数据库,采用append模式,表示将记录追加到数据库Spark的student表中
    url = "jdbc:mysql://localhost:3306/job51?useSSL=false"
    dataframe.write \
        .mode("append") \
        .jdbc(url=url, table=table, properties=props)

if __name__ == "__main__":
    #创建连接Spark环境的对象SparkSession
    spark = createSparkSession()
    #导出分析结果1
    df1 = spark.table("job51_db.m_job_post_tb")
    export_to_mysql(df1,"job_post_tb_2")
    #导出分析结果2
    df2 = spark.table("job51_db.m_post_by_companyind")
    export_to_mysql(df2, "post_by_companyind_2")
```

在运行这个任务时,需要注意两点:

(1)要确保已经启动了 Hive Metastore 服务。启动该服务的命令如下:

```
$ hive --service metastore
```

(2)执行导出前,首先在 MySQL 数据库中创建名为 job51 的数据库,SQL 语句如下:

```
mysql> create database job51;
```

8.9 项目实现:数据可视化

项目的最后一步,是分析结果的可视化展示。数据可视化的技术有很多,这里选择 Python 技术栈中的 Flask Web 框架,通过开发 Python Flask Web 项目,使用 ECharts 作为可视化组件,来展示前面的分析结果。

8.9.1 Flask 框架简介

Flask 是一个使用 Python 编写的轻量级 Web 应用框架。基于 Werkzeug WSGI(Web Server Gateway Interface)工具箱和 Jinja2 模板引擎。

Flask 较其他同类型框架更为灵活、轻便、安全且容易上手。它可以很好地结合 MVC 模式进行开发,开发人员分工合作,小型团队在短时间内就可以完成功能丰富的中小型网站或 Web 服务。另外,Flask 还有很强的定制性,用户可以根据自己的需求来添加相应的功能,在保持核心功能简单的同时可丰富与扩展其他功能,其强大的插件库可以让用户实现个性化的网站定制,开发出功能强大的网站。

和大多数的 Python 第三方模块的安装方法一样，Flask 可以直接通过 pip 来安装，安装命令如下：

```
$ pip install flask
```

注意：Anaconda3 中已经带有 Flask，如果使用 Anaconda3 安装，则不需要另外安装 Flask。

Flask 是一个微型框架，自身没有提供数据库管理、表单验证、Cookie 处理等功能，很多功能需要通过扩展才能实现。

Flask-SQLAlchemy 是 Flask 的扩展，它向应用程序添加了对 SQLAlchemy 的支持。SQLAlchemy 是一个 Python 工具包和对象关系映射器（ORM），它提供使用 Python 访问 SQL 数据库的功能。SQLAlchemy 提供了企业级持久性模式和高效、高性能的数据库访问。因为本案例开发的 Flask Web 程序需要访问 MySQL 数据库读取数据，因此需要安装该模块。

SQLAlchemy 使用 MySQL-Python 作为连接 MySQL 的默认数据库驱动。在使用 SQLAlchemy 之前要先给 Python 安装 MySQL 驱动，推荐使用 PyMySQL。PyMySQL 是 Python 中用于连接 MySQL 服务器的一个库，它遵循 Python 数据库 API 规范 v2.0，并包含了 pure-Python MySQL 客户端库。

要安装 PyMySQL，使用的命令如下：

```
$ pip install pymysql
```

安装 Flask-SQLAlchemy，使用的命令如下：

```
$ pip install flask flask-sqlalchemy
```

8.9.2　ECharts 图表库介绍

ECharts 是一个开源的纯 JavaScript 图表库，可以流畅地运行在 PC 和移动设备上，兼容当前绝大部分浏览器（IE 8/9/10/11、Chrome、Firefox、Safari 等)，底层依赖轻量级的向量图形库 ZRender，提供直观、交互丰富、可高度个性化定制的数据可视化图表，是一款非常优秀的可视化前端框架。

ECharts 提供了常规的折线图、柱状图、散点图、饼图、K 线图，用于统计的盒形图，用于地理数据可视化的地图、热力图、线图，用于关系数据可视化的关系图、旭日图，多维数据可视化的平行坐标，还有用于 BI 的漏斗图、仪表盘，并且支持图与图之间的混搭。

ECharts 最初由百度团队开源，并于 2018 年初捐赠给 Apache 基金会，成为 ASF 孵化级项目。2021 年 1 月 26 日晚，Apache 基金会官方宣布 ECharts 项目正式毕业。

Apache ECharts 图表库的使用非常简单，下面是来自 ECharts 官网的一个简单示例，在 HTML 页面中显示柱状图，代码如下：

```html
//第 8 章/echarts_demo.html

<!DOCTYPE html>
<html>
<head>
    <meta charset="utf-8">
    <title>简单 ECharts 实例</title>
    <!-- 引入 echarts.js -->
    <script src="echarts.js"></script>
</head>
<body>
    <!-- 为 ECharts 准备一个具备大小（宽和高）的 DOM -->
    <div id="main" style="width: 600px;height:400px;"></div>
    <script type="text/javascript">
        //基于准备好的 DOM，初始化 ECharts 实例
        var myChart = echarts.init(document.getElementById('main'));

        //指定图表的配置项和数据
        var option = {
            title: {
                text: '简单 ECharts 实例'
            },
            tooltip: {},
            legend: {
                data:['销量']
            },
            xAxis: {
                data: ["衬衫","羊毛衫","雪纺衫","裤子","高跟鞋","袜子"]
            },
            yAxis: {},
            series: [{
                name: '销量',
                type: 'bar',
                data: [5, 20, 36, 10, 10, 20]
            }]
        };

        //使用刚指定的配置项和数据显示图表
        myChart.setOption(option);
    </script>
</body>
</html>
```

在浏览器中打开这个网页，可以看到生成的柱状图如图 8-4 所示。

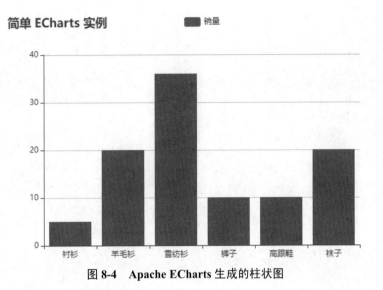

图 8-4　Apache ECharts 生成的柱状图

8.9.3　Flask Web 程序开发

接下来创建 Flask Web 服务器端应用程序。建议按以下步骤操作。

（1）创建 Flask Web 项目结构。

在合适的位置，创建项目的根目录，例如，创建一个名为 FlaskWebAppJob 的文件夹。

在 FlaskWebAppJob 目录下，再创建两个子文件夹，分别是：

① static 文件夹：用来存放 Web 静态资源，包括 CSS 样式文件、JavaScript 库文件和 images 图像文件。

② templates 文件夹：用来存放项目中用到的 Jinja2 模板文件。

此时的项目结构如图 8-5 所示。

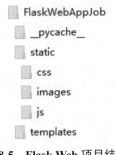

图 8-5　Flask Web 项目结构

（2）编写数据库访问工具类。

将访问数据库的代码集中到一个单独的类 DBHelper 中。为此，在项目目录下，创建一个 Python 源程序文件 dbhelper.ipynb，并编辑类 DBHelper，代码如下：

```python
#第8章/flaskwebapp/dbhelper.py

import pymysql
import pandas as pd

class DBHelper:
    #连接数据库
    def connect(self):
        try:
            self.conn = pymysql.connect(host='localhost',
                                       user='root',
                                       password='123456',
                                       database='job51',
                                       use_unicode=True,
                                       charset="utf8")
            print("连接数据库成功")
        except:
            print("连接失败")

    #统计各个岗位的招聘数量
    def get_post_by_type(self):
        self.connect()
        try:
            sql = "select post_type,post_number from post_tb;"
            df = pd.read_sql(sql, self.conn)
            return df
        finally:
            #释放连接
            self.conn.close()

    #统计指定岗位在不同行业的招聘数量
    def get_post_by_companyind(self,pt):
        self.connect()
        try:
            sql = f"""
                select companyind,post_num
                from post_of_companyind
                WHERE post_type='{pt}'
                order by post_num DESC
                limit 10;
            """
            df = pd.read_sql(sql, self.conn)
            return df
        finally:
            #释放连接
            self.conn.close()
```

注意：应将其中数据库连接参数（账号、密码）修改为读者自己的账号和密码。

然后，同样在项目目录下，创建一个 Python 主程序文件 app.ipynb，并编辑代码如下：

```python
#第8章/flaskwebapp/app.py

from flask import Flask
from flask import render_template
from dbhelper import DBHelper
import pandas as pd

app = Flask(__name__)

#路由
@app.route("/", methods=['POST', 'GET'])
def index():
    return render_template("index.html")

@app.route("/bar2")
def post_by_type():
    postByType = DBHelper().get_post_by_type()
    return render_template("post_by_type.html", df=postByType)

@app.route("/pie2/<pt>")
def post_by_companyind(pt):
    postByCompanyind = DBHelper().get_post_by_companyind(pt)
    return render_template("post_by_companyind.html", df=postByCompanyind,name=pt)

if __name__ == '__main__':
    #在本地机器上启动了 Flask 的开发服务器
    app.run(port=5000, Debug=True)
```

在上面这个程序文件中，配置路径/bar2 指向对 post_by_type()函数的调用，它将通过 DBHelper()辅助数据库访问类加载数据库中的数据，并在 templates/post_by_type.html 模板文件中进行渲染。同样，配置路径/pie2 指向对 post_by_companyind()函数的调用，它将通过 DBHelper()辅助类加载数据库中的数据，并在 templates/post_by_companyind.html 模板文件中进行渲染，其中参数 pt 代表要查询的岗位类型，其值可以是大数据、数据分析、Java 或 Python 这 4 个值之一。

8.9.4 前端 ECharts 组件开发

在 8.9.3 节中，通过 DBHelper()辅助数据库访问类加载数据库中的数据，会在 templates 目录下的模板文件中进行渲染。所谓渲染，指的是将数据按 Jinja2 模板语法填充到模板文件内的 ECharts 图表组件中。

在templates/post_by_type.html模板文件中渲染的是按类别统计的岗位数量，用ECharts中的柱状图组件可视化显示，代码如下：

```html
#第8章/flaskwebapp/templates/post_by_type.html

<!DOCTYPE html>
<html>
    <head>
        <title>柱状图</title>
        <meta charset="UTF-8">
        <!-- 引入 ECharts 文件 -->
        <script src="/static/js/echarts.min.js"></script>
    </head>
    <body class="bg-light">

    <section id="content">
        <div class="wrapper doc">
            <!-- 正文内容 -->
            <article>
                <h3>柱状图</h3>

                <!-- 在绘图前需要为 ECharts 准备一个具备大小（宽和高）的 DOM 容器 -->
                <div id="main" style="width: 100%;height:550px;"></div>

                <script type="text/javascript">
                    //基于准备好的 DOM，初始化 ECharts 实例
                    var dom1 = document.getElementById('main');
                    var myChart = echarts.init(dom1);

                    //指定图表的配置项和数据
                    var option = {
                        title: {
                            text: '各岗位招聘数量统计',
                            subtext: "一线城市",
                            textStyle: {//主标题的属性
                                color: '#C28D21',            //颜色
                                fontSize: 28,                //大小
                                fontStyle: 'oblique',        //斜体
                                fontWeight: '700',           //粗细
                                fontFamily: 'monospace'      //字体
                            },
                            subtextStyle: {//副标题的属性
                                color: '#25664A',
                                fontSize: 20                 //大小
                            }
```

```
                    },
                    tooltip: {},
                    legend: {
                        data:['岗位招聘数量统计'],
                        align: "right",
                        top: "5%", //bottom:"20%" //组件离容器的距离
                        right: "10%", //left:"10%"  //组件离容器的距离
                        width: "auto", //图例组件的宽度
                        height: "auto", //图例组件的高度
                    },
                    //x轴
                    xAxis: {
                        data: {{df["post_type"].values.tolist()|safe}}
                    },
                    //y轴
                    yAxis: {},
                    //在ECharts里, series是指: 一组数值及它们映射成的图
                    series: [{
                        name: '招聘数量',
                        type: 'bar',
                        data: {{df["post_number"].values.tolist()|safe}}
                    }],
                    grid: {
                        left: 300,
                        top: 60
                    }
                };

                //使用刚指定的配置项和数据显示图表
                myChart.setOption(option);
            </script>

        </article>
    </div>
</section>

</body>
</html>
```

注意：为了便于阅读，这里只显示了最主要的代码。完整文件内容可参考配套的源码。

上面的柱状图在浏览器中执行时，展示的结果如图 8-6 所示。

在 templates/post_by_companyind.html 模板文件中渲染的是按指定招聘岗位统计的 Top 10 行业，例如，招聘"大数据"岗位的 Top 10 行业。可以根据 url 参数来选择查询哪个岗

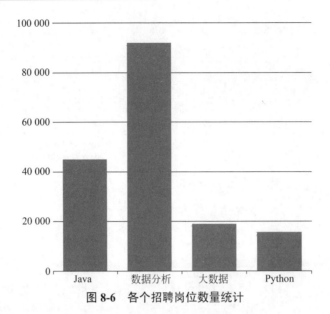

图 8-6 各个招聘岗位数量统计

位(大数据、数据分析、Java、Python 之一)的 Top 10 招聘企业。这里使用 ECharts 的饼状图进行可视化展示,代码如下:

```html
#第8章/flaskwebapp/templates/post_by_companyind.html
<!DOCTYPE html>
<html>
    <head>
        <title>饼状图</title>
        <meta charset="UTF-8">
        <!-- 引入 ECharts 文件 -->
        <script src="/static/js/echarts.min.js"></script>
    </head>

    <body class="bg-light">

    <section id="content">
        <div class="wrapper doc">
            <!-- 正文内容 -->
            <article>
                <h3>按招聘企业所属行业分析</h3>
                <!-- 在绘图前需要为 ECharts 准备一个具备大小(宽和高)的 DOM 容器 -->
                <div id="main" style="width: 100%;height:550px;"></div>

                <script type="text/javascript">
                    //基于准备好的 DOM,初始化 ECharts 实例
                    var dom1 = document.getElementById('main');
```

```
                var myChart = echarts.init(dom1);

                //指定图表的配置项和数据
                var option = {
                    title: {
                        text: '51job 招聘行业排名 Top 10',
                        subtext: '{{name}}',
                        left: 'left'
                    },
                    tooltip: {
                        trigger: 'item'
                    },
                    legend: {
                        orient: 'vertical',
                        left: 'right',
                    },
                    series: [
                        {
                            name: '行业',
                            type: 'pie',
                            radius: '50%',
                            data: [
                                    {% for list in df.values.tolist()%}
                                    {value: {{list[1]|safe}}, name: '{{list[0]|safe}}'},
                                    {% endfor %}
                            ],
                            emphasis: {
                                itemStyle: {
                                    shadowBlur: 10,
                                    shadowOffsetX: 0,
                                    shadowColor: 'rgba(0, 0, 0, 0.5)'
                                }
                            }
                        }
                    ]
                };

                //使用刚指定的配置项和数据显示图表
                myChart.setOption(option);
            </script>
        </article>
    </div>
</section>

</body>
</html>
```

注意：为了便于阅读，这里只显示了最主要的代码。完整文件内容可参考配套的源码。

在浏览器中访问这个页面，展示的结果如图 8-7 所示。

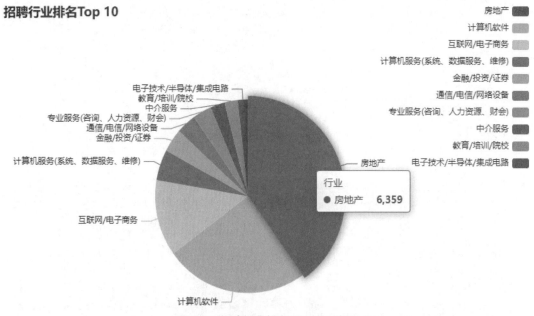

图 8-7　招聘行业排名 TOP 10 统计

图 书 推 荐

书 名	作 者
HarmonyOS 应用开发实战（JavaScript 版）	徐礼文
HarmonyOS 原子化服务卡片原理与实战	李洋
鸿蒙操作系统开发入门经典	徐礼文
鸿蒙应用程序开发	董昱
鸿蒙操作系统应用开发实践	陈美汝、郑森文、武延军、吴敬征
HarmonyOS 移动应用开发	刘安战、余雨萍、李勇军 等
HarmonyOS App 开发从 0 到 1	张诏添、李凯杰
HarmonyOS 从入门到精通 40 例	戈帅
JavaScript 基础语法详解	张旭乾
华为方舟编译器之美——基于开源代码的架构分析与实现	史宁宁
Android Runtime 源码解析	史宁宁
鲲鹏架构入门与实战	张磊
鲲鹏开发套件应用快速入门	张磊
华为 HCIA 路由与交换技术实战	江礼教
深度探索 Go 语言——对象模型与 runtime 的原理、特性及应用	封幼林
深入理解 Go 语言	刘丹冰
深度探索 Flutter——企业应用开发实战	赵龙
Flutter 组件精讲与实战	赵龙
Flutter 组件详解与实战	[加]王浩然（Bradley Wang）
Flutter 跨平台移动开发实战	董运成
Dart 语言实战——基于 Flutter 框架的程序开发（第 2 版）	亢少军
Dart 语言实战——基于 Angular 框架的 Web 开发	刘仕文
IntelliJ IDEA 软件开发与应用	乔国辉
深度探索 Vue.js——原理剖析与实战应用	张云鹏
Vue+Spring Boot 前后端分离开发实战	贾志杰
Vue.js 快速入门与深入实战	杨世文
Vue.js 企业开发实战	千锋教育高教产品研发部
Flink 原理深入与编程实战（Scala+Java）	辛立伟
Python 从入门到全栈开发	钱超
Python 全栈开发——基础入门	夏正东
Python 全栈开发——高阶编程	夏正东
Python 全栈开发——数据分析	夏正东
Python 游戏编程项目开发实战	李志远
Python 人工智能——原理、实践及应用	杨博雄 主编，于营、肖衡、潘玉霞、高华玲、梁志勇 副主编
Python 深度学习	王志立
Python 预测分析与机器学习	王沁晨
Python 异步编程实战——基于 AIO 的全栈开发技术	陈少佳
Python 数据分析实战——从 Excel 轻松入门 Pandas	曾贤志
Python 数据分析从 0 到 1	邓立文、俞心宇、牛瑶
Python Web 数据分析可视化——基于 Django 框架的开发实战	韩伟、赵盼

续表

书 名	作 者
FFmpeg 入门详解——音视频原理及应用	梅会东
Python 玩转数学问题——轻松学习 NumPy、SciPy 和 matplotlib	张骞
Pandas 通关实战	黄福星
深入浅出 Power Query M 语言	黄福星
云原生开发实践	高尚衡
云计算管理配置与实战	杨昌家
虚拟化 KVM 极速入门	陈涛
虚拟化 KVM 进阶实践	陈涛
边缘计算	方娟、陆帅冰
物联网——嵌入式开发实战	连志安
动手学推荐系统——基于 PyTorch 的算法实现（微课视频版）	於方仁
人工智能算法——原理、技巧及应用	韩龙、张娜、汝洪芳
跟我一起学机器学习	王成、黄晓辉
TensorFlow 计算机视觉原理与实战	欧阳鹏程、任浩然
分布式机器学习实战	陈敬雷
计算机视觉——基于 OpenCV 与 TensorFlow 的深度学习方法	余海林、翟中华
深度学习——理论、方法与 PyTorch 实践	翟中华、孟翔宇
深度学习原理与 PyTorch 实战	张伟振
AR Foundation 增强现实开发实战（ARCore 版）	汪祥春
ARKit 原生开发入门精粹——RealityKit + Swift + SwiftUI	汪祥春
HoloLens 2 开发入门精要——基于 Unity 和 MRTK	汪祥春
巧学易用单片机——从零基础入门到项目实战	王良升
Altium Designer 20 PCB 设计实战（视频微课版）	白军杰
Cadence 高速 PCB 设计——基于手机高阶板的案例分析与实现	李卫国、张彬、林超文
Octave 程序设计	于红博
ANSYS 19.0 实例详解	李大勇、周宝
ANSYS Workbench 结构有限元分析详解	汤晖
AutoCAD 2022 快速入门、进阶与精通	邵为龙
SolidWorks 2020 快速入门与深入实战	邵为龙
SolidWorks 2021 快速入门与深入实战	邵为龙
UG NX 1926 快速入门与深入实战	邵为龙
西门子 S7-200 SMART PLC 编程及应用（视频微课版）	徐宁、赵丽君
三菱 FX3U PLC 编程及应用（视频微课版）	吴文灵
全栈 UI 自动化测试实战	胡胜强、单镜石、李睿
pytest 框架与自动化测试应用	房荔枝、梁丽丽
敏捷测试从零开始	陈霁、王富、武夏
深入理解微电子电路设计——电子元器件原理及应用（原书第 5 版）	[美]理查德·C.耶格（Richard C. Jaeger）、[美]特拉维斯·N.布莱洛克（Travis N. Blalock）著；宋廷强译
深入理解微电子电路设计——数字电子技术及应用（原书第 5 版）	[美]理查德·C.耶格（Richard C.Jaeger）、[美]特拉维斯·N.布莱洛克（Travis N.Blalock）著；宋廷强译
深入理解微电子电路设计——模拟电子技术及应用（原书第 5 版）	[美]理查德·C.耶格（Richard C.Jaeger）、[美]特拉维斯·N.布莱洛克（Travis N.Blalock）著；宋廷强译